T0211305

# Lecture Notes in Artificial Intelligence    10583

Subseries of Lecture Notes in Computer Science

More information about this series at http://www.springer.com/series/1244

Nathalie Camelin · Yannick Estève
Carlos Martín-Vide (Eds.)

# Statistical Language and Speech Processing

5th International Conference, SLSP 2017
Le Mans, France, October 23–25, 2017
Proceedings

 Springer

*Editors*
Nathalie Camelin
University of Le Mans
Le Mans
France

Yannick Estève ⓘ
University of Le Mans
Le Mans
France

Carlos Martín-Vide ⓘ
Rovira i Virgili University
Tarragona
Spain

ISSN 0302-9743 ISSN 1611-3349 (electronic)
Lecture Notes in Artificial Intelligence
ISBN 978-3-319-68455-0 ISBN 978-3-319-68456-7 (eBook)
DOI 10.1007/978-3-319-68456-7

Library of Congress Control Number: 2017954906

LNCS Sublibrary: SL7 – Artificial Intelligence

Printed on acid-free paper

This Springer imprint is published by Springer Nature
The registered company is Springer International Publishing AG
The registered company address is: Gewerbestrasse 11, 6330 Cham, Switzerland

# Preface

These proceedings contain the papers that were presented at the 5th International Conference on Statistical Language and Speech Processing (SLSP 2017), held in Le Mans, France, during October 23–25, 2017.

The scope of SLSP deals with topics of either theoretical or applied interest discussing the employment of statistical models (including machine learning) in language and speech processing, namely:

Anaphora and coreference resolution
Authorship identification, plagiarism, and spam filtering
Computer-aided translation
Corpora and language resources
Data mining and Semantic Web
Information extraction
Information retrieval
Knowledge representation and ontologies
Lexicons and dictionaries
Machine translation
Multimodal technologies
Natural language understanding
Neural representation of speech and language
Opinion mining and sentiment analysis
Parsing
Part-of-speech tagging
Question-answering systems
Semantic role labeling
Speaker identification and verification
Speech and language generation
Speech recognition
Speech synthesis
Speech transcription
Spelling correction
Spoken dialogue systems
Term extraction
Text categorization
Text summarization
User modeling

SLSP 2017 received 39 submissions. Every paper was reviewed by three Programme Committee members. There were also a few external experts consulted. After a thorough and vivid discussion phase, the committee decided to accept 21 papers (which represents an acceptance rate of about 54%). The conference program included three invited talks and presentations of work in progress as well.

The excellent facilities provided by the EasyChair conference management system allowed us to deal with the submissions successfully and handle the preparation of these proceedings in time.

We would like to thank all invited speakers and authors for their contributions, the Program Committee and the external reviewers for their diligent cooperation, and Springer for its very professional publishing work.

July 2017

Nathalie Camelin
Yannick Estève
Carlos Martín-Vide

# Organization

## Program Committee

| | |
|---|---|
| Jon Barker | University of Sheffield, UK |
| Laurent Besacier | Grenoble Informatics Laboratory, France |
| Paul Buitelaar | National University of Ireland, Galway, Ireland |
| Felix Burkhardt | Telekom Innovation Laboratories, Germany |
| Xavier Carreras | Xerox Research Centre Europe, France |
| Francisco Casacuberta | Technical University of Valencia, Spain |
| Ciprian Chelba | Google, USA |
| Eng Siong Chng | Nanyang Technological University, Singapore |
| Jennifer Chu-Carroll | Elemental Cognition, USA |
| Doug Downey | Northwestern University, USA |
| Robert Gaizauskas | University of Sheffield, UK |
| Julio Gonzalo | National University of Distance Education, Spain |
| Keikichi Hirose | University of Tokyo, Japan |
| Gerhard Jäger | University of Tübingen, Germany |
| Gareth Jones | Dublin City University, Ireland |
| Joseph Keshet | Bar-Ilan University, Israel |
| Tomi Kinnunen | University of Eastern Finland, Finland |
| Lun-Wei Ku | Academia Sinica, Taiwan |
| Kong Aik Lee | Institute for Infocomm Research, Singapore |
| Elizabeth D. Liddy | Syracuse University, USA |
| Xunying Liu | Chinese University of Hong Kong, Hong Kong |
| Suresh Manandhar | University of York, UK |
| Carlos Martín-Vide (Chair) | Rovira i Virgili University, Spain |
| Yuji Matsumoto | Nara Institute of Science and Technology, Japan |
| Ruslan Mitkov | University of Wolverhampton, UK |
| Marie-Francine Moens | KU Leuven, Belgium |
| Seiichi Nakagawa | Toyohashi University of Technology, Japan |
| Preslav Nakov | Qatar Computing Research Institute, Qatar |
| Hermann Ney | RWTH Aachen University, Germany |
| Cécile Paris | CSIRO, Australia |
| Fuchun Peng | AISense Inc., USA |
| Pascal Perrier | Grenoble Institute of Technology, France |
| Leon Rothkrantz | Delft University of Technology, Netherlands |
| Horacio Saggion | Pompeu Fabra University, Spain |
| Murat Saraçlar | Bogaziçi University, Turkey |
| Holger Schwenk | Facebook, France |
| Brad Story | University of Arizona, USA |
| Karin Verspoor | University of Melbourne, Australia |

| Stephan Vogel | Qatar Computing Research Institute, Qatar |
| Xiaojun Wan | Peking University, China |
| Phil Woodland | University of Cambridge, UK |
| Chuck Wooters | Semantic Machines, USA |
| François Yvon | LIMSI-CNRS, France |

## Additional Reviewers

Evans, Richard
Makary, Mireille
Rohanian, Omid
Taslimipoor, Shiva

# Contents

# Invited Paper

# Author Profiling in Social Media: The Impact of Emotions on Discourse Analysis

Paolo Rosso[1(✉)] and Francisco Rangel[1,2]

[1] PRHLT Research Center, Universitat Politècnica de València, Valencia, Spain
prosso@dsic.upv.es
http://www.dsic.upv.es/~prosso
[2] Autoritas Consulting, Valencia, Spain
francisco.rangel@autoritas.es
http://www.kicorangel.com

**Abstract.** In this paper we summarise the content of the keynote that will be given at the 5th International Conference on Statistical Language and Speech Processing (SLSP) in Le Mans, France in October 23–25, 2017. In the keynote we will address the importance of inferring demographic information for marketing and security reasons. The aim is to model how language is shared in gender and age groups taking into account its statistical usage. We will see how a shallow discourse analysis can be done on the basis of a graph-based representation in order to extract information such as how complicated the discourse is (i.e., how connected the graph is), how much interconnected grammatical categories are, how far a grammatical category is from others, how different grammatical categories are related to each other, how the discourse is modelled in different structural or stylistic units, what are the grammatical categories with the most central use in the discourse of a demographic group, what are the most common connectors in the linguistic structures used, etc. Moreover, we will see also the importance to consider emotions in the shallow discourse analysis and the impact that this has. We carried out some experiments for identifying gender and age, both in Spanish and in English, using PAN-AP-13 and PAN-PC-14 corpora, obtaining comparable results to the best performing systems of the PAN Lab at CLEF.

**Keywords:** Author profiling · Graph-based representation · Shallow discourse analysis · EmoGraph

## 1 Author Profiling in Social Media

Often social media users do not explicitly provide demographic information about themselves. Therefore, due to the importance that is for marketing, but also for security or forensics, this information needs to be inferred somehow, for instance on the basis of how language is generally used among group of people that may share a more common writing style (e.g. adolescents vs. adults).

© Springer International Publishing AG 2017
N. Camelin et al. (Eds.): SLSP 2017, LNAI 10583, pp. 3–18, 2017.
DOI: 10.1007/978-3-319-68456-7_1

Studies like [8] linked the use of language with demographic traits. The authors approached the problem of gender and age identification combining function words with part-of-speech (POS) features. In [15] the authors related the language use with personality traits. They employed a set of psycho-linguistic features obtained from texts, such as POS, sentiment words and so forth. In [22] the authors studied the effect of gender and age in the writing style in blogs. They obtained a set of stylistic features such as non-dictionary words, POS, function words and hyperlinks, combined with content features, such as word unigrams with the highest information gain. They showed that language features in blogs correlates with age, as reflected in, for example, the use of prepositions and determiners.

More recently, at PAN 2013[1] and 2014[2] gender and age identification have been addressed in the framework of a shared task on author profiling in social media. Majority of approaches at PAN-AP 2013 [18] and PAN-AP 2014 [19] used combinations of style-based features such as frequency of punctuation marks, capital letters, quotations, and so on, together with POS tags and content-based features such as bag of words, dictionary-based words, topic-based words, entropy-based words, etc. Two participants used the occurrence of sentiment or emotional words as features. It is interesting to highlight the approach that obtained the overall best results using a representation that considered the relationship between documents and author profiles [14]. The best results in English were obtained employing collocations [12].

Following, in Sect. 2 we describe how discourse features can be extracted from a graph-based representation of texts, and in Sect. 3 we show the impact that considering emotions in the framework of discourse analysis may have. Finally, in Sect. 4 we draw some conclusions.

## 2   Discourse Analysis in Author Profiling

Very recently, discourse features started to be used in author profiling [23, 24]. Rhetorical Structure Theory (RST)[3] has been applied for the characterization of the writing style of authors. Features have been extracted from the discourse trees, such as the frequencies of each discourse relation per elementary discourse unit, obtaining interesting results when used in combination with other features. Unfortunately, no comparison has been made with any state-of-the-art method, for instance on the PAN-AP-13 and PAN-AP-14 corpora, and it is difficult to fully understand the impact that the use of discourse features may have on author profiling, but the preliminary results that have been obtained are quite promising.

---

[1] http://www.uni-weimar.de/medien/webis/research/events/pan-13/pan13-web/auth or-profiling.html.

[2] http://www.uni-weimar.de/medien/webis/research/events/pan-14/pan14-web/auth or-profiling.html.

[3] RST is a descriptive linguistic approach to the organization of discourse based on the linguistic theory formulated by Mann and Thompson in 1988 [11].

Our aim is instead to extract discourse features after modelling the use of language of a group of authors with a graph-based representation. These features will indicate the discourse complexity, the different structural and stylistic units the discourse is modelled in, etc. Concretely, our aim is to analyse the writing style from the perspective people combine the different POS in a text, the kind of verbs they employ, the topics they mention, the emotions and sentiments they express, etc. As Pennebaker pointed out [16], men generally use more prepositions than women and, for instance, they may use more prepositional syntagmas than women (e.g. preposition + determinant + noun + adjective). In the proposed approach, we build a graph with the different POS of authors' texts and enrich it with semantic information with the topics they speak about, the type of verbs they use, and the emotions they express. We model the text of authors of a given gender or age group as a single graph, considering also punctuation signs in order to capture how a gender or age group of authors connects concepts in sentences. Once the graph is built, we extract from the graph structure and discourse features we feed a machine learning approach with. Moreover, we will see that the way authors express their emotions depends on their age and gender. The main motivation for using a graph-based approach is its capacity to analyse complex language structures and discourses.

## 2.1   EmoGraph Graph-Based Representation

For each text of a group of authors, we carry out a morphological analysis with Freeling[4] [4,13], obtaining POS and lemmas of the words. Freeling describes each POS with an Eagle label[5]. We model each POS as a node (N) of the graph (G), and each edge (E) defines the sequence of POS in the text as directed links between the previous part-of-speech and the current one. For example, let us consider a simple text like the following:

*El gato come pescado y bebe agua.* (The cat eats fish and drinks water)

It generates the following sequence of Eagle labels:

DA0MS0 -> NCMS000 -> VMIP3S0 -> NCMS000 -> CC -> VMIP3S0 -> NCMS000 -> Fp

We model such sequence as the graph showed in Fig. 1. Due to the fact that the link VMIP3S0 -> NCMS000 is produced twice, the weight of this edge is double than the rest.

---

[4] http://nlp.lsi.upc.edu/freeling/.

[5] The Eagles group (http://www.ilc.cnr.it/EAGLES96/intro.html) proposed a series of recommendations for the morphosyntactic annotation of corpora. For Spanish, we used the Spanish version (http://nlp.lsi.upc.edu/freeling/doc/tagsets/tagset-es. html). For example in the sentence "El gato come pescado y bebe agua." (The cat eats fish and drinks water.), the word "gato" (cat) is returned as NCMS000 where NC means common noun, M means male, S means singular, and 000 is a filling until 7 chars; or the word "come" (eats) is returned as VMIP3S0 where V means verb, M means main verb (not auxiliary), I means indicative mode of the verb, P means present time, 3 means third person, S means singular, and 0 is a filling until 7 chars.

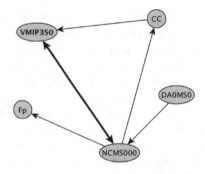

**Fig. 1.** POS Graph of "El gato come pescado y bebe agua." (The cat eats fish and drinks water)

The following step is to enrich the described graph with semantic and affective information. For each word in the text, we look for the following information:

- **Wordnet domains**[6]: If the word is a common noun, adjective or verb, we search for the domain of its lemma. We use Wordnet Domains linked to the Spanish version of the Euro Wordnet[7] in order to find domains of Spanish lemmas. If the word has one or more related topics, a new node is created for each topic and a new edge from the current Eagle label to the created node(s) is added. In the previous example, *gato* (cat) is related both to biology and animals, thus two nodes are created and a link is added from NCMS000 to each of them (NCMS000 -> biology & animals).
- **Semantic classification of verbs:** Semantic classification of (V)erbs: We search for the semantic classification of verbs. On the basis of what was investigated in [10], we have manually annotated 158 verbs with one of the following semantic categories: *(a) perception* (see, listen, smell...); *(b) understanding* (know, understand, think...); *(c) doubt* (doubt, ignore...); *(d) language* (tell, say, declare, speak...); *(e) emotion* (feel, want, love...); *(f) and will* (must, forbid, allow...). We add six features with the frequencies of each verb type. If the word is a verb we search for the semantic classification of its lemma. We create a node with the semantic label and we add an edge from the current Eagle label to the new one. For example, if the verb is a perception verb, we would create a new node named "perception" and link the node VMIP3S0 to it (VMIP3S0 -> perception).
- **Polarity of words:** If the word is a common noun, adjective, adverb or verb, we look for its polarity in a sentiment lexicon. For example, let us consider the following sentence:

*She is an incredible friend.*

It has the following sequence of Eagle labels:

PP3FS000 -> VSIP3S0 -> DI0FS0 -> NCFS000 -> AQ0CS0(->positive &negative) -> Fp

---

[6] http://wndomains.fbk.eu/.
[7] http://www.illc.uva.nl/EuroWordNet/.

The adjective node AQ0CS0 has links both to the positive and negative tags, because *incredible* could be both positive and negative depending on the context. Therefore, from a polarity viewpoint it is an ambiguous word which gives us two nodes (and two edges).

- **Emotional words:** If the word is a common noun, adjective, adverb or verb, for texts in English we look for its relationship to one emotion in Wordnet Affect[8] [25] and for texts in Spanish in the Spanish Emotion Lexicon [5]. We create a new node for each of them. See the following sentence as an example:

> *He estado tomando cursos en línea sobre temas valiosos que disfruto estudiando y que podrían ayudarme a hablar en público* (I have been taking online courses about valuable subjects that I enjoy studying and might help me to speak in public)

The representation of the previous sentence with our graph-based approach, that will call EmoGraph, is shown in Fig. 2. The sequence may be followed by starting in VAIP1S0 node. Nodes size depends on their eigenvector and nodes colour on their modularity.

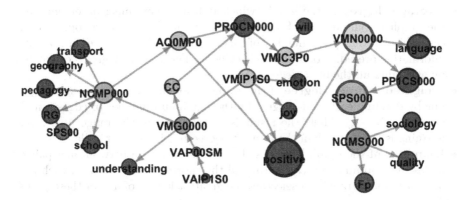

**Fig. 2.** EmoGraph of "He estado tomando cursos en línea sobre temas valiosos que disfruto estudiando y que podrían ayudarme a hablar en público" ( *"I have been taking online courses about valuable subjects that I enjoy studying and might help me to speak in public"*)

Finally, we link the last element of the sentence (e.g. Fp) with the first element of the next one, since we are also interested in how people use sentence splitters (e.g. . ; :) and any other information prone to model how people use their language.

Once the graph is built, our objective is to use a machine learning approach to model texts of gender and age groups in order to be able to classify a given text later into the right class. Therefore, we have first to extract features from

---

[8] http://wndomains.fbk.eu/wnaffect.html.

the graph. We obtain such features on the basis of graph analysis in two ways: *(a)* general properties of the graph describing the overall structure of the modelled texts; *(b)* and specific properties of its nodes and how they are related to each other, that describe how authors use language.

Following, we describe how to extract the structure-based features from the graph and what they describe from a discourse-based perspective:

- **Nodes-Edges ratio.** We calculate the ratio between the number of nodes N and the number of edges E of the graph $G = \{N, E\}$. The maximum possible number of nodes (429) is given by: *(a)* the total number of Eagle labels (247); *(b)* the total number of topics in Wordnet Domains (168); *(c)* the total number of verb classes (6); *(d)* the total number of emotions (6); *(e)* and the total number of sentiment polarities (2). The maximum possible number of edges (183, 612) in a directed graph is theoretically calculated as:

$$max(E) = N * (N - 1)$$

where N is the total number of nodes. Thus, the ratio between nodes and edges gives us an indicator of how connected the graph is, or in our case, how complicated the structure of the discourse of the user is.

- **Average degree** of the graph, which indicates how much interconnected the graph is. The degree of a node is the number of its neighbours; in our case, this is given by the number of other grammatical categories or semantic information preceding or following each node. The average degree is calculated by averaging all the node degrees.

- **Weighted average degree** of the graph is calculated as the average degree but by dividing each node degree by the maximum number of edges a node can have (N − 1). Thus, the result is transformed in the range [0, 1]. The meaning is the same than the average degree but in another scale.

- **Diameter** of the graph indicates the greatest distance between any pair of nodes. It is obtained by calculating all the shortest paths between each pair of nodes in the graph and selecting the greatest length of any of these paths. That is:

$$d = max_{n \in N} \varepsilon(N)$$

where $\varepsilon(n)$ is the eccentricity or the greatest geodesic distance between n and any other node. In our case, it measures how far one grammatical category is from others, for example how far a topic is from an emotion.

- **Density** of the graph measures how close the graph is to be completed, or in our case, how dense is the text in the sense of how each grammatical category is used in combination to others. Given a graph $G = (N, E)$, it measures how many edges are in set E compared to the maximum possible number of edges between the nodes of the set N. Then, the density is calculated as:

$$D = \frac{2 * |E|}{(|N| * (|N| - 1))}$$

- **Modularity** of the graph measures the strength of division of a graph into modules, groups, clusters or communities. A high modularity indicates that nodes within modules have dense connections whereas they have sparse connections with nodes in other modules. In our case may indicate how the discourse is modelled in different structural or stylistic units. Modularity is calculated following the algorithm described in [1].
- **Clustering coefficient** of the graph indicates the transitivity of the graph, that is, if $a$ is directly linked to $b$ and $b$ is directly linked to $c$, the probability that $a$ is also linked to $c$. The clustering coefficient indicates how nodes are embedded in their neighbourhood, or in our case, how the different grammatical categories (or semantic information such as emotions) are related to each others. For each node, the cluster coefficient (cc1) may be calculated with the Watts-Strogatzt formula [26]:

$$cc1 = \frac{\sum_{i=1}^{n} C(i)}{n}$$

Each $C(i)$ measures how close the neighbours of node i are to be a complete graph. It is calculated as follows:

$$C(i) = \frac{|\{e_{jk} : n_j, n_k \in N_i, e_{jk} \in E\}|}{k_i(k_i - 1)}$$

where $e_{jk}$ is the edge which connects node $n_j$ with node $n_k$ and $k_i$ is the number of neighbours of the node i. Finally, we calculate the global clustering coefficient as the average of all node's coefficients, excluding nodes with degree 0 or 1, following the algorithm described in [9].
- **Average path length** of the graph is the average graph-distance between all the pairs of nodes and could be calculated following [3]. It gives us an indicator on how far some nodes are from others or in our case how far some grammatical categories are from others.

Moreover, for each node in the graph, we calculate two centrality measures: betweenness and eigenvector. We use each obtained value as the weight of a feature named respectively BTW-xxx and EIGEN-xxx, where xxx is the name of the node (e.g. AQ0CS0, positive, enjoyment, animal and so on):

- **Betweenness centrality** measures how important a node is by counting the number of shortest paths of which it is part of. The betweenness centrality of a node x is the ratio of all shortest paths from one node to another node in the graph that pass through x. We calculate it as follows:

$$BC(x) = \sum_{i,j \in N - \{n\}} \frac{\sigma_{i,j}(n)}{\sigma_{i,j}}$$

where $\sigma_{i,j}$ is the total number of shortest paths from node i to node j, and $\sigma_{i,j}(n)$ is the total number of those paths that pass through n. In our case, if one node has a high betweenness centrality means that it is a common element used for link among parts-of-speech, for example, prepositions, conjunctions, or even verbs or nouns. This measure may give us an indicator of what the most common links in the linguistic structures used by authors are.

- **Eigenvector centrality** of a node measures the influence of such node in the graph [2]. Given a graph and its adjacency matrix $A = a_{n,t}$ where $a_{n,t}$ is 1 if a node n is linked to a node t, and 0 otherwise, we can calculate the eigenvector centrality score as:

$$x_n = \frac{1}{\lambda} \sum_{t \in M(n)} x_t = \frac{1}{\lambda} \sum_{t \in G} a_{n,t} x_t$$

where $\lambda$ is a constant representing the greatest eigenvalue associated with the centrality measure, $M(n)$ is a set of the neighbours of node n and $x_t$ represents each node different to $x_n$ in the graph. This measure may give us an indicator of what are the grammatical categories with the most central use in the authors' discourse, for example nouns, verbs, adjectives, etc.

## 2.2   Experiments with PAN-AP-13 and PAN-AP-14 Corpora

Below we present the results that have been obtained for gender and age identification with a Support Vector machine with a Gaussian Kernel on the PAN-AP-13 and PAN-AP-14 corpora.

We carried out the experiments with the Spanish partition of the PAN-AP-13 social media corpus. In Table 1 the results for gender identification are shown. The proposed graph-based approach obtained competitive results with respect to the two best performing systems (with no statistically significant difference). In Table 2 EmoGraph shows a better performance than the system that was

**Table 1.** Results in accuracy for gender identification in PAN-AP-13 corpus (Spanish)

| Ranking | Team | Accuracy |
|---------|------|----------|
| 1 | Santosh | 0.6473 |
| 2 | **EmoGraph** | 0.6365 |
| 3 | Pastor | 0.6299 |
| 4 | Haro | 0.6165 |
| 5 | Ladra | 0.6138 |
| 6 | Flekova | 0.6103 |
| 7 | Jankowska | 0.5846 |
| ... | ... | |
| 17 | Baseline | 0.5000 |
| ... | ... | |
| 22 | Gillam | 0.4784 |

**Table 2.** Results in accuracy for age identification in PAN-AP-13 corpus (Spanish)

| Ranking | Team | Accuracy |
|---------|------|----------|
| 1 | **EmoGraph** | 0.6624 |
| 2 | Pastor | 0.6558 |
| 3 | Santosh | 0.6430 |
| 4 | Haro | 0.6219 |
| 5 | Flekova | 0.5966 |
| ... | ... | |
| 19 | Baseline | 0.3333 |
| ... | ... | |
| 21 | Mechti | 0.0512 |

ranked first at the shared task was obtained for age identification (10s, 20s and 30s), although statistically with no significant difference (t-Student test).

We studied what topics the different group of authors wrote about in the corpus (we removed the most frequent topics[9] because not so informative being at the top of the domain hierarchy). We obtained the topics with the help of Wordnet Domains. The corresponding word clouds are shown in Figs. 3, 4 and 5 for females in each age group (10s, 20s and 30s), and in Figs. 6, 7 and 8 for males in the same age groups. Younger people tend to write more about many different disciplines such as physics, linguistics, literature, metrology, law, medicine, chemistry and so on, maybe due to the fact that this is the stage of life when people mostly speak about their homework. Females seem to write more about chemistry or gastronomy, and males about physics or law. Both write about music and play. On the contrary of what one could might think, 10s females write about

**Fig. 3.** Top domains for 10s females in PAN-AP-13 corpus

**Fig. 4.** Top domains for 20s females in PAN-AP-13 corpus

**Fig. 5.** Top domains for 30s females in PAN-AP-13 corpus

---

[9] E.g. biology, quality, features, psychological, economy, anatomy, period, person, transport, time and psychology.

**Fig. 6.** Top domains for 10s males in PAN-AP-13 corpus

**Fig. 7.** Top domains for 20s males in PAN-AP-13 corpus

**Fig. 8.** Top domains for 30s males in PAN-AP-13 corpus

sexuality whereas males do not, and the contrary for commerce (shopping). As they grow up, both females and males show more interest in buildings (maybe due to the fact that they look for flats to rent), animals, gastronomy, medicine, and about religion, although in a highest rate among males.

With respect to the use of verb types, we were interested in investigating what kind of actions (verbs) females and males mostly refer to and how this changes over time. Figure 9 illustrates that males use more *language* verbs (e.g. tell, say, speak...), whereas females use more *emotional* verbs (e.g. feel, want, love...) conveying more verbal emotions than males.

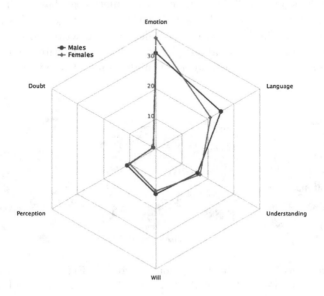

**Fig. 9.** Use of verb types per gender in PAN-AP-13 corpus

Moreover, we analysed the evolution of the use of verbs over the age. Figures 10 and 11 show the evolution through 10s, 20s and 30s. The use of *emotional* verbs decreases over years, although we can assert that females use more *emotional* verbs than males in any stage of life. The contrary happens with verbs of *language*. Verbs of *understanding* (e.g. know, understand, think...) seem to increase for males and remain stable for females, but it has to be said that females started using more verbs of understanding already in the early age at a similar ratio than males do later. Similarly, verbs of *will*[10] (e.g. must, forbid, allow...) increase for both genders, but at a higher rate for males.

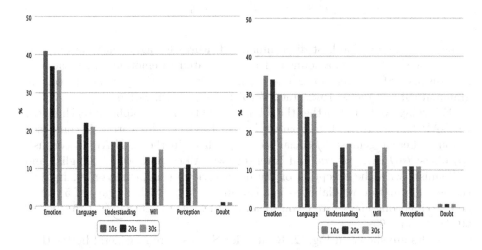

**Fig. 10.** Evolution in the use of verb types for females in PAN-AP-13 corpus

**Fig. 11.** Evolution in the use of verb types for males in PAN-AP-13 corpus

Finally, we analyse the most discriminative features for the identification of gender and age on the basis of information gain [27]. Table 3 shows the top 20 features over 1100. Betweenness ($BTW$-$xxx$) and eigenvector ($EIGEN$-$xxx$) features are among the top features. We can identify a higher number of *eigen* features (mainly for verbs, nouns and adjectives) in gender identification in comparison to the higher number of *betweenness* features (mainly prepositions or punctuation marks) in age identification. This means that features describing the important nodes in the discourse provide more information to gender identification, whereas features describing the most common links in the discourse provide more information to the age identification. In other words, the selection of the position in the discourse for words such as nouns, verbs or adjectives, which mainly give the meaning of the sentence, is the best discriminative features for gender identification, whereas the selection of connectors such as prepositions, punctuation marks

---

[10] "Verbs of will": verbs that suggest interest or intention of doing things (such as *must, forbid, allow*). Verbs of will do not have any relationship with *will* as the auxiliary verb for the future in English.

**Table 3.** Most discriminating features for gender and age identification

| Ranking | Gender | Age | Ranking | Gender | Age |
|---|---|---|---|---|---|
| 1 | Punctuation-semicolon | Words-length | 11 | BTW-NC00000 | EIGEN-SPS00 |
| 2 | EIGEN-VMP00SM | Pron | 12 | BTW-Z | BTW-NC00000 |
| 3 | EIGEN-Z | BTW-SPS00 | 13 | EIGEN-DA0MS0 | Punctuation-exclamation |
| 4 | EIGEN-NCCP000 | BTW-NCMS000 | 14 | BTW-Fz | Emoticon-happy |
| 5 | Pron | Intj | 15 | BTW-NCCP000 | BTW-Fh |
| 6 | Words-length | EIGEN-Fh | 16 | EIGEN-AQ0MS0 | Punctuation-colon |
| 7 | EIGEN-NC00000 | BTW-PP1CS000 | 17 | SEL-disgust | Punctuation |
| 8 | EIGEN-administration | EIGEN-Fpt | 18 | EIGEN-DP3CP0 | BTW-Fpt |
| 9 | Intj | EIGEN-NC00000 | 19 | EIGEN-DP3CS0 | EIGEN-DA0FS0 |
| 10 | SEL-sadness | EIGEN-NCMS000 | 20 | SEL-anger | Verb |

or interjections are the best discriminative features for age identification. It is important to notice the amount of features related to emotions (SEL-sadness, SEL-disgust, SEL-anger) for gender identification and the presence of certain grammatical categories (Pron, Intj, Verb) for age identification.

Following, we tested further the robustness of the EmoGraph approach on the PAN-AP-14 corpus, both in Spanish and in English. This corpus is composed of four different genres: *(i)* social media (such as in the PAN-AP-13 corpus); *(ii)* blogs; *(iii)* Twitter; *(iv)* and hotel reviews. All corpora were in English and in Spanish, with the exception of the hotel reviews (in English only). In 2014 the age information was labelled in a continuous way (without gaps of 5 years), and the following classes were considered: *(i)* 18–24; *(ii)* 25–34; *(iii)* 35–49; *(iv)* 50–64; *(v)* and 65+.

Results are shown in Fig. 12. Results for Spanish are in general better than for English. This may be due to the higher variety of the morphological information obtained with Freeling for Spanish. In fact, Freeling obtains 247 different annotations for Spanish whereas it obtains 53 for English. For example, in the Spanish version the word "cursos" (courses) for the given example in Fig. 2 is returned as NCMP000 where NC means common noun, M means male, P means plural, and 000 is a filling until 7 chars; in the English version, the word "courses" is annotated as NNS.

With respect to the results obtained in the PAN-AP-13 corpus for Spanish, the results for age are lower due to the higher number of classes (3 classes in 2013 vs. 5 continuous ones in 2014). Results for Twitter and blogs are better than for social media and reviews. This is due to the quality of the annotation, in fact both blogs and Twitter corpora were manually annotated, ensuring that the reported gender and age of each author was true. On the contrary, in social media and reviews what the authors reported was assumed to be true. Furthermore, in blogs and also in Twitter there were enough texts per author in order to obtain a better profile. In fact, although in Twitter each tweet is short (as much 140 characters), we had access to hundreds of tweets per author. The worst results were obtained for the reviews. Besides the possibility of deceptive information regarding age and gender in the reviews corpus, it is important to know that reviews were bounded to the hotel domain and just to two kinds of emotions: complain or praise.

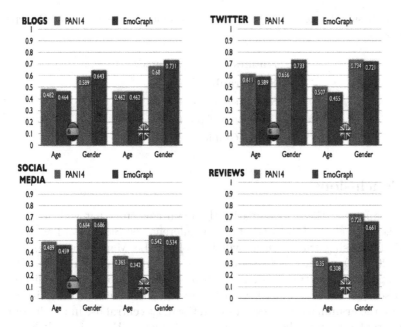

**Fig. 12.** Results in accuracy in PAN-AP-14 corpus: EmoGraph vs. the best team

# 3   The Impact of Emotions

In order to understand further the impact of emotions in our graph-based representation of texts, we carried out a further experiment with another corpus, the EmIroGeFB [17] corpus of Facebook comments in Spanish, that we previously annotated with the Ekmans's six basic emotions [6]. We compared the proposed approach where emotions are taken into account with other variations of the graph-based representation that take into account some of the structure and discourse features:

- **Simple Graph:** a graph built only with the grammatical category of the Eagle labels (the first character of the Eagle label), that is, verb, noun, adjective and so on;
- **Complete Graph:** a graph built only with the complete Eagle labels, but without topics, verbs classification and emotions;
- **Semantic Graph:** a graph built with all the features described above (Eagle labels, topics and verbs classification) but without emotions.

Results for gender identification are shown in Table 4. The best results were obtained when also emotions that were used in the discourse were considered in the graph-based approach.

**Table 4.** Results for gender identification in accuracy on the EmIroGeFB corpus (in Spanish)

| Features | Accuracy |
|---|---|
| **EmoGraph** | 0.6596 |
| Semantic Graph | 0.5501 |
| Complete Graph | 0.5192 |
| Simple Graph | 0.5083 |

## 4   Conclusions

In this paper we tried to summarise the main concepts that will be addressed in the keynote at the 5th International Conference on Statistical Language and Speech Processing (SLSP) that will be held in Le Mans, France in October 23–25, 2017. Our aim was to show that with a graph-based representation of texts is possible to extract discourse features that describe how complicated the discourse is, how the discourse is modelled in different structural or stylistic units, what are the grammatical categories with the most central use in the discourse of a demographic group, where in the discourse emotion-bearing words have been used, etc. *Eigen* features describing the important nodes in the discourse (e.g. the position in the discourse of words such as nouns, verbs or adjectives, which mainly give the meaning of the sentence) showed to help in gender identification, whereas *betweenness* features describing the most common links in the discourse (e.g. connectors such as prepositions, punctuation marks or interjections) helped more in age identification.

A more complete description of the EmoGraph graph-based approach and the experiments carried out on the PAN-AP-13 and PAN-AP-14 can be found in [21] and in [20].

**Acknowledgments.** We thank the SLSP Conference for the invitation for giving the keynote on Author Profiling in Social Media. The research work described in this paper was partially carried out in the framework of the SomEMBED project (TIN2015-71147-C2-1-P), funded by the Spanish Ministry of Economy, Industry and Competitiveness (MINECO).

## References

1. Blondel, V.D., Guillaume, J.L., Lambiotte, R., Lefebvre, E.: Fast unfolding of communities in large networks. J. Stat. Mech.: Theory Exp. **2008**(10), 10008 (2008)
2. Bonacich, P.: Factoring and weighting approaches to clique identification. J. Math. Soc. **2**(1), 113–120 (1972)
3. Brandes, U.: A faster algorithm for betweenness centrality. J. Math. Soc. **25**(2), 163–177 (2001)
4. Carreras, X., Chao, I., Padró, L., Padró, M.: FreeLing : an open-source suite of language analyzers. In: Proceedings of the 4th International Conference on Language Resources and Evaluation (LREC 2004) (2004)

5. Díaz Rangel, I., Sidorov, G., Suárez-Guerra, S.: Creación y evaluación de un diccionario marcado con emociones y ponderado para el español. Onomazein **29**, 23 (2014). (in Spanish)
6. Ekman, P.: Universals and cultural differences in facial expressions of emotion. In: Symposium on Motivation, Nebraska, pp. 207–283 (1972)
7. Forner, P., Navigli, R., Tufis, D. (eds.): CLEF 2013 Evaluation Labs and Workshop, Working Notes Papers, September 2013, Valencia, Spain, vol. 1179, pp. 23–26. CEUR-WS.org (2013)
8. Koppel, M., Argamon, S., Shimoni, A.: Automatically categorizing written texts by author gender. Literay Linguist. Comput. **17**(4), 401–412 (2003)
9. Latapy, M.: Main-memory triangle computations for very large (sparse (power-law)) graphs. Theor. Comput. Sci. (TCS) **407**(1–3), 458–473 (2008)
10. Levin, B.: English Verb Classes and Alternations. University of Chicago Press, Chicago (1993)
11. Mann, W.C., Thompson, S.A.: Rhetorical structure theory: toward a functional theory of text organization. Text-Interdiscip. J. Study Discourse **8**(3), 243–281 (1988)
12. Meina, M., Brodzinska, K., Celmer, B., Czokow, M., Patera, M., Pezacki, J., Wilk, M.: Ensemble-based classification for author profiling using various features notebook for PAN at CLEF 2013. In: Forner et al. [7]
13. Padró, L., Stanilovsky, E.: FreeLing 3.0: towards wider multilinguality. In: Proceedings of the Language Resources and Evaluation Conference (LREC 2012) (2012)
14. Lopez-Monroy, A.P., Montes-Gomez, M., Jair Escalante, H., Villasenor-Pineda, L., Villatoro-Tello, E.: INAOEs participation at PAN13: author profiling task. Notebook for PAN at CLEF 2013. In: Forner et al. [7]
15. Pennebaker, J.W., Mehl, M.R., Niederhoffer, K.: Psychological aspects of natural language use: our words, our selves. Annu. Rev. Psychol. **54**, 547–577 (2003)
16. Pennebaker, J.W.: The Secret Life of Pronouns: What Our Words Say About Us. Bloomsbury Press, London (2011)
17. Rangel, F., Hernández, I., Rosso, P., Reyes, A.: Emotions and irony per gender in Facebook. In: Proceedings of the Workshop on Emotion, Social Signals, Sentiment & Linked Open Data (ES3LOD), LREC-2014, Reykjavik, Iceland, 26–31 May 2014, pp. 68–73 (2014)
18. Rangel, F., Rosso, P., Koppel, M., Stamatatos, E., Inches, G.: Overview of the author profiling task at PAN 2013. In: Forner et al. [7]
19. Rangel, F., Rosso, P., Chugur, I., Potthast, M., Trenkmann, M., Stein, B., Verhoeven, B., Daelemans, W.: Overview of the 2nd author profiling task at PAN 2014. In: Cappellato, L., Ferro, N., Halvey, M., Kraaij, W. (eds.) Notebook Papers of CLEF 2014 LABs and Workshops, vol. 1180, pp. 951–957. CEUR-WS.org (2014)
20. Rangel, F., Rosso, P.: On the multilingual and genre robustness of EmoGraphs for author profiling in social media. In: Mothe, J., Savoy, J., Kamps, J., Pinel-Sauvagnat, K., Jones, G.J.F., SanJuan, E., Cappellato, L., Ferro, N. (eds.) CLEF 2015. LNCS, vol. 9283, pp. 274–280. Springer, Cham (2015). doi:10.1007/978-3-319-24027-5_28
21. Rangel, F., Rosso, P.: On the impact of emotions on author profiling. Inf. Process. Manag. **52**(1), 73–92 (2016)
22. Schler, J., Koppel, M., Argamon, S., Pennebaker, J.W.: Effects of age and gender on blogging. In: AAAI Spring Symposium: Computational Approaches to Analyzing Weblogs, AAAI, pp. 199–205 (2006)

23. Soler-Company, J. Wanner, L.: Use of discourse and syntactic features for gender identification. In: The Eighth Starting Artificial Intelligence Research Symposium. Collocated with the 22nd European Conference on Artificial Intelligence, pp. 215–220 (2016)
24. Soler-Company, J., Wanner, L.: On the relevance of syntactic and discourse features for author profiling and identification. In: 15th Conference of the European Chapter of the Association for Computational Linguistics (EACL), Valencia, Spain, pp. 681–687 (2017)
25. Strapparava, C., Valitutti, A.: WordNet affect: an affective extension of WordNet. In: Proceedings of the 4th International Conference on Language Resources and Evaluation, Lisboa, pp. 1083–1086 (2004)
26. Watts, D.J., Strogatz, S.H.: Collective dynamics of 'small-world' networks. Nature **393**(6684), 409–410 (1998)
27. Yang, Y., Pedersen, J.O.: A comparative study on feature selection in text categorization. In: Proceedings of the Fourteenth International Conference on Machine Learning, ICML, pp. 412–420 (1997)

# Language and Information Extraction

# Neural Machine Translation by Generating Multiple Linguistic Factors

Mercedes García-Martínez$^{(\boxtimes)}$, Loïc Barrault, and Fethi Bougares

LIUM, Le Mans University, Le Mans, France
{mercedes.garcia_martinez,Loic.Barrault,Fethi.Bougares}@univ-lemans.fr

**Abstract.** Factored neural machine translation (FNMT) is founded on the idea of using the morphological and grammatical decomposition of the words (factors) at the output side of the neural network. This architecture addresses two well-known problems occurring in MT, namely the size of target language vocabulary and the number of unknown tokens produced in the translation. FNMT system is designed to manage larger vocabulary and reduce the training time (for systems with equivalent target language vocabulary size). Moreover, we can produce grammatically correct words that are not part of the vocabulary. FNMT model is evaluated on IWSLT'15 English to French task and compared to the baseline word-based and BPE-based NMT systems. Promising qualitative and quantitative results (in terms of BLEU and METEOR) are reported.

**Keywords:** Machine translation · Neural networks · Deep learning · Factored representation

## 1 Introduction and Related Works

In contrast to the traditional phrased-based statistical machine translation [12] that automatically translates subparts of the sentences, standard Neural Machine Translation (NMT) systems use the sequence to sequence approach at word level and consider the entire input sentence as a unit for translation [2,5,25].

Recently, NMT showed better accuracy than existing phrase-based systems for several language pairs. Despite these positive results, NMT systems still face several challenges. These challenges include the high computational complexity of the softmax function which is linear to the target language vocabulary size (Eq. 1).

$$p_i = e^{o_i} / \sum_{r=1}^{N} e^{o_r} \text{ for } i \in \{1, \ldots, N\} \tag{1}$$

where $o_i$ are the outputs, $p_i$ their softmax normalization and $N$ the total number of outputs.

In order to solve this issue, a standard technique is to define a *short-list* limited to the $s$ most frequent words where $s << N$. The major drawback of this technique is the growing rate of unknown tokens generated at the output. Another work around has been proposed in [11] by carefully organising the

© Springer International Publishing AG 2017
N. Camelin et al. (Eds.): SLSP 2017, LNAI 10583, pp. 21–31, 2017.
DOI: 10.1007/978-3-319-68456-7_2

batches so that only a subset $K$ of the target vocabulary is possibly generated at training time. This allows the system to train a model with much larger target vocabulary without substantially increasing the computational complexity. Another possibility is to define a structured output layer (SOUL) to handle the words not appearing in the shortlist. This allows the system to always apply the softmax normalization on a layer with reduced size [14]. The problem of unknown words was addressed making use of the alignments produced by an unsupervised aligner [16]. The unknown generated words are substituted in a post-process step by the translation of their corresponding aligned source word or copying the source word if no translation is found. The translation of the source word is made by means of a dictionary.

Other recent work have used subword units instead of words. In [24], some unknown and rare words are encoded as subword units with the Byte Pair Encoding (BPE) method. Authors show that this can also generates words unseen at training time. As an extreme case, the character-level neural machine translation has been presented in several works [6,7,15] and showed very promising results. The character-level NMT architectures are composed of many layers, to deal with the long distance dependencies, increasing aggressively the computational complexity of the training process. In [22] has been shown that character-level decoders outperform subwords units using BPE method when processing unknown words, but they perform worse when extracting morphosyntactic information about the sentences, due to the long distances.

Among other previous works, our work can be seen as a continuation of [9]. Several works have used factors as additional information for the input words in neural language modelling with interesting results [1,18,26]. More recently, factors have also been integrated into a word-level NMT system as additional linguistic input features [23]. Unlike these previous works, we are considering factors as translation unit. We refer to *factors* as some linguistic annotations at word level, *e.g.* the Part of Speech (POS) tag, number, gender, etc. The advantages of using factors as translation unit are two-fold: reducing the output vocabulary size and allowing to generate surface forms which are never seen in the training data.

Factors were first introduced for NMT at output side in [9] where two factored synchronous symbols are simultaneously generated. Authors presented an investigation of the architecture of their factored NMT system to show that better results are obtained using a feedback of the two generated outputs concatenation.

Our work is different from previous efforts in that we consider only the best type of feedback for the network. We also introduce an additional factor about the case information (lowercase, uppercase or in capitals) and evaluate using a different translation test. Moreover, we apply an unknown words (*unk*) replacement technique using the alignments of the attention mechanism to replace the generated unknown words in target side. For that, we make use of an unigram dictionary to find the translation of the source word corresponding to the generated *unk*.

We compare this architecture to the state of the art BPE approach and the classic word-level NMT approach on the English to French dataset from IWSLT'15 evaluation campaign. We provide, in addition a quantitative and qualitative study about the obtained results.

The remainder of this paper is organized as follows: Sect. 2 describes the attention-based NMT system and Sect. 3 its extension using the factored approach. In Sect. 4, we describe the experiments and the obtained results. Finally, Sect. 5 concludes the paper and presents the future work.

## 2  Neural Machine Translation

The standard NMT model consists of a sequence to sequence encoder-decoder of two recurrent neural networks (RNN), one used by the encoder and the other by the decoder. The source language sequence is mapped into an embedded dimension in the encoder and the decoder maps the representation back to a target language sequence.

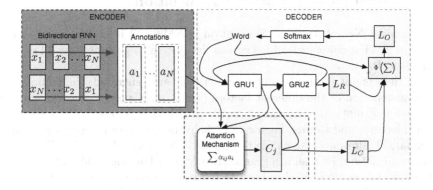

**Fig. 1.** Attention-based NMT system (Color figure online)

The architecture includes a bidirectional RNN encoder (see left part of Fig. 1) equipped with an attention mechanism [2]. Each input sentence word $x_i$ ($i \in 1 \ldots N$ with $N$ the source sequence length) is encoded into an annotation $a_i$ by concatenating the hidden states of a forward and a backward RNN provided by a gated recurrent unit (GRU) [5] to control the flow of information. These annotations $a_1 \ldots a_N$ represents the whole sentence with a focus on the word being processed. One difference from the architecture of [2] is that the decoder contains a conditional GRU [8] which consists of two GRUs interspersed with the attention mechanism (see right top part of the Fig. 1). The first GRU combines the embedding of the previous decoded token and the previous hidden state in order to generate an intermediate representation which is an input of the attention mechanism and the second GRU. The attention mechanism (bottom yellow part of the Fig. 1) computes a source context vector $C_j$ as a convex combination of annotation vectors, where the weights of each annotation are computed locally using a feed-forward network. These weights can be used to align the target words with the source positions. The second GRU generates the hidden state of the conditional GRU by looking at the output of the first GRU and the context vector $C_j$. The decoder RNN takes as input the embedding of the previous output word (feedback of the network) in the first GRU, the context vector $C_j$ in the second GRU and its hidden state. The output layer $L_O$ is connected to the network through a hyperbolic tangent sum operation $\Phi(\sum)$ which takes as input the embedding of the previous output word as well as the context vector and the output of the decoder from the second GRU (both adapted with a linear transformation, respectively, $L_C$ and $L_R$). Finally, the output probabilities for each word in the target vocabulary are computed with a *softmax* function. The word with the highest probability is the translation output at each timestep. The encoder and the decoder are trained jointly to maximize the conditional probability of the reference translation.

## 3    Factored Neural Machine Translation

The Factored Neural Machine Translation (FNMT) [9] is an extension of the standard NMT architecture which allows the system to generate several output symbols at the same time.

For the sake of simplicity, only two symbols are generated: the lemma and the concatenation of the different factors (verb, tense, person, gender, number and case information). The target words are then represented by a factored output: lemmas and factors. Factors may help the translation process providing grammatical information to enrich the output. The task of this work is English to French translation, English is a grammatically poor language and factors do not help for its translation, this has been tested in previous experiments. Therefore, we apply the factors only in the target side when translating to French which is a grammatically rich language. In the example shown in Fig. 3, from the verbal form in French *devient*, we obtain the lemma *devenir* and its factors *VP3#SL* (**V**erb, in **P**resent, **3**rd person, no gender (**#**), **S**ingular and **L**owercased form). Moreover, we can see the word *interéssant* with the lemma *interéssant* and factors *Adj##MSL* (**Adj**ective, no tense (**#**) and no person (**#**), **M**asculine gender, **S**ingular number and **L**owercased form). The morphological analyser MACAON toolkit [17] is used to obtain the lemma and factors for each word taking into account its context with nearly 100% accuracy. The first entry is used in the few cases that MACAON proposes multiple words (e.g. same word written in two forms).

The FNMT architecture is presented in Fig. 2. The encoder and attention mechanism of Fig. 1 remain unchanged. However, the decoder has been modified to get multiple outputs. The hidden state of the conditional GRU (cGRU) is shared to produce simultaneously several outputs. The output from the layer $L_O$ has been diversified to two *softmax* layers, one to generate the lemma and

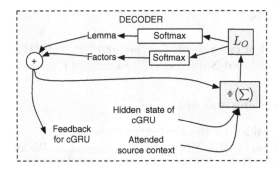

**Fig. 2.** Detailed view of the decoder of the Factored NMT system

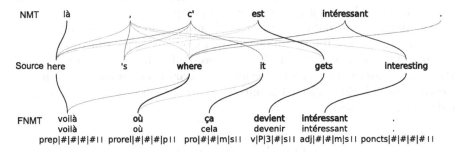

**Fig. 3.** Examples of NMT and FNMT outputs aligned against the source sentence

the other to generate the factors. An additional design decision is related to the decoder feedback. Contrary to the word based model, where the feedback is naturally the previous word (see Fig. 1), we have multiple choices where multiple outputs are generated for each decoding time-step. We have decided to use the concatenation of the embeddings of both generated symbols based on the work [9].

The FNMT model may lead to sequences with a different length, since lemmas and factors are generated simultaneously but separately (each sequence ends after the generation of the end of sequence <eos> token). To avoid this, the sequences length is decided based on the lemmas stream length (i.e. the length of the factors sequence is constrained to be equal to the length of the lemma sequence). This is motivated by the fact that the lemmas contain most of the information of the final surface form (word).

Once we obtain the factored outputs from the neural network, we need to combine them to obtain the surface form (word representation). This operation is also performed with the MACAON tool, which, given a lemma and some factors, provides the word. Word forms given by MACAON toolkit have a 99% success rate. In the cases (e.g. name entities) that the word corresponding to the lemma and factors is not found, the system outputs the lemma itself.

# 4   Experiments

We performed a set of experiments for Factored NMT (FNMT) and compared
them with the word-based NMT and BPE-based NMT systems.

## 4.1   Data Processing and Selection

The systems are trained on the English to French (EN-FR) Spoken Language
Translation task from IWSLT 2015 evaluation campaign[1]. We applied data selection using modified Moore-Lewis filtered by XenC [21] to obtain a sub part
of the available parallel corpora (news-commentary, united-nations, europarl,
wikipedia, and two crawled corpora). The Technology Entertainment Design
(TED) [4] corpus has been used as in-domain corpus.

   We preprocess the data to convert html entities and filter out the sentences
with more than 50 words for both source and target languages. Finally, we obtain
a corpus of 2M sentences with 147k unique words for the English side and 266k
unique words for the French side. French vocabulary is bigger than English since
French is more highly inflected language. Table 1 shows training, development
and testing sets statistics.

**Table 1.** Datasets statistics

| Data | Corpus name | Datasets | # Sents | # Words EN-FR |
|------|-------------|----------|---------|---------------|
| Training | train15 | data selection | 2M | 147–266k |
| Development | dev15 | dev10 + test10 + test13 | 3.6k | 7.3–8.9k |
| Testing | test15 | test11 + test12 | 1.9k | 4.5–5.4k |

## 4.2   Training

Models are trained using NMTPY [3], an NMT toolkit in Python based on
Theano[2]. The following hyperparameters have been chosen to train the systems.
The embedding and recurrent layers have the dimensions 620 and 1000, respectively. The batch size is set to 80 sentences and the parameters are trained
using the Adadelta [27] optimizer. We clipped the norm of the gradient to be
no more than 1 [20] and initialize the weights using *Xavier* [10]. The systems
are validated on dev15 dataset using early stopping based on BLEU [19]. The
vocabulary size of the source language is set to 30K. The output layer size of the
baseline NMT system is set to 30K. For the sake of comparability and consistency, the same value (30k) is used for the lemma output of the FNMT system.
This 30K FNMT vocabulary includes 17k lemmas obtained from the original
NMT vocabulary (30k word level gives 17k lemmas when all the derived forms

---

[1] https://sites.google.com/site/iwsltevaluation2015.
[2] https://github.com/lium-lst/nmtpy.

of the verbs, nouns, adjectives, etc. are discarded) increased with additional new lemmas to fit the 30K desired value. The factors have 142 different units in their vocabulary. When it comes to combining the lemmas and the factors vocabulary, the system is able to generate 172K different words, using the external linguistic resources, which is 6 times bigger than a standard word-based NMT vocabulary.

For BPE systems, bilingual vocabulary has been built using source and target language applying the joint vocabulary BPE approach. In order to create comparable BPE systems, we set the number of merge operations for the BPE algorithm (the only hyperparameter of the method) as 30K minus the number of character according to the paper [24]. Then, we apply a total of 29388 merge operations to learn the BPE models on the training and validation sets. During the decoding process, we use a beam size of 12 as used in [2].

### 4.3   Quantitative Results

The Factored NMT system aims at integrating linguistic knowledge into the decoder in order to overcome the restriction of having a large vocabulary at target side. We first compare our system with the standard word-level NMT system. For the sake of comparison with state of the art systems, we have built a subword system using the BPE method. Subwords were calculated at the input and the output side of the neural network as described in [24]. The results are measured with two automatic metrics, the most common metric for machine translation BLEU and METEOR [13]. We evaluate on test15 dataset from the IWSLT 2015 campaign and results are presented in Table 2.

**Table 2.** Results on IWSLT test15. %BLEU and %METEOR performance of NMT and FNMT systems with and without UNK replacement (UR) are presented. For each system we provide the number of generated UNK tokens in the last column

| Model | %METEOR↑ | %BLEU↑ | | | #UNK |
|---|---|---|---|---|---|
| | Word | Word | Lemma | Factors | |
| NMT/+UR | 62.21/63.38 | 41.80/42.74 | 45.10 | 51.80 | 1111 |
| BPE | 62.87 | 42.37 | 45.96 | 53.31 | 0 |
| FNMT/+UR | **64.10/64.81** | **43.42/44.15** | **47.18** | **54.24** | **604** |

As we can see from the Table 2 results, the FNMT system obtains better %BLEU and %METEOR scores compared to the state of the art NMT and BPE systems. An improvement of about 1 %BLEU point is achieved compared to the best baseline system (BPE). This improvement is even bigger (**1.4** %BLEU point) when UNK replacement is applied to both systems. In a quest to better understand the reasons of this improvement, we also computed the %BLEU scores of each output level (lemmas and factors) for FNMT. Theses scores are presented in Table 2. The lemma and factors scores of NMT and BPE systems are obtained through a decomposing of their word level output into lemma and factors. We observe yet again that FNMT systems gives better score at both lemma and factors level. Replacement of unknown words has been performed using the alignments extracted from the attention mechanism. We have replaced the generated UNK tokens by translating its highest probability aligned source word. We see an improvement of around 1 point %BLEU score in both NMT and FNMT systems.

The last column of Table 2 shows, for each system, the number of generated UNK tokens. As shown in the table our FNMT system produces half of the UNK tokens compared to the word-based NMT system. This tends to prove that the Factored NMT system effectively succeed in modelling more words compared to the word based NMT system augmenting the generalization power of our model and preserving manageable output layer sizes. Though we can see that BPE system does not produce UNK tokens, this is not reflected in the scores. Indeed, this can be due to the possibility of generation of incorrect words using BPE units in contrast to the FNMT system.

### 4.4   Qualitative Analysis

The strengths of FNMT are considered under this qualitative analysis. We have studied and compared the translation outputs of NMT at word-level and BPE-level with the ones of FNMT systems. Two examples are presented in Fig. 3 and Table 3.

**Table 3.** Examples of translations with NMT, BPE and FNMT systems (without unknown words replacement)

| Src | we in medicine , I think , are baffled | | | | | | | |
|---|---|---|---|---|---|---|---|---|
| Ref | Je pense que en médecine nous sommes dépassés | | | | | | | |
| NMT | Nous | , | en | médecine | , | je | pense | , | sont | UNK |
| BPE | nous | , | en | médecine | , | je | pense | , | sommes | b@@ af@@ és |
| FNMT | nous | , | en | médecine | , | je | pense | , | sont | déconcertés |
| Lemmas | lui | , | en | médecine | , | je | penser | , | être | déconcerter |
| Factors | pro-1-p-l | pct-l | prep-l | nc-f-s-l | pct-l | cln-1-s-l | v-P-1-s-l | pct-l | v-P-3-p-l | vppart-K-m-p-l |

The reference translation of the source sentence presented in Fig. 3 is *"mais voilà où ça devient intéressant"*. As we can see, contrary to the baseline NMT system, the FNMT system matches exactly the reference and thus produces the correct translation. An additional interesting observation is that the alignment provided by the attention mechanism seems to be better defined and more helpful when using factors. Also, one can notice the difference between the attention distributions made by the systems over the source sentence. The NMT system first translated "here" into "là", added a coma, and then was in trouble for translating the rest of the sentence, which is reflected by the rather fuzzy attention weights. The FNMT system had better attention distribution over of the source sentence in this case.

Table 3 shows another example comparing NMT, BPE and FNMT systems. The NMT system generated an unknown token (UNK) when translating the English word *"baffled"*. We observe that BPE translates *"baffled"* to *"bafs"* which does not exist in French. This error probably comes from the shared vocabulary between the source and target languages creating an incorrect word very similar to its aligned source tokens. FNMT translates it to *"dconcerts"* which is a better translation than in the reference. One should note that it is not generated by the unknown word replacement method. However, for this particular example, an error on the factors leads to the word *"sont"* instead of *"sommes"*, resulting in lower automatic scores for FNMT output.

## 5    Conclusion

In this paper, the Factored NMT approach has been further explored. Factors based on linguistic *a priori* knowledge have been used to decompose the target words. This approach outperforms a strong baseline system using subword units computed with byte pair encoding. Our FNMT system is able to model an almost 6 times bigger word vocabulary with only a slight increase of the computational cost. By these means, the FNMT system is able to halve the generation of unknown tokens compared to word-level NMT. Using a simple unknown word replacement procedure involving a bilingual dictionary, we are able to obtain even better results (+0.8 %BLEU compared to previous best system).

Also, the use of external linguistic resources allows us to generate new word forms that would not be included in the standard NMT system *shortlist*. The advantage of this approach is that the new generated words are controlled by the linguistic knowledge, that avoid producing incorrect words, as opposed to actual systems using BPE. We demonstrated the performance of such a system on an inflected language (French). The results are very promising for use with highly inflected languages like Arabic or Czech.

**Acknowledgments.** This work was partially funded by the French National Research Agency (ANR) through the CHIST-ERA M2CR project, under the contract number ANR-15-CHR2-0006-01.

## References

1. Alexandrescu, A.: Factored neural language models. In: HLT-NAACL (2006)
2. Bahdanau, D., Cho, K., Bengio, Y.: Neural machine translation by jointly learning to align and translate. CoRR abs/1409.0473 (2014)
3. Caglayan, O., García-Martínez, M., Bardet, A., Aransa, W., Bougares, F., Barrault, L.: NMTPY: a flexible toolkit for advanced neural machine translation systems. arXiv preprint arXiv:1706.00457 (2017)
4. Cettolo, M., Girardi, C., Federico, M.: WIT[3]: web inventory of transcribed and translated talks. In: Proceedings of the 16th Conference of the European Association for Machine Translation (EAMT), Trento, Italy, pp. 261–268, May 2012

5. Cho, K., van Merrienboer, B., Gülçehre, Ç., Bougares, F., Schwenk, H., Bengio, Y.: Learning phrase representations using RNN encoder-decoder for statistical machine translation. CoRR abs/1406.1078 (2014)

6. Chung, J., Cho, K., Bengio, Y.: A character-level decoder without explicit segmentation for neural machine translation. CoRR abs/1603.06147 (2016)

7. Costa-Jussà, M.R., Fonollosa, J.A.R.: Character-based neural machine translation. CoRR abs/1603.00810 (2016)

8. Firat, O., Cho, K.: Conditional gated recurrent unit with attention mechanism (2016). github.com/nyu-dl/dl4mt-tutorial/blob/master/docs/cgru.pdf

9. García-Martínez, M., Barrault, L., Bougares, F.: Factored neural machine translation architectures. In: Proceedings of the International Workshop on Spoken Language Translation, IWSLT 2016, Seattle, USA (2016). http://workshop.2016. iwslt.org/downloads/IWSLT_2016_paper_2.pdf

10. Glorot, X., Bengio, Y.: Understanding the difficulty of training deep feedforward neural networks. In: Proceedings of the International Conference on Artificial Intelligence and Statistics (AISTATS 2010). Society for Artificial Intelligence and Statistics (2010)

11. Jean, S., Cho, K., Memisevic, R., Bengio, Y.: On using very large target vocabulary for neural machine translation. CoRR abs/1412.2007 (2014)

12. Koehn, P., Hoang, H., Birch, A., Callison-Burch, C., Federico, M., Bertoldi, N., Cowan, B., Shen, W., Moran, C., Zens, R., Dyer, C., Bojar, O., Constantin, A., Herbst, E.: Moses: open source toolkit for statistical machine translation. In: Proceedings of the 45th Annual Meeting of the ACL on Interactive Poster and Demonstration Sessions, ACL 2007, pp. 177–180. Association for Computational Linguistics, Stroudsburg (2007)

13. Lavie, A., Agarwal, A.: METEOR: an automatic metric for mt evaluation with high levels of correlation with human judgments. In: Proceedings of the Second Workshop on Statistical Machine Translation, StatMT 2007, pp. 228–231. Association for Computational Linguistics, Stroudsburg (2007)

14. Le, H.S., Oparin, I., Messaoudi, A., Allauzen, A., Gauvain, J.L., Yvon, F.: Large vocabulary SOUL neural network language models. In: INTERSPEECH (2011). sources/Le11large.pdf

15. Ling, W., Trancoso, I., Dyer, C., Black, A.W.: Character-based neural machine translation. CoRR abs/1511.04586 (2015)

16. Luong, T., Sutskever, I., Le, Q.V., Vinyals, O., Zaremba, W.: Addressing the rare word problem in neural machine translation. CoRR abs/1410.8206 (2014). http:// arxiv.org/abs/1410.8206

17. Nasr, A., Béchet, F., Rey, J.F., Favre, B., Roux, J.L.: MACAON, an NLP tool suite for processing word lattices. In: Proceedings of the ACL-HLT 2011 System Demonstrations, pp. 86–91 (2011)

18. Niehues, J., Ha, T.L., Cho, E., Waibel, A.: Using factored word representation in neural network language models. In: Proceedings of the First Conference on Machine Translation, pp. 74–82. Association for Computational Linguistics, Berlin, August 2016

19. Papineni, K., Roukos, S., Ward, T., Zhu, W.J.: BLEU: a method for automatic evaluation of machine translation. In: Proceedings of the 40th Annual Meeting on Association for Computational Linguistics, ACL 2002, Stroudsburg, PA, USA, pp. 311–318 (2002)

20. Pascanu, R., Mikolov, T., Bengio, Y.: Understanding the exploding gradient problem. CoRR abs/1211.5063 (2012)

21. Rousseau, A.: XenC: an open-source tool for data selection in natural language processing. Prague Bull. Math. Linguist. **100**, 73–82 (2013)
22. Sennrich, R.: How grammatical is character-level neural machine translation? Assessing MT quality with contrastive translation pairs. CoRR abs/1612.04629 (2016)
23. Sennrich, R., Haddow, B.: Linguistic input features improve neural machine translation. CoRR abs/1606.02892 (2016)
24. Sennrich, R., Haddow, B., Birch, A.: Neural machine translation of rare words with subword units. In: Proceedings of the 54th Annual Meeting of the Association for Computational Linguistics. Long Papers, vol. 1, pp. 1715–1725. Association for Computational Linguistics (2016)
25. Sutskever, I., Vinyals, O., Le, Q.V.: Sequence to sequence learning with neural networks. CoRR abs/1409.3215 (2014)
26. Wu, Y., Yamamoto, H., Lu, X., Matsuda, S., Hori, C., Kashioka, H.: Factored recurrent neural network language model in TED lecture transcription. In: IWSLT (2012)
27. Zeiler, M.D.: ADADELTA: an adaptive learning rate method. arXiv preprint arXiv:1212.5701 (2012)

# Analysis and Automatic Classification of Some Discourse Particles on a Large Set of French Spoken Corpora

Denis Jouvet[1,2,3](✉), Katarina Bartkova[4,5], Mathilde Dargnat[4,5], and Lou Lee[4,5]

[1] Inria, 54600 Villers-lès-Nancy, France
[2] Université de Lorraine, LORIA, UMR 7503, 54600 Villers-lè-Nancy, France
denis.jouvet@loria.fr
[3] CNRS, LORIA, UMR 7503, 54600 Villers-lès-Nancy, France
[4] Université de Lorraine, ATILF, UMR 7118, 54063 Nancy, France
{katarina.bartkova,mathilde.dargnat}@univ-lorraine.fr,
lou.lee4@etu.univ-lorraine.fr
[5] CNRS, ATILF, UMR 7118, 54063 Nancy, France

**Abstract.** In French, quite a number of words and expressions are frequently used as discourse particles in spoken language, especially in spontaneous speech. The semantic load of these words or expressions differ whether they are used as discourse particles or not. Therefore, the correct identification of their discourse function remains of great importance. In this paper the distribution of the discourse function (or not discourse function), and of the detailed discourse functions of some of these words, is studied on a large set of French corpora ranging from prepared speech (e.g. storytelling and broadcast news) to spontaneous speech (e.g. interviews and interactions between people). The paper is focused on a subset of discourse particles that are recurrent in the considered corpora. The discourse function of a few thousand occurrences of these words have been manually annotated. A statistical analysis of the functions of the words is presented and discussed with respect to the types of spoken corpora. Finally, some statistics with respect to a few prosodic correlates of the discourse particles are presented, as well as some results of automatic classification and detection of the word function (discourse particle or not) using prosodic features.

**Keywords:** Discourse particles · French language · Prosodic parameters · Discourse function statistics · Discourse particle detection

## 1 Introduction

In French, some words and expressions are frequently used as *Discourse Particles* (DPs) in spoken language. The ongoing study aims at investigating the correlation between the main semantico-pragmatic values of the DPs and their prosodic features (pause, position in prosodic group, duration...). To not be biased by a

© Springer International Publishing AG 2017
N. Camelin et al. (Eds.): SLSP 2017, LNAI 10583, pp. 32–43, 2017.
DOI: 10.1007/978-3-319-68456-7_3

single type of speech data, our study is based on a large variety of speech corpora that range from storytelling and prepared speech to highly spontaneous speech resulting from interactions between people.

Studies of discourse markers, including DPs, have flourished in the last twenty years, but they most often address only the semantic, pragmatic and sometimes syntactic components of linguistic description, from a synchronic as well as diachronic point of view. However, prosodic considerations remain peripheral or quite general (see for instance [1–5]).

Using the term DP raises problems about terminology and categorization. Several terms coexist (discourse or pragmatic markers, discourse or modal particles, phatic connectives, etc.) and they are not always interchangeable. DPs are frequently defined in contrast to discourse markers (connectives) or to modal particles [6–8]. In this paper, a DP is defined as a functional category [9], whose lexical members, in addition to being a DP, have more traditional uses (conjunction, adverb, interjection, adjective, etc.). Semantically, analyzing DPs raises the complex problem of the referential status of those items, in particular of their indexical [10,11] and procedural values [12,13]. In short, a DP is an invariable linguistic item that functions at the discourse level: it conveys deictic information available only at utterance time. The information content can concern utterance interpretation, epistemic state and affective mood of the speaker or management of interaction.

The three items studied here behave differently, but all exist as DPs and non-DPs in French. *"Alors"* (then, what's up...) can be a temporal anaphoric adverb, a discourse connective or a DP. *"Bon"* (well, all right, OK...) can be an adjective, a noun or a DP. *"Donc"* (therefore, well...) can be a discourse connective or a DP. As DP, they present most of the prototypical properties listed in the scientific literature (see [1,6,14,15] for some general approaches).

For conducting the study, we rely on a large variety of speech corpora coming from the ESTER2 speech recognition evaluation campaign [16] and from the ORFEO project [17]. This amounts to several millions of time aligned words. Occurrences of the selected words (*"alors"*, *"bon"*, *"donc"*) have been chosen at random and then manually annotated.

The paper is organized as follows. Section 2 presents the speech corpora and the annotations. Section 3 discusses the DP vs. non-DP usage of the words with respect to the various speech corpora. Section 4 focuses on some frequent detailed discourse functions. Finally, Sect. 5 presents some information on a few prosodic correlates of the DPs, and Sect. 6 discusses some automatic classification experiments.

## 2  Speech Corpora and Annotations

The study is based on a large set of French speech corpora, of various degree of spontaneity, coming from the ESTER2 evaluation campaign [16] and the ORFEO project [17].

**Storytelling. FRE** (FREnch oral narrative corpus, [18]) is a corpus of oral storytelling in French.

**News (Prepared Speech). EST** (ESTER 2, [16]) is a corpus of French broadcast news collected from various radio channels. It contains mainly prepared speech and a few interviews.

**Interviews, Dialogues, and Conversations. CFP** (*Corpus de Français Parlé Parisien* – French spoken in Paris, [19,20]) contains interviews about Paris and its suburbs. **COR** (French part of the C-ORAL-ROM project – integrated reference corpora for spoken romance languages, [21,22]) contains dialogues and conversations as well as some more formal speech. **CRF** (*Corpus de référence du français parlé* – reference corpus for spoken French, [23,24]) contains speech recorded from speakers with various education levels. **TUF** refers to the French part of TUFS speech corpus [25]. And **VAL** refers to a part of the Valibel speech database [26].

**Interactions. CLA** refers to a part of the CLAPI corpus (*Corpus de LAngue Parlée en Interaction* – Corpus of spoken language in interaction, [27]). **FLE** (a part of the FLEURON corpus, [28]) corresponds to interactions between students and other speakers (such as university staff, professors ... ). **TCO** (TCOF: *Traitement de Corpus Oraux en Français* – processing French oral corpora, [29]) consists of interactions between speakers. **OFR** (OFROM: *Corpus Oral de français de Suisse Romande* – Speech corpus from French-speaking Switzerland, [30,31]) contains data recorded during interactions and interviews. **DEC** (DECODA corpus, [32]) contains anonymized dialogs recorded from calls to the Paris transport authority (RATP) call-center. Finally, **HUS** is a speech corpus containing recordings of working meetings.

All the corpora have been recorded in France, except VAL (recorded in Belgium) and OFR (recorded in Switzerland).

Except for ESTER2, which is not part of the ORFEO project, we have used the automatic speech-text alignments carried on in the ORFEO project. Table 1 reports the number of words in the alignments for each corpus. Globally, for the 13 corpora, more than 5 million word occurrences have been speech-text aligned. Also, Table 1 displays for each selected word its frequency of occurrence in each corpus; this vary from 0.05% for the word "*donc*" in the FRE corpus up to 1.61% for the word "*donc*" in the CLA corpus.

In each corpus, a subset of occurrences of the words "*alors*", "*bon*" and "*donc*" has been selected at random, and manually annotated by listening to a speech segment spanning the considered occurrence (15 words before and 15 words after). The manual annotation consists in indicating whether the occurrences correspond to DP functions or non-DP functions. For DP functions, a finer annotation is made to detail the pragmatic function (e.g., concluding, rephrasing, expressing emotion, (re)introducing etc.). Incorrect data (e.g. too bad speech-text alignment) have been discarded from detailed manual annotations. Table 1 indicates for each word, the number of items annotated (either as DP or as non-DP).

**Table 1.** Counts and statistics for the three studied words, for the various corpora.

| Corpus | | Story | News | Interviews, conversations | | | | | Interactions, ... | | | | | |
|---|---|---|---|---|---|---|---|---|---|---|---|---|---|---|
| | | FRE | EST | CFP | COR | CRF | TUF | VAL | CLA | FLE | TCO | OFR | DEC | HUS |
| Number of words (millions) | | 0.14 | 1.82 | 0.41 | 0.22 | 0.38 | 0.58 | 0.25 | 0.02 | 0.03 | 0.36 | 0.29 | 0.65 | 0.17 |
| Articulation rate (pho./sec.) | | 11.9 | 13.7 | 13.1 | 13.0 | 12.8 | 14.8 | 13.5 | 14.9 | 13.9 | 13.9 | 12.9 | 13.3 | 15.1 |
| "alors" (what's up, then, ...) | Freq. (%) | 0.56 | 0.16 | 0.38 | 0.39 | 0.36 | 0.24 | 0.33 | 0.23 | 0.49 | 0.38 | 0.45 | 0.79 | 0.40 |
| | Nb. annot. | 98 | 172 | 87 | 86 | 91 | 84 | 77 | 35 | 73 | 71 | 91 | 66 | 79 |
| | DP (%) | 24 | 55 | 79 | 77 | 68 | 67 | 71 | 63 | 93 | 75 | 84 | 79 | 89 |
| "bon" (all right, well, ...) | Freq. (%) | 0.11 | 0.06 | 0.37 | 0.31 | 0.52 | 0.48 | 0.38 | 0.49 | 0.30 | 0.45 | 0.23 | 0.38 | 0.45 |
| | Nb. annot. | 88 | 181 | 75 | 89 | 80 | 83 | 79 | 78 | 69 | 75 | 91 | 82 | 66 |
| | DP (%) | 59 | 58 | 87 | 80 | 90 | 75 | 90 | 39 | 61 | 93 | 86 | 84 | 82 |
| "donc" (therefore, well, ...) | Freq. (%) | 0.05 | 0.24 | 0.72 | 0.68 | 0.87 | 0.52 | 0.41 | 0.32 | 1.61 | 0.76 | 0.71 | 0.80 | 0.91 |
| | Nb. annot. | 70 | 191 | 84 | 76 | 90 | 82 | 88 | 60 | 89 | 79 | 95 | 85 | 68 |
| | DP (%) | 67 | 78 | 75 | 83 | 80 | 85 | 89 | 67 | 93 | 91 | 87 | 88 | 90 |

## 3 Discourse Particle or Not

DPs do not contribute to propositional content, but they add some pragmatic function for ongoing discourse and elaborate the meaning of the utterance [33]. The three words studied here have a 'traditional' grammatical or lexical meaning, but can also convey a 'pragmatic' function when used as a DP. Non-DP "alors" is either an adverb of time (Table 2, Ex. 1) or a discourse connective. When "alors" is a DP, it no longer has its traditional meaning or function. As a DP, "alors" (re)introduces a topic, expresses speaker's emotions, attracts the interlocutors' attention, or structures the speech flow, sometimes in correlation with the cognitive process, etc. In Table 2, Ex. 2, it expresses a hesitation, not a consecutive, nor a temporal meaning. In the same way, the basic role of "bon" is an adjective; however, when "bon" is a DP it can be used to connect two discourse units.

An interesting distributional tendency of words used as DPs is observed with respect to the type of corpus. As shown in Table 1, the frequency of DPs in the spontaneous speech (interviews or interactions) is significantly higher than in the prepared speech (storytelling or broadcast news). This is of no surprise if

**Table 2.** Examples of non-DP and DP usages for the word "alors".

| Ex. 1 | ... la question que tout le monde se posait **alors** était les ventes de ces nains de jardin refléteraient elles ... |
|---|---|
| Non-DP | ... the question that everyone was asking **then** was would the sales of these garden dwarves reflect ... |
| Ex. 2 | ... il a dit qu'il avait qu'il avait dix-huit, dix-neuf euh **alors** euh presque dix-neuf ans ... |
| DP | ... he said that he was that he was eighteen, nineteen ah **then** ah almost nineteen ... |

we accept that the main characteristics of DPs are pragmatic/deictic functions, showing speaker's intentions or emotions rather than actually conveying a lexical or a grammatical meaning.

The word "*alors*" is, originally, an anaphoric adverb of time ('then' or 'at that time') as well as a discourse connective. As it can be seen in Table 1, most of the "*alors*" in storytelling are non-DPs (only 24% are DPs). The narrative nature of these corpora can explain this distribution: "*alors*" is one of the favorite markers to make narration progress. However, more than 50% of the "*alors*" are DPs in broadcast news, and the percentage increases in interviews (around 70%) and gets even larger for interactions (up to 93% for FLE). The highest number of DPs "*alors*" are observed in the FLE, OFR, and HUS corpora. This can be explained by the fact that these corpora contain a high number of interactions between two or more speakers, and therefore a high number of turn-takings and hesitations.

As, for the word "*bon*", a significantly greater number of DPs are also found in spontaneous speech than in prepared speech. FRE (storytelling) and EST (broadcast news) have rather low rates of DPs (59% and 58%) compared to the other corpora that have over 80% of DP rates.

The word "*donc*" exhibits less difference between the various types of speech (spontaneous and prepared), though it has a slightly higher number of DPs in the spontaneous speech. Moreover, "*donc*" is the most frequent DP observed in prepared speech among the three DPs studied in this paper.

## 4 Discourse Particle Function

The DPs have been further annotated with respect to their most frequent and prominent pragmatic meanings, based on specific studies and on our annotation experience.

Six pragmatic functions were identified for "*alors*" (hesitation, introduction, re-introduction, conclusion, interaction, addition); six pragmatic functions for "*bon*" (conclusion, transition-confirmation, transition-dialogue, transition-incision, interruption, emotion); and five pragmatic functions for "*donc*" (re-introduction, introduction, conclusion, interaction, addition). Some example of DP pragmatic functions for the word "*alors*" are displayed in Table 3. Each DP has also a 'complex' pragmatic function when the word occurs along with one or more other DPs. This complex function is necessary as the meaning of DPs occurring in such contexts is different from the one they have when they occur alone (e.g. "*bon bah*" (well), "*mais bon*" (but OK), "*enfin bon*" (anyway), "*bon alors*" (well then), "*donc voilà*" (here we are), etc.)

The frequency of usage of the various DP pragmatic functions has been studied, and only the usage frequencies for the most frequent pragmatic functions are reported in Table 4, along with the number of word occurrences that were labelled 'DP' in each set of data. As it can be observed, the pragmatic functions of DPs depend on the type of corpus, whether it is prepared speech, interview, or interactions between speakers.

**Table 3.** Examples of DP pragmatic functions for the word *"alors"*.

| DP-introduction | *... la les forces régulières les forces loyalistes vont mettre le paquet sur bouaké [pause] **alors** la question qui qui se pose à la mi journée c'est de savoir qui ...*<br><br>... the regular forces the loyalist forces will provide full backing on bouaké [pause] **then** the question arising at midday is to know ... |
|---|---|
| DP-conclusion | *... en achetant tout simplement des produits vous savez étiquetés satisfait ou remboursé **alors** c'est une gestion mais ça marche il l'a prouvé il a rempli son frigo ...*<br><br>... by simply buying products you know labeled satisfied or refunded **then** it is a management but it works he proved it he has filled its fridge ... |
| DP-interaction | [Speaker1] *... et vous pensez l'avoir perdu où madame?/*[Speaker2] ***alors** euh j'ai deux endroits possibles alors je sais que je l'ai passé au à le au métro ...*<br><br>[speaker1] ... and you think you have lost it where madam?/[speaker2] **so** uh there are two possible places then I know I used it at at station ... |

**Table 4.** Statistics for the main discursive functions.

| Word | DP function | Story | News | Interviews | Interactions | *total* |
|---|---|---|---|---|---|---|
| *"alors"* | Nb. times DP | 23 | 95 | 308 | 341 | 767 |
| | Conclusion | 4% | 7% | 12% | 27% | 18% |
| | Hesitation | 4% | 20% | 12% | 6% | 10% |
| | Introduction | 4% | 71% | 33% | 26% | 34% |
| | Reintroduction | 35% | 0% | 22% | 24% | 21% |
| *"bon"* | Nb. times DP | 52 | 104 | 341 | 343 | 840 |
| | Complex | 33% | 11% | 35% | 43% | 35% |
| | Trans.-confirm | 27% | 26% | 19% | 17% | 20% |
| | Trans.-incision | 10% | 22% | 13% | 10% | 13% |
| *"donc"* | Nb. times DP | 47 | 149 | 346 | 414 | 956 |
| | Addition | 13% | 20% | 23% | 26% | 23% |
| | Conclusion | 19% | 36% | 31% | 28% | 30% |
| | Reintroduction | 21% | 26% | 34% | 33% | 32% |

The DP *"alors"* in spontaneous speech corpora show more variety with respect to their pragmatic functions, compared to its usage in storytelling. The highest usage percentage of the DP *"alors"* in storytelling has the 'reintroduction' function, and in prepared speech the 'introduction' function.

A significant number of complex DPs are found for *"bon"*, especially in spontaneous speech. This shows that *"bon"* is very often combined with other DPs, and in that case the meaning is not necessarily compositional. Less of 'complex'

DPs "*bon*" are found in prepared speech (broadcast news) compared to all the other corpora. A more formal – and less emotional – language used in broadcast news can explain this fact. Further studies are needed for complex DPs with a finer-grained analysis of their actual meanings or functions. Moreover, there are three subcategories of the pragmatic 'transition' function for the DP "*bon*": confirmation, when the speaker agrees with his interlocutor; dialogue, for simple transition between two speakers; and incision, when the speaker wants to add more information or details. The amount of transition-confirmation functions of DPs is reduced when the spontaneity degree increases.

As, for the DP "*donc*", the functions 'addition', 'conclusion' and 'reintroduction' have very similar frequency in our data.

## 5   Analysis of a Few Prosodic Correlates

In [34], prosodic correlates of a few words that can be used as discourse particles have been analyzed, but using data mainly from prepared speech. Here, as mentioned in Sect. 2, we consider a much larger set of speech corpora spanning various speaking styles (from storytelling to highly spontaneous speech). We report and discuss here statistics on a few prosodic correlates.

The prosodic annotation has been carried on automatically. The presence (or not) of a pause before or after the word results from the analysis of the force speech-text alignments. The segmentation of the speech stream into intonation groups is obtained with the ProsoTree software [35], which relies on F0 slope inversions as described in [36], and locates intonation group boundaries using information based on F0 slope values, pitch level and vowel duration.

**Pauses Before the Word.** Table 5 displays the percentage of occurrences of pauses before the considered word, when used as DP or as non-DP. For the word "*alors*", there is no difference in pause occurrences between its DP and non-DP functions in storytelling style. However, in the three remaining styles there are significantly fewer pauses in non-DP than in DP functions. As far as the word "*bon*" is concerned, pauses in non-DP functions are very few in storytelling style while their number remains significantly lower than in DP functions in the other styles too. The word "*donc*" has approximately the same frequency of pause occurrences in DP and non-DP functions with, however, a slightly higher number of pauses in interaction data when DP.

**Pauses After the Word.** With respect to the occurrences of pauses after the word (Table 6), in general, there are more pauses occurring after the DP functions of the studied words than after the non-DP functions. When a pause occurs after the word "*bon*", there are substantial differences in storytelling (high number of pause after DPs) while in the other styles the number of pauses is either very similar between DPs and non-DPs or only slightly higher in DPs. For the word "*alors*" the highest differences are found in "interview" and "interaction" styles (higher number of pauses after DPs). The "interaction" style is also the one where the greatest difference is found for the word "*donc*" (also more pauses after DPs).

**Table 5.** Occurrences of pauses before the word.

| Word | DP/non-DP | Story | Prepared | Interviews | Interactions |
|------|-----------|-------|----------|------------|--------------|
| "alors" | DP | 82% | 79% | 63% | 62% |
| | Non-DP | 82% | 51% | 42% | 38% |
| "bon" | DP | 42% | 54% | 34% | 42% |
| | Non-DP | 3% | 10% | 7% | 14% |
| "donc" | DP | 34% | 31% | 52% | 59% |
| | Non-DP | 45% | 32% | 51% | 38% |

**Table 6.** Occurrences of pauses after the word.

| Word | DP/non-DP | Story | Prepared | Interviews | Interactions |
|------|-----------|-------|----------|------------|--------------|
| "alors" | DP | 18% | 17% | 26% | 25% |
| | Non-DP | 12% | 20% | 9% | 13% |
| "bon" | DP | 49% | 36% | 34% | 30% |
| | Non-DP | 3% | 21% | 22% | 31% |
| "donc" | DP | 12% | 20% | 25% | 24% |
| | Non-DP | 9% | 17% | 20% | 8% |

**Table 7.** Position of the word in the intonation group.

| Word | | Position | Story | Prepared | Interviews | Interactions |
|------|--|----------|-------|----------|------------|--------------|
| "alors" | DP | Alone | 82% | 65% | 82% | 77% |
| | | First | 18% | 26% | 18% | 20% |
| | Non-DP | Alone | 89% | 61% | 66% | 58% |
| | | First | 11% | 30% | 32% | 39% |
| "bon" | DP | Alone | 67% | 78% | 63% | 66% |
| | | Last | 29% | 9% | 23% | 17% |
| | Non-DP | Alone | 38% | 40% | 41% | 55% |
| | | Last | 50% | 43% | 45% | 42% |
| "donc" | DP | Alone | 83% | 61% | 75% | 76% |
| | | Last | 12% | 18% | 12% | 11% |
| | Non-DP | Alone | 68% | 59% | 74% | 64% |
| | | Last | 18% | 10% | 10% | 18% |

**Position in Intonation Group.** According to Table 7, the studied words occur more often alone in prosodic groups when they are used as DPs than when non-DPs. The highest differences are observed for the word "alors" and "bon" while no substantial difference is observed for the word "donc". On the other hand, in non-DP functions, "alors" occurs more frequently in first position in prosodic groups while *"bon"* is more frequently in last position. In the interview and interaction styles, the word *"alors"* is more frequently alone when DP. The word *"bon"*, when DP, is more frequently alone in all considered styles while, when non-DP, it is more frequent in last position. Finally, for the word *"donc"* there is a noteworthy difference between DP and non-DP functions in storytelling (more DPs alone than when non-DPs) while in the other styles there is either only a slight difference ('interaction') or no significant difference is found.

## 6   Automatic Classification and Detection

In the reported experiments, prosodic correlates are used to automatically classify word occurrences as DP or non-DP, and a neural network (NN) approach is used. For each of the three words, experiments are conducted using the Keras toolkit [37]. 60% of the data are used for training the NN parameters, 10% for validation, and the remaining 30% are used for evaluating performance.

First experiments are conducted using prosodic features computed over the considered words and its neighbors (a few words before and after). The prosodic features include absolute and normalized values of the duration and energy of the last vowel of the words, F0 values at the end of the words and their slopes, the presence and the duration of pauses, . . . The best classification results with these prosodic parameters are obtained by taking into account features associated to sequences of five to nine words centered over the considered word. As reported in Table 8, this leads to a correct classification rates ranging from 69% (for *"alors"*) to 82% (for *"bon"*). With respect to DP detection, the F1-measure ranges from 78% (for *"alors"*) to 88% (for *"bon"*).

Another set of experiments have been conducted by considering only the F0 values (computed with the RAPT [38] approach of the SPTK toolkit [39]) over a time window centered over the considered word. Best results are obtained by considering a 3 to 5 second window. Classification and detection results are reported in Table 9. The correct classification rate ranges from 64% (for *"alors"*) to 73% (for *"donc"*). With respect to DP detection, the F1-measure ranges from 75% (for *"alors"*) to 84% (for *"donc"*).

The results obtained with the F0 curve are almost as good as those achieved with the prosodic parameters (which include more information, as for example the durations of the last vowel of the words, pauses, . . . ). Further classification experiments will consider combining these two sets of features.

**Table 8.** Automatic classification and detection results using prosodic features.

|  | Classification correct | DP detection | | |
|---|---|---|---|---|
|  |  | Recall | Precision | F1-measure |
| *"alors"* | 69% | 81% | 75% | 78% |
| *"bon"* | 82% | 90% | 86% | 88% |
| *"donc"* | 71% | 79% | 84% | 81% |

**Table 9.** Automatic classification and detection results using fundamental frequency values.

| | Classification correct | DP detection | | |
|---|---|---|---|---|
| | | Recall | Precision | F1-measure |
| *"alors"* | 64% | 79% | 71% | 75% |
| *"bon"* | 69% | 84% | 76% | 80% |
| *"donc"* | 73% | 87% | 81% | 84% |

# 7  Conclusion

In this paper, we have analyzed and discussed the distribution of discourse functions for three very frequent French words that can be used as discourse particles (DP) or not discourse particles (non-DP), over a large set of speech corpora. These corpora exhibit different speaking styles ranging from storytelling to highly spontaneous speech corresponding to oral interactions between speakers, and including intermediate styles such as prepared speech (from broadcast news) and spontaneous speech from interviews, dialogues and conversations.

For the three words (*"alors"*, *"bon"* and *"donc"*) considered in this study, a noticeable increase of their usage as a DP is observed from (1) storytelling (lowest percentage of DP usage), to (2) prepared speech, then to (3) interviews and conversations, and finally to (4) highly spontaneous speech observed in oral interactions between speakers (highest percentage of DP usage). A detailed study of the DP pragmatic function also show that the pragmatic usage vary across the corpora, and seems dependent on the spontaneity degree of the data.

Prosodic correlates of the words vary whether they are used as DPs or non-DPs. Moreover, in many cases, the distribution of the prosodic correlates also varies with respect to the spontaneity degree of the speech data. Automatic classification tests show that prosodic parameters (over a few words window) as well as the F0 curve (over a few second windows) carry significant information with respect to DP vs. non-DP function.

**Acknowledgments.** This work has been carried out in the framework of the Prosod-Corpus operation supported by the CPER LCHN (*Contrat Plan Etat Région "Langues, Connaissances et Humanités Numériques"*). Some experiments presented in this paper have been carried out using the Grid'5000 testbed, supported by a scientific interest group hosted by Inria and including CNRS, RENATER and several Universities as well as other organizations (see https://www.grid5000.fr).

# References

1. Aijmer, K.: Understanding Pragmatic Markers. A Variational Pragmatic Approach. Edinburgh UP, Edinburgh (2006)
2. Bartkova, K., Bastien, A., Dargnat, M.: How to be a discourse particle? In: Speech Prosody 2016, Boston, USA, pp. 859–863 (2016)

3. Degand, L., Fagard, B.: Alors between discourse and grammar: the role of syntactic position. Funct. Lang. **18**, 19–56 (2011)
4. Hansen, M.B.M.: Particles at the Semantics-Pragmatics Interface: Synchronic and Diachronic Issues. Elsevier, Amsterdam (2008)
5. Wichmann, A., Simon-Vandenbergen, A.-A., Aijmer, K.: How prosody reflects semantic change: a synchronic case study of of course. In: Davidse, K., Vandelanotte, L., Cuyckens, H. (eds.) Subjectification, Intersubjectification and Grammaticalization, pp. 103–154. Mouton de Gruyter, Berlin (2010)
6. Brinton, L.J.: Pragmatic Markers in English. Grammaticalization and Discourse Functions. De Gruyter, Berlin (1996)
7. Degand, L., Cornillie, B., Pietrandrea, P. (eds.): Discourse Markers and Modal Particles: Categorization and Description. John Benjamins, Amsterdam (2013)
8. Dostie, G.: Pragmaticalisation et marqueurs discursifs. De Boeck/Duculot, Liège (2004)
9. Hansen, M.B.M.: The Function of Discourse Particles. Benjamins, Amsterdam (1998)
10. Ducrot, O.: Le Dire et le dit. Editions de Minuit, Paris (1984)
11. Kleiber, G.: Sémiotique de l'interjection. Langue française **161**, 10–23 (2006)
12. Sperber, D., Wilson, D.: Relevance: Communication and Cognition. Blackwell, Oxford (1986)
13. Blakemore, D.: Semantic Constraints on Relevance. Blackwell, Oxford (1987)
14. Denturck, E.: Ètude des marqueurs discursifs - L'exemple de "quoi". Master Diss., Gent University (2008)
15. Fernandez-Vest, J.: Les particules énonciatives dans la construction du discours. Presses Universitaires de France, Paris (1994)
16. Galliano, S., Gravier, G., Chaubard, L.: The ESTER 2 evaluation campaign for rich transcription of French broadcasts. In: INTERSPEECH 2009, 10th Annual Conference of the International Speech Communication Association, Brighton, UK, pp. 2583–2586 (2009)
17. ORFEO project: http://www.projet-orfeo.fr/
18. French oral narrative: http://frenchoralnarrative.qub.ac.uk
19. CFPP2000: http://cfpp2000.univ-paris3.fr/
20. Branca-Rosoff, S., Fleury, S., Lefeuvre, F., Pires, M.: Discours sur la ville. Présentation du Corpus de Français Parlé Parisien des années 2000 (CFPP 2000)
21. C-ORAL-ROM: http://lablita.dit.unifi.it/corpora/descriptions/coralrom/
22. Cresti, E., do Nascimento, F. B., Moreno-Sandoval, A., Veronis, J., Martin, P., Choukri, K.: The C-ORAL-ROM CORPUS. A multilingual resource of spontaneous speech for romance languages. In: LREC 2004, 4th International Conference on Language Resources and Evaluation, Lisbon, Portugal (2004)
23. CRFP: http://www.up.univ-mrs.fr/delic/corpus/index.html
24. Delic team: Autour du Corpus de référence du français parlé. Recherches sur le français parlé, no. 18, Publications de l'université de Provence, 265 p. (2004)
25. TUFS:      http://www.tufs.ac.jp/ts/personal/ykawa/art/2014_Waseda_Corpus_TUFS.pdf
26. Valibel: http://www.uclouvain.be/81834.html
27. CLAPI: http://clapi.ish-lyon.cnrs.fr/
28. FLEURON: https://apps.atilf.fr/fleuron2/
29. TCOF: http://www.cnrtl.fr/corpus/tcof/
30. OFROM: http://www.unine.ch/ofrom

31. Avanzi, M., Béguelin, M.-J., Diémoz, F.: Présentation du corpus OFROM - corpus oral de français de Suisse romande. Université de Neuchâtel, Switzerland (2012–2015)
32. Bechet, F., Maza, B., Bigouroux, N., Bazillon, T., El-Beze, M., De Mori, R., Arbillot, E.: DECODA: a call-centre human-human spoken conversation corpus. In: LREC 2012, 8th International Conference on Language Resources and Evaluation, Istanbul, Turkey (2012)
33. Stede, M., Schmitz, B.: Discourse particles and discourse functions. Mach. Transl. 15(1–2), 125–147 (2000)
34. Dargnat, M., Bartkova, K., Jouvet, D.: Discourse particles in French: prosodic parameters extraction and analysis. In: SLSP 2015, International Conference on Statistical Language and Speech Processing, Budapest, Hungary (2015)
35. Bartkova, K., Jouvet, D.: Automatic detection of the prosodic structures of speech utterances. In: Železný, M., Habernal, I., Ronzhin, A. (eds.) SPECOM 2013. LNCS, vol. 8113, pp. 1–8. Springer, Cham (2013). doi:10.1007/978-3-319-01931-4_1
36. Martin, P.: Prosodic and rhythmic structures in French. Linguistics 25, 925–949 (1987)
37. Keras: https://keras.io/
38. Talkin, D.: A robust algorithm for pitch tracking (RAPT). In: Kleijn, W.B., Paliwal, K.K. (eds.) Speech Coding and Synthesis, pp. 495–518. Elsevier, Amsterdam (1995)
39. SPTK: http://sp-tk.sourceforge.net/

# Learning Morphology of Natural Language as a Finite-State Grammar

Javad Nouri and Roman Yangarber[✉]

Department of Computer Science, University of Helsinki, Helsinki, Finland
{javad.nouri,roman.yangarber}@cs.helsinki.fi

**Abstract.** We present algorithms that learn to segment words in morphologically rich languages, in an unsupervised fashion. Morphology of many languages can be modeled by finite state machines (FSMs). We start with a baseline MDL-based learning algorithm. We then formulate well-motivated and general linguistic principles about morphology, and incorporate them into the algorithm as heuristics, to constrain the search space. We evaluate the algorithm on two highly-inflecting languages. Evaluation of segmentation shows gains in performance compared to the state of the art. We conclude with a discussion about how the learned model relates to a morphological FSM, which is the ultimate goal.

**Keywords:** Unsupervised morphology induction · Minimum description length principle · MDL · Finite-state automata

## 1 Introduction

We present work on unsupervised *segmentation* for languages with complex morphology. Our ultimate research question is to explore to what extent morphological structure can be induced without supervision, from a large body of unannotated text (a corpus). This has implications for the question whether the morphological system is somehow *"inherently encoded"* in language, as represented by the corpus.

The paper is organized as follows: we state the morphology induction problem (Sect. 2), review prior work (Sect. 3), present our models (Sects. 4 and 5), present the evaluation and experiments (Sect. 6), and finally discuss future work.

## 2 Morphological Description

We focus on highly-inflecting languages. In the experiments, we use Finnish and Turkish, which belong to different language families (Uralic and Turkic) and exhibit different morphological phenomena.

Finnish is considered to be agglutinative, although it has some fusional phenomena—morpho-phonological alternation. It has complex derivation and inflection, and productive compounding. Derivation and inflection are achieved

© Springer International Publishing AG 2017
N. Camelin et al. (Eds.): SLSP 2017, LNAI 10583, pp. 44–57, 2017.
DOI: 10.1007/978-3-319-68456-7_4

via suffixation; prefixes do not exist. Multiple stems in the compound may be inflected, e.g., *kunnossapidon* = "of keeping (smth.) in usable condition":

| *kunno* + | *ssa* # | *pido* + | *n* |
|-----------|---------|----------|-----|
| condition + | Iness. # | keeping + | Gen. |

where # is a compound boundary, + is a morpheme boundary, *Iness.* marks *inessive* case (presence in a location, or being in a state), and *Gen.* marks the *genitive*. The morph *kunno* is a "weak" allomorph of the stem *kunto*; the weakening is conditioned by the closed syllable environment, i.e., the following double consonant -*ss*-.

Turkish is similar to Finnish: agglutination, no prefixation, minimal compounding.

The ultimate goal in the future is to model aspects of morphology, including classification of morphemes into morphological categories, and capturing *allomorphy*—morpho-phonological alternation.[1] However, at present, as in most prior work, we address the problem of *segmentation* only, to try to establish a solid baseline. Once we have a good way of segmenting words into morphs, we plan to move to modeling the more complex morphological phenomena.[2]

## 3  Prior Work

Interest in unsupervised morphology induction has surged since 2000. Detailed surveys are found, e.g., in [15,19], and in proceedings from a series of "Morpho-Challenge" workshops between 2005 and 2010 [17,18,20]. Our approach is founded on the Minimum Description Length principle (MDL) as a measure of model quality, see e.g., [14]. Linguistica, [11], also uses MDL, combined with the idea of a *signature*—set of affixes that belong to a morphological paradigm; e.g., suffixes for a certain class of nouns form one signature, etc. Our models also group morphs into different morphological *classes*, though using a different approach (Sect. 4).

Finite state machines (FSMs) are used in [12], which are similar to our FSMs; however, the approach there is less general, since it uses heuristics unsuitable for languages with very rich morphology.

The MDL-based Morfessor and its variants, e.g., [3,5], are closely related to our work. Unlike [4], we do not posit morphological classes *a priori*, but allow the model to learn the classes and distributions of morphs automatically. Word embedding, modeling semantic relations between words, have been used with orthographic features of words to learn the morphology of languages as "morphological chains" in [21].

---

[1] In the example above, the morph *kunno*- is an allomorph of *kunto*; which allomorph appears in a given word is determined by its phonological environment.

[2] Note, we do not claim that the problem must be sub-divided this way. As others before us, we begin with segmentation because it is more tractable.

Evaluation of morphological learning is a complex challenge in itself [23, 26]; prior work on this topic is described in detail in Sect. 6.

The Morpho-Challenges have seen attempts to model aspects of morphology which we do not address: using analogical reasoning, handling *ablaut*-type morphology (as in German, and Semitic languages), etc. Beyond segmentation, modeling allomorphy has been attempted, e.g., by [24], but the performance of the proposed algorithms on segmentation so far falls short of those that do not model allomorphy. Research on induction of morphological structure is driven in part by the observation that children learn it at a very early age,[3] which makes acquisition by machine a fascinating challenge.

## 4   The Learning Algorithm: SMORPH

Morphological systems for many languages are modeled by a finite-state machine (FSM), where each state corresponds to a morphological *class*. The seminal approach of *Two-Level Morphology*, [16], represents morphological grammar by a FSM. We can associate a state with, e.g., a set of verb stems that belong to a certain inflectional paradigm; or the set of suffixes in a certain noun paradigm, etc. The edges in the FSM define the *morphotactics* of the language—the permissible ordering among the states.

The data $\mathcal{D}$ (the *corpus*) is a large list of words in the given language. For every word $w \in \mathcal{D}$, SMORPH tries to learn the most probable sequence of states that generates $w$: it treats the problem as finding a model that produces the most probable **segmentation** for each word $w$ into "morphs"—fragments of $w$—and the **classification**—assignment of the morphs to classes (classes are identified with the states).[4]

The learning algorithm searches for the best model in a certain model class. Thus, the full description of the algorithm must specify a. the *model class* (Sect. 4.1), b. the *objective function*: a way of assigning a score—the cost—to each model (Sect. 4.3), and c. a *search strategy* for optimizing the objective across the model class (Sect. 4.2).

### 4.1   Morphology Models

We begin with a hidden Markov model (HMM), with a set of hidden *states/classes* $\{C_i\}$. States emit morphs with certain *emission* probabilities, and *transition* probabilities between states. To code each word $w$, the model starts at the initial state $C_0$. From state $C_i$, it can transition to another state $C_j$ with a certain probability $P_{tr}(C_j|C_i)$ and emit a morph—a segment of $w$ from $C_j$. The probability of emitting a morph $\mu$ from state $C_j$ is denoted by $P_{em}(\mu|C_j)$. Once the entire $w$ is emitted, the model transitions to the final state $C_F$ and emits a special end-of-word symbol #.

---

[3] This relates to the *poverty of stimulus* claim, [2], about human ability to learn complex systems from very limited data.

[4] Note that we evaluate only the *segmentation*, (Sect. 6).

Ideally, the states will correspond to "true" morphological classes; for example, nouns may fall into different declension paradigms and each paradigm would be assigned its own class/state in the model.

Probabilities $P_{tr}$ and $P_{em}$ are determined by counting how words are segmented into morphs, and which states emit which morphs. Many segmentations are possible for the given corpus. We need a way to choose the best segmentation for $\mathcal{D}$—the best parameters for the model. To approach this model selection problem via MDL we define a *code-length* for the data $\mathcal{D}$, which is the number of bits required to encode it.

## 4.2   Search

---

**Algorithm 1.** Baseline search algorithm, using Expectation-Maximization [10]

---

**Input**: Data: a large list $\mathcal{D}$ of words in the language
1  Initialize: create a **random** segmentation and classification—split all words in $\mathcal{D}$ into morphs randomly, and assign morphs to classes randomly;
2  **repeat**
3      *Compute Parameters*: (E-step) based on the current segmentation and classification, compute all emission and transition costs;
4      *Re-segment*: (M-step) given the newly computed parameter values find the best segmentation for all words in the corpus (using Dynamic Programming, Sect. 4.6)
5  **until** *convergence in cost*;

---

Convergence is determined by the MDL code-length of the complete model (*cost*), defined in Sect. 4.3. We fix the number of classes $K$,[5] and begin with a random segmentation and classification (assignment of morphs to classes). We then greedily re-segment each word $w \in \mathcal{D}$, minimizing the MDL code-length of the model plus the data.

## 4.3   The Objective: Two-Part MDL Cost

Finding the best segmentation and classification can be viewed as the problem of *compressing* $\mathcal{D}$. In MDL 2-part coding, we try to minimize the cost of the *complete* data: the cost of the model plus the cost of the data given the model. In our case, this means summing the costs of coding the Lexicon, the transitions and the emissions:

$$L(\mathcal{D}) = L(M) + L(\mathcal{D}|M) = L(Lex) + L(Tr) + L(Em)$$

The cost of the **Lexicon**: $L(Lex)$, is the number of bits needed to encode each class, morph by morph, *irrespective of the order* of the morphs in the class:

$$L(Lex) = \sum_{i=1}^{K} \left[ \sum_{\mu \in C_i} L(\mu) - \log |C_i|! \right] \tag{1}$$

---

[5] Ideally, $K$ should be large, to give the model sufficient expressiveness.

where $K$ is the number of classes, $\mu$ ranges over all morphs in class $C_i$, $L(\mu)$ is the code length of a morph $\mu$ (Eq. 2), and $|C_i|$ is the number of morphs in class $C_i$. The term $-\log|C_i|!$ accounts for the fact $C_i$ is a set, and we do not need to code the morphs in $C_i$ in any particular order.

The code length $L(\mu)$ of morph $\mu$ is computed similarly to [5], as the number of bits needed to encode $\mu$:

$$L(\mu) = (|\mu| + 1) \cdot \log(|\Sigma| + 1) \tag{2}$$

where $|\mu|$ is the number of symbols in $\mu$, and $|\Sigma|$ is the size of the alphabet; one is added to $|\Sigma|$ to account for one special *morph-boundary* symbol.[6]

**Transitions:** Given the lexicon, we code the paths of class transitions from $C_0$ to $C_F$, from word start to word finish, using Bayesian Marginal Likelihood (ML), as introduced in [9]. In ML, we treat each transition $(C_iC_j)$ in the data as an "event" to be coded. If in a set of events $\mathcal{E} = \{E_j\}$, each $E_j$ has a corresponding "prior" count $\alpha_j$ and count of observed occurrences $O_j$, the cost of coding $\mathcal{E}$ is:

$$L(\mathcal{E}) = -\sum_j \log \Gamma(O_j + \alpha_j) + \sum_j \log \Gamma(\alpha_j) + \log \Gamma \sum_j (O_j + \alpha_j) - \log \Gamma \sum_j \alpha_j \tag{3}$$

where the summations range over the set $\mathcal{E}$.[7] We use uniform priors, $\alpha_j = 1, \forall j$. Thus $\log \Gamma(\alpha_j) = 0$, and the second term is always zero in this equation.

To compute the cost of all transitions $L(Tr)$, we apply Eq. 3 to each class $C_i$, as $i$ ranges from 0 to $K$; the set of "events" $\mathcal{E}$ is the set of all classes $C_j$, which are the targets of transitions from $C_i$:

$$\sum_{i=0}^{K} \left[ -\sum_{j=1}^{K+1} \log \Gamma\Big(f(C_iC_j) + 1\Big) + \log \Gamma \left( \sum_{j=1}^{K+1} \Big[f(C_iC_j) + 1\Big] \right) - \log \Gamma(K) \right]$$

**Emissions:** We code the emissions $L(Em)$ analogously:

$$\sum_{i=1}^{K} \left[ -\sum_{\mu \in C_i} \log \Gamma\Big(f(\mu, C_i) + 1\Big) + \log \Gamma \left( \sum_{\mu \in C_i} \Big[f(\mu, C_i) + 1\Big] \right) - \log \Gamma(|C_i|) \right]$$

where $f(C_i, C_j)$ is the count of transitions from $C_i$ to $C_j$, and $f(\mu, C)$ is the number of times morph $\mu$ was emitted from class $C$. The cost $L(Em)$ is computed by applying Eq. 3, where the set of "events" $\mathcal{E}$ is now the set of emissions $(\mu, C_i)$ of all morphs $\mu$ from $C_i$.

---

[6] Another way to code $L(\mu)$ is to account for the *frequencies* of the symbols of the alphabet.

[7] For additional explanation about this coding scheme please refer to the original paper.

## 4.4   Input

For each language, we pre-process a large text by collecting *distinct* words from the text. We do not model the *text*—i.e., the distribution of words in the text—but the *language*, i.e., the observed distinct words, as is done in recent prior work. In this paper, *corpus* refers to the list of *distinct* words.[8]

## 4.5   Initialization

The initial segmentation and classification is obtained by randomly placing morph boundaries in each input word $w$, independently, according to a Bernoulli distribution,[9] and by randomly assigning the morphs to one of the classes in the Lexicon.

## 4.6   Dynamic Programming Re-segmentation

We compute the most probable segmentation of every word $w$ in the corpus into morphs, at iteration $t$, given a set of transitions and emissions from iteration $t - 1$.

We apply a Viterbi-like dynamic programming (DP) search for every word $w$, to compute the most probable path through the HMM, given $w$, *without* using the segmentation of $w$ at iteration $t - 1$. Standard Viterbi would only give us the best class assignment *given* a segmentation. Here, the search algorithm fills in the DP matrix starting from the leftmost column toward the right.

**Notation:** $\sigma_a^b$ is a *substring* of $w$, from position $a$ to position $b$, inclusive. We number positions starting from 1. The shorthand $\sigma^b \equiv \sigma_1^b$ is a prefix of $w$ up to $b$, and $\sigma_a \equiv \sigma_a^n$ is a suffix, when $|w| = n$. A single **morph** $\mu_a^b$ lies between positions $a$ and $b$ in $w$. Note that $\sigma_a^b$ is just a sub-string, and may contain several morphs, or cut across morph boundaries. In the cell $(i, j)$, marked $X$, in the DP matrix, we compute the cost of the HMM being in state $C_i$ and having emitted the prefix up to the $j$-th symbol of $w$, $L(C_i|\sigma^j)$. This cost is computed as the minimum over the following expressions, using values computed previously and already available in columns to the left of $\sigma^j$:

$$L(C_i|\sigma^j) = \min_{q,b} \left( L(C_q|\sigma^b) + L(C_i|C_q) + L(\mu_{b+1}^j|C_i) \right) \qquad (4)$$

This says that the best way of getting to state $C_i$ and emitting $w$ up to $\sigma^j$ is to come from some state $C_q$ having emitted $w$ up to $\sigma^b$, then jump from $C_q$ to $C_i$,

---

[8] This is different from some of the earlier work, e.g., [3,5], and agrees with [25], who observe "that training on word types...give similar scores, while...training on word tokens, is significantly worse." We do not claim that there is no useful information in the distribution of words for learning morphology; however, the current models do not utilize it.

[9] With parameter $\rho$. In the current experiments we used $\rho = 0.20$ and 0.25. This is similar to the approach in [6], where they used a Poisson distribution with a fixed parameter.

and emit $\mu_{b+1}^j$ as a *single morph* from $C_i$. $L(C_i|C_q)$ is the cost of transition from state $C_q$ to $C_i$. This can be calculated using the Bayesian Marginal Likelihood code-length formula as:

$$L(C_i|C_q) = \Delta L = L(Tr \cup t) - L(Tr) = -\log \frac{f(C_q C_i) + 1}{\sum_{k=0}^{K-1} \sum_{j=1}^{K} \left( f(C_k C_j) + 1 \right)}$$

where $t$ is the transition $C_q \to C_i$. The cost of emitting morph $\mu$ from class $C_i$ is:

$$L(\mu|C_i) = \begin{cases} -\log \frac{f(\mu, C_i) + 1}{f(C_i) + |C_i|} & \text{if } \mu \in C_i \\ -\log \frac{|C_i|}{\left( f(C_i) + |C_i| \right) \left( f(C_i) + |C_i| + 1 \right)} + L(\mu) & \text{if } \mu \notin C_i \end{cases}$$

The second case is for when $\mu$ is not emitted from $C_i$ yet and does not exist in its lexicon and $L(\mu)$ is the cost of adding $\mu$ to the lexicon.

In Eq. 4, the minimum is taken over all states $q = 0, 1, \ldots, K$, including the initial state $C_0$, and over all columns $b$ that precede column $j$: $b = j - 1, \ldots, 0$. Here $L(\mu_{b+1}^j|C_i)$ is the cost of emitting the $\mu_{b+1}^j$ from $C_i$, for some $b < j$. For the empty string, $\sigma^0 \equiv \epsilon$, we set $L(C_0|\sigma^0) \equiv 0$ for the initial state, and $L(C_q|\sigma^0) \equiv \infty$ for $q \neq 0$.

The transition to the final state $C_F$ is computed in the rightmost column of the matrix, marked #, using the transition from the last morph-emitting state—in column $\sigma^n$—to $C_F$. (State $C_F$ always emits the word boundary #). Thus, the cost of the best path to generate $w$ is:

$$L(w) = \min_{q=1,\ldots,K} L(C_q|\sigma^n) + L(C_F|C_q) + L(\#|C_F)$$

where the last factor $L(\#|C_F)$ is always 0. In addition to storing $L(C_i|\sigma^j)$ in cell $(i, j)$ of the matrix, we store also the "best" (least expensive) state $q$ and the column $b$ from which we arrived at this cell. These values, the previous row and column, allow us to backtrack through the matrix at the end, to reconstruct the lowest-cost—most probable—path through the HMM.

## 5    Enhancements to Baseline Model

We next present enhancements to the baseline algorithm described in Sect. 4, which yield improvements in performance.

**Simulated Annealing:** The greedy search for the best segmentation for all words in the corpus quickly converges to local—far from global—optima. To avoid local optima, we use simulated annealing, with temperature $T$ varying between fixed starting and ending values, $T_0$ and $T_F$, and a geometric cooling schedule, $\alpha$. In Eq. 4, rather than using min to determine the *best* cell $(q, b)$ from which to jump to $X$ in the DP matrix, we select a candidate cell at random, depending on its cost, from a gradually narrowing window.[10]

---

[10] We use a standard approach to simulated annealing, for details please see [1].

This ensures that the model does not always greedily choose the best solution, and enables it to initially make random jumps to avoid local optima.

Next, as mentioned in the abstract, we introduce heuristics that constrain the search, based on simple yet general linguistic principles.

**1. Directionality of the FSM:** The FSM must be *directional*: the morphotactics of any language specify exactly the order in which morphological classes may follow one another. In many Indo-European languages, e.g., a word can have some prefixes, then a stem, then suffixes—always in a fixed order. Further, different kinds of suffixes have strict ordering among them— e.g., derivation precedes inflection.[11]

To enforce directionality, in Sect. 4.6, we constrain the DP matrix so that the preceding state $q$ in Eq. 4 ranges only from 0 up to $i - 1$, rather than up to $K$. Since the states are ordered, this blocks transitions from a later state to an earlier one.

**2. Natural Classes of Morphs:** As a general principle, morph classes fall into two principal kinds: stems vs. affixes. We arbitrarily fix some range of states in the beginning to be *prefix* states, followed by a range of *stem* states, followed by *suffix* states.[12] A simple heuristic based on this is that the HMM must pass through at least one stem state during the DP search.

**3. Bulk Re-segmentation:** An important linguistic principle is that stems and affixes have very *different properties*. First, stem classes are usually *open*—i.e., potentially very large, whereas all affix classes are necessarily *closed*—very limited. This is reflected, e.g., in borrowing: one language may borrow any number of stems from another freely, but it is extremely unlikely to borrow a suffix.

Second, in general a randomly chosen affix is typically expected to occur *much* more frequently in the corpus than a random stem.[13]

Based on this principle, we introduce another heuristic to guide the search: after the normal re-segmentation step, we check all classes for "bad" morphs that violate this principle: very frequent morphs in stem classes, and very rare morphs in affix classes.[14] With a certain probability $\pi(T)$ which depends only on the simulated annealing temperature,[15] all words that contain a bad morph

---

[11] A problem for the directionality heuristic is compounding, where, e.g., in Finnish, the FSM can jump back to a stem class, even after some suffix classes, as seen in the examples in Sect. 2. We will model compounding in the future via a special "restart" state in the grammar. Despite this, even for Finnish, with its heavy compounding, the directional models still perform better than non-directional ones.

[12] Here we divide 15 available states as: 2 classes for prefixes, 6 for stems, and 7 for suffixes.

[13] This is true in general (though a language may have some exceptionally rarely used affix, which might happen to be less frequent than a very frequent stem).

[14] We model this via two hyper-parameters: $s_{max}$ for maximum tolerated count of a stem, and $a_{min}$ for minimum frequency of an affix. In the experiments, we set both to 100.

[15] For example, $\pi(T)$ can be $e^{-T}$.

are removed from the model *in bulk* (from the lexicon, and their transition and emission counts), and re-segmented afresh. When $T$ is high, $\pi(T)$ is small; as T cools, $\pi(T) \to 1$.

# 6   Evaluation

Evaluation of morphology discovery is challenging, specifically since morphological analysis is limited to *segmentation*—here, as well as in most prior work— because in general it is not possible to posit definitively "correct" segmentation boundaries, which by definition ignore allomorphy.

An evaluation based on probabilistic sampling is suggested in [23]. Another scheme is suggested in the papers about the HUTMEGS "Gold-standard" evaluation corpus for Finnish, [6,7]. However, these approaches to evaluation are problematic, in that they ignore the issue of **consistency** of the segmentation.

[7] observe correctly that positing a single "proper" morphological analysis for a word $w$ is not possible, in general. A motivating example is English $w = tries$: it can be analyzed as *tri+es* or *trie+s*. In actuality, $w$ has two morphemes, which can have more than one allomorph—a stem {*try-/tri-*} or {*try-/trie-*}, and 3rd person suffix {*-s/-es*} or {*-s*}. Restricting morphological analysis to segmentation makes the problem ill-defined: it is not possible to posit a "proper" way to place the morpheme boundary and then to expect an automatic system to discover that particular way. HUTMEGS proposes "fuzzy" morpheme boundaries, to allow the system free choice within the bounds of the fuzzy boundary—as long as the system splits the word somewhere inside the boundary, it is not penalized.

Recent work has called into question the validity of evaluation schemes that disregard the *consistency* of segmentations across the data, considering such schemes as too permissive: the system should commit to one way of segmenting similar words—its chosen "theory"—and then *consistently* segment according to its theory—and should be penalized for violating its own theory by placing the boundaries *differently* in similar words. We follow the evaluation scheme in [22], which provides for gold-standard segmentations and an evaluation algorithm so as to enforce consistency while giving the learning algorithm maximal benefit of the doubt.

Consider again the example of *tries*. In the suggested gold standard, one annotates the segmentation neither as *tri-es* nor as *trie-s*, but as $tri \overset{X}{\cdot} e \overset{X}{\cdot} s$— the special markers (dots) indicate that this segmentation is "ambiguous"—can be handled in more than one way. The label $X$ identifies this particular kind of ambiguity. (The gold-standard defines a separate set of labels for each language.) In this case, the definition of $X$ states that it accepts two *"theories"*: {10, 01} as valid—i.e., a morpheme boundary in the first position (10), or in the second (01), but not both (not 11; also not 00). Similar words: *cries, dries, flies* are then annotated similarly in the gold-standard.

Suppose the words *tries, flies,* and *applies* are in the gold standard, annotated with $X$ as above, and the model segments them as: *trie-s, flie-s,* but *appli-es*. We conclude that A. its preferred theory for handling $X$ is to put the boundary in the second position (2 out of 3), and B. when it segmented *appli-es* it placed the boundary incorrectly: it violated its own preferred theory (defined by the majority of its choices). Thus its accuracy will be 2/3. The label $X$ is used to co-index possible "ambiguous" boundaries in the gold standard, for a particular type of ambiguity. Annotators use these labels consistently across the evaluation corpus.

## 6.1   Experiments

We test on data from Finnish and Turkish. For each language, the corpus consists of about 100,000 distinct word forms, from texts of novels. Approximately 1000 words extracted at random from the data were annotated for the gold-standard by at least two annotators with knowledge of morphology and native proficiency.[16]

**Ablation Studies:** Shown in Table 1, confirm the gains in performance yielded by the heuristics. The enhancements to the basic algorithm are simulated annealing (SA), directionality (Dir), natural classes (NC) and bulk re-segmentation (Bu). Without SA, performance drops substantially. With SA, the table shows the gain obtained from adding all possible combinations of the heuristics. It might seem that adding directionality reduces the scores of the model, however, adding heuristics built upon directionality help the model achieve better results compared to the non-directional model.[17]

**Comparison Studies:** Are shown in Fig. 1, for Finnish and Turkish. Each point in the plots represents a single run of an algorithm. The coordinates of each point are its recall and precision; the accuracy of each run is in its label.

For comparison, we ran Morfessor CatMAP [8], on the same data, since it currently obtains the best performance over all Morfessor variants, as explained in [13]. FlatCat performs better than CatMAP with *semi-supervised* learning, but falls short of CatMAP performance in the unsupervised setting. CatMAP has a perplexity threshold parameter, $b$, "which indicates the point where a

**Table 1.** Ablation studies—Finnish

| SA | Dir | NC | Bu | Recall | Precis. | F-1 | Accuracy |
|----|-----|-----|-----|--------|---------|-------|----------|
| − | − | − | − | 30.83 | 70.79 | 42.96 | 75.94 |
| + | − | − | − | 34.30 | 79.48 | 47.92 | 78.06 |
| + | + | − | − | 33.93 | 77.80 | 47.25 | 77.70 |
| + | + | + | − | 34.77 | 73.32 | 47.17 | 77.65 |
| + | + | + | + | **36.37** | **83.83** | **50.73** | **79.80** |

---

[16] Disagreements between annotators were resolved. All annotated gold-standard data, for Finnish and Turkish, will be made publicly available with this paper on-line.

[17] To save space, we show results for Finnish only; other languages follow similar patterns (included in final paper).

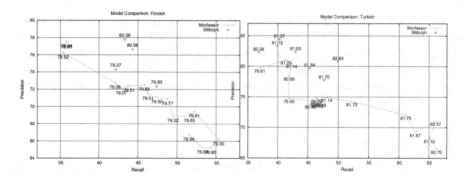

**Fig. 1.** Precision vs. recall: A—Finnish data, B—Turkish data (Color figure online)

morph is as likely to be a prefix as a non-prefix" [8]. This parameter trades off recall for precision; the more data there is, the higher $b$ should be. As $b$ grows, words are split less, giving higher precision but lower recall. Running Morfessor with varying $5 \leq b \leq 800$, yields the red line in the plots. SMORPH also has hyper-parameters, which we test in the experiments. Probability $\rho$ of a morph boundary between two adjacent symbols during the initial random segmentation, 0.20–0.25; the number of classes $K$, 15; assignment of classes to prefix, stem and suffix kinds, $s_{max}$ and $a_{min}$.

The blue points in the plots correspond to runs of SMORPH, with different settings of the hyper-parameters,[18] which can be optimized further, e.g., on a development corpus.

The runs of SMORPH show an improvement in terms of recall and precision over Morfessor CatMAP: the blue points lie above the red curve. For example, *at a given level of recall*, SMORPH reaches higher precision. For Finnish, the gain in precision is 2–8%; for Turkish, 2–7%. Conversely, *at a given level of precision*, SMORPH reaches higher recall; for very large $b$, Morfessor reaches higher recall, but generally at a loss in precision. SMORPH and Morfessor obtain similar accuracy values, though at a fixed level of recall SMORPH has higher accuracy. More fine-grained effects of the hyper-parameters on performance are to be explored and investigated in future work.

**Qualitative Evaluation of Classification:** An important feature of SMORPH is that the morph *classes* it learns are of a high quality. Manual inspection confirms that the classes group together morphs of a similar nature: noun-stem classes separate from verb stems, classes of affixes of similar kinds, etc.; hence the high precision.

As is natural in MDL, if some affixes appear frequently together, they will eventually be learned as a *single* affix; this explains lower recall. However,

---

[18] The parameters have not been tuned jointly; we started with values for the parameters as above, and checked the effect of varying them independently. Choice of the parameter values is driven by the observed total MDL cost (i.e., with no reference to the gold-standard evaluations).

this problem may be addressable as a post-processing step, after learning is complete.[19] (To be explored in future work.) Of course, evaluating the classes quantitatively is difficult, hence we evaluate quantitatively only the segmentations.

## 6.2   Error Analysis

A *non-directional* model trained on Finnish is shown in Fig. 2—with only 5 classes for clearer visualization. Each node shows the 8 most frequent morphs emitted from it, as well as the number of distinct morphs ($|Lex|$) and emission frequency ($freq$). Probabilities of transition between states are shown on the edges. (Edges with probability <0.02 are omitted for clarity.) The model has learned to often emit stems from states $S_1$ and $S_5$, and suffixes from $S_2$, $S_3$ and $S_4$. As expected, stem states are much larger than suffix states. They exhibit different properties: $S_1$ and $S_5$ have much flatter distribution, while $S_3$ and $S_4$ have spiked distributions: a few morphs with very high frequencies, and many with very low frequencies.

Further, $S_1$ has mostly verbal stems, whereas $S_5$ mostly nominal ones; $S_4$ is heavy on nominal suffixes, while $S_2$ has mostly verbal ones.[20]

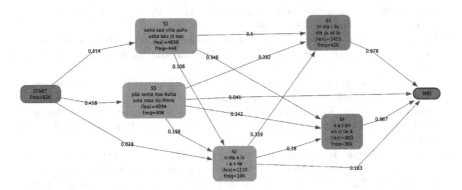

**Fig. 2.** Visualization of a model trained on Finnish data (with only 5 states)

## 7   Conclusions and Current Work

We have presented an algorithm for segmentation of a large corpus of words, which improves upon the state of the art. There are several important differences between SMORPH and Morfessor models. SMORPH tries to approach the

---

[19] Hence it is easier to recover from recall errors (false negatives) than from precision errors (false positives), and thus they are not equally important in this setting. Note that we do not consider F-score in the evaluation, but rather follow both recall and precision. F-score favors points where recall and precision are as near as possible. For example, whereas Morfessor trades off precision for recall to achieve a higher F-score, we do not consider it a benefit.

[20] This shows that the model indeed begins to resemble a FSM that we hope to achieve.

problem in a systematic way by grouping the discovered morphs into classes that respect general linguistic principles: directionality of morphotactics and natural differences between stems vs. affixes. It starts from a random initial model with no prior assumptions about the language, and learns to segment the data by optimizing a two-part cost function and Bayesian Marginal Likelihood, different from coding schemes used in prior work. The model is evaluated using a scheme, which avoids some of the problems in earlier evaluations.

To assure replicability, all gold-standard segmentations and code are made publicly available. Future improvements include those mentioned above, learning the optimal number of classes automatically, which should be reflected in the code-length, modeling compounding and allomorphy.

**Acknowledgments.** This research was supported in part by the FinUgRevita Project, No. 267097, of the Academy of Finland. We thank Hannes Wettig for his contributions to this work.

# References

1. Černý, V.: Thermodynamical approach to the traveling salesman problem: an efficient simulation algorithm. J. Optim. Theory Appl. **45**(1), 41–51 (1985)
2. Chomsky, N.: Rules and Representations. Basil Blackwell, Oxford (1980)
3. Creutz, M.: Unsupervised segmentation of words using prior distributions of morph length and frequency. In: Proceedings of 41st Meeting of ACL, Sapporo, Japan (2003)
4. Creutz, M.: Induction of a simple morphology for highly-inflecting languages. In: Proceedings of ACL SIGPHON, Barcelona, Spain (2004)
5. Creutz, M., Lagus, K.: Unsupervised discovery of morphemes. In: Proceedings of Workshop on Morphological and Phonological Learning, Philadelphia, PA, USA (2002)
6. Creutz, M., Lagus, K., Lindén, K., Virpioja, S.: Morfessor and Hutmegs: unsupervised morpheme segmentation for highly-inflecting and compounding languages. In: Proceedings of 2nd Baltic Conference on Human Language Technologies, Tallinn, Estonia (2005)
7. Creutz, M., Lindén, K.: Morpheme segmentation gold standards for Finnish and English. Technical report A77, HUT (2004)
8. Creutz, M., Lagus, K.: Inducing the morphological lexicon of a natural language from unannotated text. In: Proceedings of the International and Interdisciplinary Conference on Adaptive Knowledge Representation and Reasoning (AKRR-05), Espoo, Finland (2005)
9. Dawid, A.: Statistical theory: the prequential approach. J. Roy. Stat. Soc. A **147**(2), 278–292 (1984)
10. Dempster, A., Laird, N., Rubin, D.: Maximum likelihood from incomplete data via the EM algorithm. J. Roy. Stat. Soc. **39**(B), 1–38 (1977)
11. Goldsmith, J.: Unsupervised learning of the morphology of a natural language. ACL **27**(2), 153–198 (2001)
12. Goldsmith, J., Hu, Y.: From signatures to finite state automata. In: Midwest Computational Linguistics Colloquium, Bloomington, IN (2004)

13. Grönroos, S.A., Virpioja, S., Smit, P., Kurimo, M.: Morfessor FlatCat: an HMM-based method for unsupervised and semi-supervised learning of morphology. In: Proceedings of COLING 2014, the 25th International Conference on Computational Linguistics, Dublin, Ireland (2014)
14. Grünwald, P.: The Minimum Description Length Principle. MIT Press, Cambridge (2007)
15. Hammarström, H., Borin, L.: Unsupervised learning of morphology. Comput. Linguist. **37**(2), 309–350 (2011)
16. Koskenniemi, K.: Two-level morphology: a general computational model for word-form recognition and production. Ph.D. thesis, University of Helsinki, Finland (1983)
17. Kurimo, M., Creutz, M., Lagus, K. (eds.): Proceedings of the PASCAL Challenge Workshop on Unsupervised Segmentation of Words into Morphemes. PASCAL European Network of Excellence, Venice (2006)
18. Kurimo, M., Turunen, V., Varjokallio, M.: Overview of morpho challenge 2008. In: Peters, C., et al. (eds.) CLEF 2008. LNCS, vol. 5706, pp. 951–966. Springer, Heidelberg (2009). doi:10.1007/978-3-642-04447-2_127
19. Kurimo, M., Virpioja, S., Turunen, V., Lagus, K.: Morpho-challenge 2005–2010: evaluations and results. In: Proceedings of the 11th Meeting of the ACL Special Interest Group on Computational Morphology and Phonology. Association for Computational Linguistics, Uppsala (2010)
20. Kurimo, M., Virpioja, S., Turunen, V.T.: Proceedings of the Morpho Challenge 2010 Workshop. Technical report, TKK-ICS-R37, Aalto University, School of Science and Technology, Department of Information and Computer Science, Espoo, Finland (2010)
21. Narasimhan, K., Barzilay, R., Jaakkola, T.: An unsupervised method for uncovering morphological chains. Trans. Assoc. Comput. Linguist. **3**, 157–167 (2015)
22. Nouri, J., Yangarber, R.: A novel evaluation method for morphological segmentation. In: Proceedings of LREC 2016: The Tenth International Conference on Language Resources and Evaluation, Portorož, Slovenia (2016)
23. Spiegler, S., Monson, C.: EMMA: a novel evaluation metric for morphological analysis. In: Proceedings of the 23rd International Conference on Computational Linguistics. Association for Computational Linguistics (2010)
24. Virpioja, S., Kohonen, O., Lagus, K.: Unsupervised morpheme analysis with allomorfessor. In: Peters, C., Di Nunzio, G.M., Kurimo, M., Mandl, T., Mostefa, D., Peñas, A., Roda, G. (eds.) CLEF 2009. LNCS, vol. 6241, pp. 609–616. Springer, Heidelberg (2010). doi:10.1007/978-3-642-15754-7_73
25. Virpioja, S., Kohonen, O., Lagus, K.: Evaluating the effect of word frequencies in a probabilistic generative model of morphology. In: Proceedings of the NODALIDA Conference (2011)
26. Virpioja, S., Turunen, V.T., Spiegler, S., Kohonen, O., Kurimo, M.: Empirical comparison of evaluation methods for unsupervised learning of morphology. TAL **52**(2), 45–90 (2011). http://www.atala.org/IMG/pdf/2-Virpioja-TAL52-2-2011.pdf

# Incorporating Coreference to Automatic Evaluation of Coherence in Essays

Michal Novák[✉], Kateřina Rysová, Magdaléna Rysová, and Jiří Mírovský

Charles University, Faculty of Mathematics and Physics,
Institute of Formal and Applied Linguistics,
Malostranské náměstí 25, 11800 Prague 1, Czech Republic
{mnovak,rysova,magdalena.rysova,mirovsky}@ufal.mff.cuni.cz

**Abstract.** The paper contributes to the task of automated evaluation of surface coherence. It introduces a coreference-related extension to the EVALD applications, which aim at evaluating essays produced by native and non-native students learning Czech. Having successfully employed the coreference resolver and coreference-related features, our system outperforms the original EVALD approaches by up to 8% points. The paper also introduces a dataset for non-native speakers' evaluation, which was collected from multiple corpora and the parts with missing annotation of coherence grade were manually judged. The resulting corpora contains sufficient number of examples for each of the grading levels.

**Keywords:** Evaluation of coherence · Acquisition corpora processing · Coherence · Anaphora · Coreference · Discourse

## 1 Introduction

The task of automated evaluation of coherence can be defined as assigning grades to essays based on the level of their surface coherence. The essays can be written by native as well as non-native speakers of a given language. A system that addresses this problem may be particularly beneficial for teachers and may facilitate their work in grading the school essays. At the same time, it might also be useful for students themselves who can easily and quickly verify the level of their writing skills. They can also use the automatic evaluation as a practical tool for learning and practicing writing a coherent piece of text.

We focus on possibilities of automated evaluation of surface text coherence in students' essays in Czech. Particularly, we experiment with the EVALD application, which has been recently released in two variants: EVALD 1.0 [14] and EVALD 1.0 for Foreigners [15] (both described in [17]). EVALD 1.0 for native speakers evaluates the texts according to the scale commonly used in Czech schools, i.e. it assigns grades 1–5 (excellent–fail). EVALD 1.0 for Foreigners uses six grades A1–C2 (beginner–mastery) according to the internationally used scale established by Common European Framework of Reference for Languages, CEFR). The EVALD applications have so far focused only on discourse and lexical phenomena [16]. For instance, they take into account diversity and frequency

© Springer International Publishing AG 2017
N. Camelin et al. (Eds.): SLSP 2017, LNAI 10583, pp. 58–69, 2017.
DOI: 10.1007/978-3-319-68456-7_5

of discourse connectives. However, other aspects of coherence such as coreference relations or topic-focus articulation has been so far neglected.

This paper introduces a coreference-related extension of EVALD. We integrate a coreference resolver to the processing pipeline and extend the feature set to utilize its output. Furthermore, we design features that focus on pronouns but do not require a coreference resolver to be run.

The original EVALD for Foreigners as proposed in [16] uses a dataset with a skewed distribution over the CEFR levels, which often prevents it from predicting a low-populated level. We attempt to rectify this shortcoming by merging examples from multiple corpora, thus creating a much richer dataset.

The structure of the paper is as follows. Section 2 presents an original EVALD application, its data preprocessing stage and a discourse-based feature set it applies. The coreference-based extensions including the coreference resolver and the new set of features are introduced in Sect. 3. Datasets employed in the experiments and their construction is described in Sect. 4. Finally, in Sect. 5, the proposed approaches are evaluated and compared with a couple of baselines, before the paper concludes in Sect. 7.

## 2 Evaluator of Discourse

EVALD (EVALuator of Discourse) is a software application that employs machine learning methods to automatically rate the quality of surface coherence in Czech texts, based on the methodology and experiments first reported in [16] and later in [17]. It exists in two versions optimized for two different groups of users: one for Czech native speakers (EVALD 1.0, [14]), one for learners of Czech as a foreign language (EVALD 1.0 for Foreigners, [15]).

In the present paper, we use an updated experimental setting and an enlarged set of features (as compared with [16]). This section elaborates on the two main components of the application: the data preprocessing pipeline, and the original feature set extracted from this preprocessed data. Eventually, the final EVALD application to rate a level of surface coherence is built by using the Random Forest algorithm for machine learning.

### 2.1 Data Preprocessing

The preprocessing pipeline carries out an automatic linguistic analysis on the raw texts. It enriches them with morphological, surface and deep syntax information. Deep syntax (*tectogrammatical*) representation follows the theory of Functional Generative Description [21] and attempts to resemble manually annotated structures from the Prague Dependency Treebank [2]. A sentence on the tectogrammatical layer is structured as a dependency tree consisting only of content words and possibly reconstructed expressions that are elided on the surface. It enables representing pro-drops, which is essential for modeling anaphoric and coreference relations in Czech. Furthermore, this representation layer accommodates annotation of discourse connectives and relations.

All these preprocessing steps were performed in the current version of Treex [12], which is a modular system for natural language processing, using a predefined scenario for Czech text analysis. The scenario includes tens of individual steps, most notably tokenization, sentence segmentation, morphological analysis and part-of-speech tagging by the MorphoDiTa tool [23], dependency parsing by MST parser adjusted to Czech [8]. The scenario proceeds with transformation of the surface tree to a tectogrammatical tree: auxiliary nodes are made hidden, pro-drops are reconstructed, semantic roles are assigned, etc.

The preprocessing stage of the original EVALD application is concluded with a discourse parser. It focuses on discourse relations marked by explicit connectives, employing the approach described in [11]. It is a lexically based approach inspired by the annotation of the Penn Discourse Treebank 2.0 [13] and aims at capturing local discourse relations (between clauses, sentences, or short spans of texts). During processing, it first addresses intra-sentential discourse relations, followed by recognition of inter-sentential relations. The latter procedure utilizes lists of common Czech inter-sentential connectives and their most frequent discourse types (senses) extracted from the Prague Discourse Treebank 2.0 [18, PDiT 2.0].

## 2.2 Original Feature Sets

The original feature set as introduced in [16] embraces discourse-related and lexical features. In fact, for experimental purpose we distinguish its two subsets: *surface* and *advanced* features.

*Surface Features.* This set consists of features that only use tokenization and sentence segmentation. No advanced part of the text analysis such as syntactic or discourse parsing is exploited.

The lexical part of the surface set includes features such as the number of tokens per sentence, the normalized number of different lemmas, Yule's index of lemmas repetition [26], and Simpson's index of lemmas diversity [22].

One of the discourse-oriented surface features uses a list of 49 most frequent discourse connectives extracted from the discourse annotation in the PDiT 2.0, complemented by a few informal variants that are likely to appear in texts written by non-native speakers (e.g. *teda* as an informal variant of *tedy* [*so, therefore*]). The feature counts number of occurrences of these connective words in the tested text, without trying to distinguish their connective and non-connective usages, and normalizes the count by the number of sentences. Another two features count the number of coordinating and subordinating connective words separately.

*Advanced Features.* Information comes from the rich linguistic preprocessing producing automatically parsed trees and associated discourse relations (see Sect. 2.1).

Some of the features measure the frequency of predicate-less sentences, some of them the frequency of discourse relations – intra-sentential, inter-sentential as well as all together.

Another group of features questions variety of used discourse relations. It contains 11 features measuring a ratio of relations with a given connective as well as features expressing the ratio of 4 major types of discourse relationship (temporal, contingency, contrast and expansion). Finally, it includes a total number of unique connectives used in the whole text.

# 3   Coreference-Related Extension to EVALD

The original EVALD applications take advantage of discourse-related and lexical features only. This paper introduces coreference-related features. We designed these features by focusing on potentially coreferential expressions, mostly pronouns, and collecting their relative frequencies. Further features also inspect the coreference links reconstructed in the text, which would not be possible without a coreference resolution system.

First, we introduce the coreference resolution system, which enriches the preprocessing pipeline described in Sect. 2.1. After that, we proceed with presenting the new features to extend the original feature sets from Sect. 2.2.

## 3.1   Coreference Resolution System

Our extension to EVALD uses information acquired by the state-of-the-art coreference resolver for Czech – Treex Coreference Resolver [7, Treex CR]. As its name suggests, like the EVALD preprocessing stage it is an integral part of the Treex framework [12]. Although it is available for multiple languages, we utilize only its Czech version in this work.

Similarly to EVALD, Treex CR also operates on the text analyzed to the tectogrammatical representation. Tectogrammatics not only allows for exploiting the rich linguistic annotation available on this layer, but it also facilitates addressing zero anaphora. Addressing zero anaphora is central in Czech especially for subject pronouns, which are rarely expressed on the surface.

Treex CR consists of a sequence of models optimized by a supervised machine learning method. Every model targets a different type of coreferential expressions. Namely for Czech, Treex CR comprises a model for: (1) relative pronouns, (2) reflexive pronouns including the reflexive possessive pronoun "svůj", (3) 3rd person zero subjects, personal and possessive pronouns.

Models were trained on a dataset extracted from the training portion of the Prague Dependency Treebank 3.0 [2], containing more than 38,000 sentences and 652,000 words. Table 1 shows the performance of around 7-times smaller testset

**Table 1.** Counts of selected anaphoric expressions and performance of Treex CR on them measured in terms of F-score. The type *ZsPP3* denotes 3rd person zero subjects, personal and possessive pronouns.

|             | Relative | Reflexive | ZsPP3 | All    |
|-------------|----------|-----------|-------|--------|
| # instances | 1,075    | 579       | 1,950 | 3,604  |
| Treex CR    | 78.40    | 76.19     | 61.31 | **68.46** |

extracted from the same corpus. The F-score measuring how good the resolver is in finding any of the pronoun's antecedents is around 70% in total for all the selected expression categories.

## 3.2 Anaphora-Related Features

This work proposes two sets of additional features to the original EVALD applications. A common property of both sets is that they aim at describing how successful the essay's writer is in maintaining coreference relations. On the other hand, the main difference between the two sets lies in whether they use the information obtained by the CR system or not.

*Pronoun Features.* Pronouns are one of the most prominent part-of-speech categories used to retain coherence in a text. The reason is that they frequently appear as anaphoric occurrences, i.e. their interpretation often depends on previous context. It is pronouns and their distribution in the text that is in focus of the first set of features. We do not employ information coming from the CR system here, though. The aggregated frequencies thus may capture both anaphoric and non-anaphoric occurrences of pronouns.

The set includes a relative frequency of pronouns among all words as well as among nouns and pronouns together. Same ratios are also computed for each of the 21 pronoun subtypes. The subtype is specified within the positional part-of-speech tag assigned by the MorphoDiTa tagger [23]. In addition, a proportion of every subtype to all pronouns is included as a feature. Analogously, similar counts are aggregated over the tectogrammatical tree. Unlike the surface features above, these are able to capture also zero subjects, which appear massively in Czech.

Besides the quantity of pronouns' usage, we measure also its quality, i.e. how wide the repertoire of used pronouns is. Again, we compute a relative frequency of the number of different pronoun lemmas. Furthermore, another group of features focus on pronouns or zeros filling the semantic role of Actor in a sentence and measure what is the distribution of lemmas at this position. Czech learners may prefer using overt pronouns here, despite their low proportion compared to zeros in stylistically well-written texts. Similarly, an excessive use of the Czech demonstrative pronoun *"to"* is a typical sign of informal style or spoken language.

*Coreference Features.* The second set of new features takes advantage of the output of the CR system introduced in Sect. 3.1.

Quantitative features collect the count of coreference chains (i.e. the mentions of the same entity), and links and normalize them by the text length. A distribution of coreferential chains by their length forms another group of features. At the same time, the proportion of intra- and inter-sentential coreference links is recorded as a feature.

Qualitative features measure the variety of expressions forming the coreferential chains, focusing on lemmas and part-of-speech subtypes of the expressions.

# 4    Sources of Data

To build and test the EVALD applications, we created two datasets that consist of text authored by learners for whom Czech is a native language (*L1 dataset*), or a foreign language (*L2 dataset*). We compiled these datasets from the content of three language acquisition corpora: MERLIN corpus [3], CzeSL-SGT/AKCES 5 [19], and Skript2012/AKCES 1 [20].

*MERLIN.* The original corpus contains altogether 2,286 writing samples by non-native speakers (students) of Czech, German and Italian. The texts have been collected from the CEFR-related test provided by institutions that offer internationally recognized language exams in accordance with the high standards defined by The Association of Language Testers in Europe.[1] Namely, the total number of 441 Czech texts has been provided by ÚJOP Institute in Prague.

All the texts are rated with respect to various criteria, including coherence/cohesion. The Czech texts consist of tests for overall examination levels A2–B1. Hence, the majority of coherence ratings stays at these levels and the A1 and C1 grades account only for less than 3% of the texts.

*CzeSL-SGT/AKCES 5.* It is another corpus of texts written by non-native speakers of Czech. It contains 8,617 texts that were created in courses of Czech for foreigners from 2009 to 2013. The texts come from the authors that speak 54 different first languages.

Almost all the texts are labeled with an overall language proficiency CEFR level of the writer. Nevertheless, unlike the Merlin corpus, no annotation of coherence levels is provided in CzeSL-SGT.

*Skript2012/AKCES 1.* This corpus comprises texts written by native speakers of Czech, particularly, the students' essays created during the lessons of Czech language at elementary and high schools. The original corpus contains 1,694 texts.

Most of the texts are labeled with overall grades 1–5. Similarly to the CzeSL-SGT corpus, no specific coherence level annotation is available.

The L1 dataset was formed solely from Skript2012, because as far as we know it is the only corpus of Czech essays produced by native students. We included only the essays that are labeled with an overall grade. However, we had to manually judge all of them on the level of coherence.

A core of the L2 dataset is constituted by the MERLIN corpus containing explicit coherence level annotation. However, due to the fact that it comprises only tests for certain examination levels, we do not consider it representative. Unlike other authors (e.g., [10,24]), who excluded the under-represented levels A1 and C2, we decided to keep all levels by merging multiple data sources. Less populated levels were thus supplied with the texts from the CzeSL-SGT corpus. As it has no coherence levels annotated, we selected a sample of CzeSL-SGT texts with less populated overall proficiency levels (A1, A2, and C1), manually assessed the level of coherence there, and added them to the L2 dataset.

---

[1] http://www.alte.org.

**Table 2.** Basic statistics of the collected datasets.

| L1 dataset | 1 | 2 | 3 | 4 | 5 | | Total |
|---|---|---|---|---|---|---|---|
| # documents | 484 | 149 | 121 | 239 | 125 | | 1,118 |
| # sentences | 20,986 | 4,449 | 2,913 | 3,382 | 939 | | 32,669 |
| # tokens | 301,238 | 65,684 | 40,054 | 43,797 | 11,379 | | 462,152 |
| L2 dataset | A1 | A2 | B1 | B2 | C1 | C2 | Total |
| # documents | 174 | 176 | 171 | 157 | 105 | 162 | 945 |
| # sentences | 1,802 | 2,179 | 2,930 | 2,302 | 1,498 | 10,870 | 21,581 |
| # tokens | 15,555 | 21,750 | 27,223 | 37,717 | 21,959 | 143,845 | 268,049 |

To collect examples also for the C2 level, we exploit the fact that this level is generally characterized as a near-native language competence. That is, the authors of such texts are likely to have developed so high degree of proficiency in Czech that their texts are hardly distinguishable from the texts produced by native speakers. Therefore, we filled the C2 level with a sample of texts from the Skript2012 corpus.

Our L2 dataset merged from three different corpora differs from the dataset used in previous experiments with EVALD in [16]. Their dataset contained only MERLIN essays, with levels A1 and C1 low-populated and C2 not represented at all. Therefore, the performance scores of EVALD that we present in Sect. 5 are not directly comparable with their results.

The basic statistics of the L1 and L2 datasets are highlighted in Table 2.

## 5    Experiments

Extensions to EVALD applications as proposed in Sect. 3 are evaluated in three variants of additional feature sets: (1) pronoun features only, (2) coreference features only, (3) both pronoun and coreference features. We compare these settings with two baselines that follow a design introduced in Sect. 2, namely the baseline (1) with surface features only, and (2) with both surface and advanced features.

All these variants are evaluated by 10-fold cross-validation. However, the performance cannot be measured directly by accuracy as we arbitrarily adjusted the dataset to make the levels more balanced (see Sect. 4). Moreover, the process of how the source corpora were collected does not convince us that the distribution of grades there reflects their real distribution over the population. We thus assume that new essays to be evaluated come from a uniform distribution over all possible grades.

One way to ensure that the test data fold is uniformly distributed is to sample from it so that each of the class has the same size as the smallest one. A shortcoming of such approach is that many valuable data instances are thrown away. Aggregated over all folds, such limited test L1 and L2 data consists of 475

**Table 3.** Performance of the new feature sets compared to the original feature sets. It is measured by F-score on complete dataset, and by exact (e-Acc) and one-level tolerance (1-Acc) accuracy on the balanced dataset.

|  | L1 dataset | | | L2 dataset | | |
|---|---|---|---|---|---|---|
|  | F | e-Acc | 1-Acc | F | e-Acc | 1-Acc |
| Surface | 40.1 | 42.1 | 72.4 | 47.6 | 48.5 | 74.7 |
| Surface+advanced | 44.9 | 46.1 | 80.8 | 51.3 | 55.5 | 82.5 |
| +pronoun | 45.9 | 48.2 | **83.0** | 58.6 | 62.3 | **86.8** |
| +coref | 45.2 | 47.0 | 81.3 | 54.7 | 58.7 | 85.2 |
| +pronoun+coref | **46.0** | **49.5** | **83.0** | **59.0** | **63.3** | 85.5 |

and 600 instances, which account for 43% and 64% of all available instances. We used this limited testset to calculate the *exact accuracy* score.

Alternatively, macro-averaging of class-oriented scores has a similar balancing effect without a need to reduce the test data. The F-score $F_c$ is first calculated for every class $c$ in the dataset. The scores are the averaged over all classes to produce the *macro-averaged F-score*: $F = \frac{1}{|C|} \sum_{c \in C} F_c$.

Results of evaluation are highlighted in Table 3. They exhibit consistent improvement of the coreference-related extension of EVALD over the baselines in terms of all these measures. Baseline methods are outperformed by all the three new settings. Nevertheless, the combination of all the presented feature sets seems to perform the best. Evaluating each of the two new additional feature sets without the other one suggests that the pronoun features are more valuable.

The new feature sets also appear to have a stronger effect on the L2 dataset, i.e. in evaluation of non-native speakers' essays. It may be attributed to the quality of these essays, which is supposed to vary more widely than it is for native speaker's essays.

Besides the two metrics presented above, we also score the methods with the *one-level tolerance accuracy* in Table 3. Like the exact accuracy, this variant is also calculated on the balanced data. However, a positive point is counted even if a method predicts a class adjacent to the true one. Such score is justified by the fact that even a human judge has often difficulties to determine the evaluation grade precisely. According to this score, the baselines are again surpassed by all the new settings, achieving over 80% in accuracy on both datasets. Nevertheless, this time the pronoun feature set alone reaches the superior results.

Figure 1 shows F-scores of all the variants measured for individual grades in the L1 (left) and L2 dataset (right). For both datasets, the evaluator seems to perform best for extreme grades, which sounds reasonable in the light of the scale nature of the evaluation grades. On the other hand, note that in the L2 dataset the performance for the neighboring grades of the extreme ones (i.e., A2 and C1) is the worst. These neighboring grades are in fact often misclassified as the A1 and C2 levels, respectively.

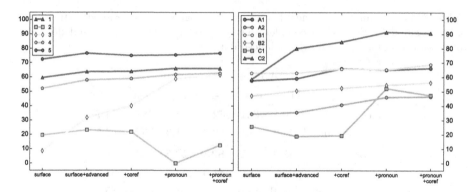

**Fig. 1.** F-scores for individual grades in the L1 (left) and L2 dataset (right).

Inspecting the performance while adding new features, the score improves only slightly for most of the grades. The overall improvement should thus be mostly attributed to performance jumps for only a few grades, namely, the grade 3 for the L1 dataset, and C1 and C2 levels for the L2 dataset. The positive effect of coreference-related features for the C2 level may be a consequence of apparently longer documents (see Table 2), which are supposed to contain more coreferential relations. Nevertheless, the performance jumps for the other grades would seem to be difficult to justify without inspecting the data manually.

## 6   Related Work

The aim of the present work is to judge texts and assign grades to them, both with respect to the level of its coherence/cohesion. We do this on texts written by native Czech speakers as well as by learners of Czech as a foreign language. Hence, the task is analogous to the tasks of *Automated essay scoring (AES)* and *Proficiency Level Classification (PLC)*.

The goal of AES is to automatically assign grades to essays written in an educational setting. The research on AES traces back to the prehistory of automatic processing, with the first program called Project Essay Grade[TM] [9]. Over the years multiple even commercial applications to address this task have been developed (see [5] for an overview).

Due to using discourse structure and stylistic features for AES, the e-rater® [1] is particularly related to our work. In detail, they exploit a linear discourse structure associated with the text by a specialized tool [4]. Analogous to our system, this tool takes advantage of discourse marker words and structures, and syntactic structures such as subordinating and relative clauses.

In a recent work by Zupanc and Bosnić [27], a wide repertoire of coherence features is employed for AES. These features are not linguistically motivated, they rather address statistical properties of coherent texts. Nevertheless, apart from the score, this system also provides a comprehensive feedback on errors. The preprocessing pipeline for this task integrates also two coreference resolution systems. In addition, just like in our work, among the variety of machine-learning methods they had tested, the Random Trees method appeared to perform the best.

A CR system is also utilized in the AES system of Wonowidjojo et al. [25] to preprocess the data subsequently modeled by Latent Semantic Analysis.

Proficiency Level Classification can be considered as a special case of AES applied on text written by non-native speakers of a language (L2). The following review of the related works focuses on works that share these common features: the L2 language is other than English, and the texts are graded by CEFR levels.

Works of Hancke and Meurers [6], Vajjala and Lõo [24], and Pilán et al. [10] address PLC on German, Estonian, and Swedish texts, respectively. Like us, all these works encountered the issue of few data for especially marginal level A1 and C2. Whereas we filled these levels from another corpora, [10,24] decided to remove under-represented levels from training and evaluation. In [6], this issue seems to be ignored.

Except for the German one, all works report scores that treat the evaluation as if the test data were uniformly distributed over the grades. Accuracy on balanced set with each level reduced to the size of the minimal one reached 77% for Estonian. Their macro-averaged F-score is even higher – 78.5%. The Swedish system performed 72% in F-score, which improved by 2.5% points after applying domain adaptation with additional data coming from coursebooks. All the scores were calculated by 10-fold cross-validation. Compared to these scores, the performance of our classifier falls behind by 10–15% points. Nonetheless, the reader should keep in mind that our task is to evaluate solely coherence of the essays, which might be viewed as more difficult task than overall evaluation.

# 7   Conclusion

This work contributed to the task of automated evaluation of coherence.

The paper introduced a coreference-related extension to the EVALD applications, which aim at evaluating essays in Czech produced by native and non-native speakers of this language. We successfully integrated the Treex CR system for coreference resolution and designed a set of features investigate the level of retaining coreference relations in the essays. Measured by exact accuracy on a balanced dataset, the extended version outperformed the original EVALD applications by 3 and 5% points in native and non-native settings, respectively.

The paper also introduced a dataset for evaluation of surface coherence in texts produced by non-native speakers of Czech. The dataset was collected from multiple corpora and the parts with missing annotation of coherence grade were manually judged. The resulting corpora thus contains sufficient number of examples for each of the CEFR levels.

In the future work, we would like to extend the evaluator with features related to topic-focus articulation. Apart from the discourse and coreference relations, this is the third main component that strongly influences surface coherence of a text.

**Acknowledgment.** The authors acknowledge support from the Ministry of Culture of the Czech Republic (project No. DG16P02B016 *Automatic Evaluation of Text Coherence in Czech*). This work has been using language resources developed, stored and distributed by the LINDAT/CLARIN project of the Ministry of Education, Youth and Sports of the Czech Republic (project LM2015071).

# References

1. Attali, Y., Burstein, J.: Automated essay scoring with e-rater® V.2. J. Technol. Learn. Assess. **4**(3), 1–31 (2006)
2. Bejček, E., Hajičová, E., Hajič, J., Jínová, P., Kettnerová, V., Kolářová, V., Mikulová, M., Mírovský, J., Nedoluzhko, A., Panevová, J., Poláková, L., Ševčíková, M., Štěpánek, J., Zikánová, Š.: Prague Dependency Treebank 3.0. ÚFAL MFF UK, Prague (2013)
3. Boyd, A., Hana, J., Nicolas, L., Meurers, D., Wisniewski, K., Abel, A., Schöne, K., Štindlová, B., Vettori, C.: The MERLIN corpus: learner language and the CEFR. In: Proceedings of the 9th International Conference on Language Resources and Evaluation (LREC 2014), pp. 1281–1288. European Language Resources Association, Reykjavík (2014)
4. Burstein, J., Marcu, D., Knight, K.: Finding the WRITE stuff: automatic identification of discourse structure in student essays. IEEE Intell. Syst. **18**(1), 32–39 (2003)
5. Dikli, S.: An overview of automated scoring of essays. J. Technol. Learn. Assess. **5**(1), 1–36 (2006)
6. Hancke, J., Meurers, D.: Exploring CEFR classification for German based on rich linguistic modeling. Learner Corpus Research 2013, Book of Abstracts, Bergen, Norway, pp. 54–56 (2013)
7. Novák, M.: Coreference resolution system not only for Czech. In: ITAT 2017: Information Technologies-Applications and Theory (Proceedings). CreateSpace Independent Publishing Platform, Martinské Hole (2017)
8. Novák, V., Žabokrtský, Z.: Feature engineering in maximum spanning tree dependency parser. In: Matoušek, V., Mautner, P. (eds.) TSD 2007. LNCS (LNAI), vol. 4629, pp. 92–98. Springer, Heidelberg (2007). doi:10.1007/978-3-540-74628-7_14
9. Page, E.B.: The use of the computer in analyzing student essays. Int. Rev. Educ. **14**(2), 210–225 (1968)
10. Pilán, I., Volodina, E., Zesch, T.: Predicting proficiency levels in learner writings by transferring a linguistic complexity model from expert-written coursebooks. In: COLING 2016, 26th International Conference on Computational Linguistics, Proceedings of the Conference: Technical Papers, pp. 2101–2111. ACL, Osaka (2016)
11. Poláková, L., Mírovský, J., Nedoluzhko, A., Jínová, P., Zikánová, Š., Hajičová, E.: Introducing the Prague discourse Treebank 1.0. In: Proceedings of the Sixth International Joint Conference on Natural Language Processing, pp. 91–99. Asian Federation of Natural Language Processing, Nagoya (2013)
12. Popel, M., Žabokrtský, Z.: TectoMT: modular NLP framework. In: Loftsson, H., Rögnvaldsson, E., Helgadóttir, S. (eds.) NLP 2010. LNCS (LNAI), vol. 6233, pp. 293–304. Springer, Heidelberg (2010). doi:10.1007/978-3-642-14770-8_33
13. Prasad, R., Dinesh, N., Lee, A., Miltsakaki, E., Robaldo, L., Joshi, A., Webber, B.: The Penn discourse Treebank 2.0. In: Proceedings of the Sixth International Conference on Language Resources and Evaluation (LREC 2008), pp. 2961–2968. European Language Resources Association, Marrakech (2008)

14. Rysová, K., Mírovský, J., Novák, M., Rysová, M.: EVALD 1.0. ÚFAL MFF UK, Prague (2016)
15. Rysová, K., Mírovský, J., Novák, M., Rysová, M.: EVALD 1.0 for Foreigners. ÚFAL MFF UK, Prague (2016)
16. Rysová, K., Rysová, M., Mírovský, J.: Automatic evaluation of surface coherence in L2 texts in Czech. In: Proceedings of the 28th Conference on Computational Linguistics and Speech Processing ROCLING XXVIII, pp. 214–228. National Cheng Kung University, The Association for Computational Linguistics and Chinese Language Processing (ACLCLP), Taipei (2016)
17. Rysová, K., Rysová, M., Mírovský, J., Novák, M.: Automatic evaluation of discourse in Czech - software applications EVALD 1.0 and EVALD 1.0 for foreigners. In: Recent Advances in Natural Language Processing 2017. RANLP 2017 Organising Committee/ACL, Varna (2017)
18. Rysová, M., Synková, P., Mírovský, J., Hajičová, E., Nedoluzhko, A., Ocelák, R., Pergler, J., Poláková, L., Pavlíková, V., Zdeňková, J., Zikánová, Š.: Prague Discourse Treebank 2.0. ÚFAL MFF UK, Prague (2016)
19. Šebesta, K., Bedřichová, Z., Šormová, K., et al.: AKCES 5 (CzeSL-SGT). ÚTKL FF UK, Prague (2014)
20. Šebesta, K., Goláňová, H., Letafková, J., et al.: AKCES 1. ÚTKL FF UK, Prague (2016)
21. Sgall, P., Hajičová, E., Panevová, J., Mey, J.: The Meaning of the Sentence in its Semantic and Pragmatic Aspects. Springer, Heidelberg (1986)
22. Simpson, E.H.: Measurement of diversity. Nature **163**, 688 (1949). doi:10.1038/163688a0
23. Straková, J., Straka, M., Hajič, J.: Open-source tools for morphology, lemmatization, POS tagging and named entity recognition. In: Proceedings of 52nd Annual Meeting of the Association for Computational Linguistics: System Demonstrations, pp. 13–18. Association for Computational Linguistics, Baltimore (2014)
24. Vajjala, S., Lõo, K.: Automatic CEFR level prediction for Estonian learner text. In: Proceedings of the Third Workshop on NLP for Computer-Assisted Language Learning at SLTC 2014, no. 107, pp. 113–127. Linköping University Electronic Press, Linköping (2014)
25. Wonowidjojo, G., Hartono, M.S., Frendy, Suhartono, D., Asmani, A.B.: Automated essay scoring by combining syntactically enhanced latent semantic analysis and coreference resolution. In: 6th International Workshop on Computer Science and Engineering, Tokyo, Japan, pp. 580–584 (2016)
26. Yule, G.U.: The Statistical Study of Literary Vocabulary. Cambridge University Press, Cambridge (1944)
27. Zupanc, K., Bosnić, Z.: Automated essay evaluation with semantic analysis. Knowl.-Based Syst. **120**(3), 118–132 (2017)

# Graph-Based Features for Automatic Online Abuse Detection

Etienne Papegnies[1,2(✉)], Vincent Labatut[1], Richard Dufour[1],
and Georges Linarès[1]

[1] LIA – EA 4128, University of Avignon, Avignon, France
[2] Nectar de Code, Barbentane, France
{etienne.papegnies,vincent.labatut,richard.dufour,
georges.linares}@univ-avignon.fr

**Abstract.** While online communities have become increasingly important over the years, the moderation of user-generated content is still
performed mostly manually. Automating this task is an important step
in reducing the financial cost associated with moderation, but the majority of automated approaches strictly based on message content are highly
vulnerable to intentional obfuscation. In this paper, we discuss methods
for extracting conversational networks based on raw multi-participant
chat logs, and we study the contribution of graph features to a classification system that aims to determine if a given message is abusive.
The conversational graph-based system yields unexpectedly high performance, with results comparable to those previously obtained with a
content-based approach.

**Keywords:** Text categorization · Abuse detection · Online communities · Moderation

## 1 Introduction

The widespread availability of Internet access allows users from around the
world to congregate into online communities. With the ever-increasing number of users, online communities have become important places to trade ideas,
and have acquired a great socio-economical importance.

However, because of the anonymity provided by the medium, an online community is often confronted with users that display abusive behaviors. For community maintainers, it can be important to act on this issue through the use
of moderation, because failure to do so can poison the community, trigger user
exodus, and expose the administrators to legal jeopardy. Moderation is the application of sanctions when users are judged to violate the community rules. When
done by humans, this work is expensive and companies have a vested interest
in automating the process. One can distinguish two types of automated systems assisting in moderation: (1) an automated flagging system that raises some

© Springer International Publishing AG 2017
N. Camelin et al. (Eds.): SLSP 2017, LNAI 10583, pp. 70–81, 2017.
DOI: 10.1007/978-3-319-68456-7_6

messages to the attention of moderators; and (2) a fully automated system that detects abusive messages and executes sanctions on users that are breaking the community rules.

In this work, we consider the classification problem of automatically determining if a message from a user is abusive or not. For this purpose, we propose an original approach aiming at exploring a range of graph-based features extracted from online textual conversations. We first extract various types of *conversational networks*, *i.e.* graphs where vertices represent users and where edges correspond to supposed message-based interactions between them. We then process a number of graph-theoretical measures that characterize these networks in different ways. A classifier is then trained and tested on a corpus of chat logs originating from the community of the French massively multiplayer on-line game *SpaceOrigin*[1]. We finally conduct a qualitative study to analyze the impact of each graph-based feature on the automatic abusive message classification performance.

The rest of this paper is organized as follows. In Sect. 2, we review related work on abuse detection and network extraction from raw conversation logs. In Sect. 3, we describe the method proposed to extract conversational networks, and the topological features that we compute for the resulting graphs. In Sect. 4, our dataset is presented as well as the overall experimental setup for the classification task. A discussion and a qualitative study of our results is also provided. Finally, we summarize our contributions in Sect. 5 and present some perspectives.

## 2  Related Work

This section is a brief review of the literature focusing on the most relevant works relating to two aspects of the problem at hand. First, in Subsect. 2.1, we review general works regarding the detection of online abuse. Second, in Subsect. 2.2, we explore previously used techniques to extract network structures from raw conversation data.

### 2.1  Abuse Detection

One can distinguish two main categories of works related to abuse detection: those using the content of the exchanged messages and those focusing on their context. Some works also propose to combine both categories.

From the content-based point-of-view, the work initiated by Spertus in [19] was a first attempt to create a classifier for hostile messages. This is relevant to us, because abusive messages often contain hostility. They use static rules to extract linguistic markers for each message: Imperative Statement, Profanity, Condescension, Insult, Politeness and Praise. These are then used as features in a binary classifier. They obtain good results, except in specific cases like hostility through sarcasm. However, the limitation of this approach is that its application to another language is a difficult task, requiring to transpose it to other grammar rules and idioms.

---

[1] https://play.spaceorigin.fr/.

Cheng *et al.* [5] note that a word tagged as offensive in a message is not a definite indication that the message is offensive, i.e. while "You are stupid." is clearly offensive, "This is stupid. xD" is not. Lack of context can be somewhat mitigated by looking at word $n$-grams instead of unigrams (i.e. single words).

Dinakar *et al.* [8] use *tf-idf* features, a static list of badwords and of widely used sentences containing verbal abuse to detect cyberbullying in Youtube comments. Again, their model showed good results, except when sarcasm was used.

In [4], Chavan and Shylaja review machine learning approaches to detect aggressive messages in on-line social networks. They show that Pronoun Occurrence, usually neglected in text classification, is important, and use Skip-Gram features to mitigate the context issues.

Content-based text classification usually makes for a good baseline. Content features are inexpensive to compute. However, such methods have severe limitations: for instance, abuse can be spread over a succession of messages. Some messages can reference a shared history between two users. Even more common are users that are voluntarily obfuscating message content to work around badwords detection. Indeed, abusers can bypass automatic systems by making the abusive content difficult to detect [11]: for instance, they can intentionally modify the spelling of a forbidden word.

Because the reactions of other users to an abuse case are completely beyond the control of the abuser, some works consider the content of messages *around* the targeted message.

For instance, Yin *et al.* [21] use features derived from the neighboring phrases of a given message to detect harassment on the Web. Their goal is to spot conversations going off-topic, and use that as an indicator. Their approach shows good results when used against multi-participant chat logs, and they note that sentiment features seem to constitute mostly noise due to the high misspelling rate.

In [6], Cheng *et al.* propose to focus on building user behavior models. For this purpose, they perform a comprehensive study of antisocial behavior in online discussion communities. Their work provides insight into the devolution of abusive users over time in a community, regarding both the quality of their contributions and their reactions towards other members of the community. A critical result of the analysis is that instances of antisocial messages usually generate a bigger response from the community, compared to normal messages.

Balci and Salah [1] make use of user features to detect abuse in the community of an online game. These features include information such as gender, number of friends, financial investment, avatars, and general rankings. The goal is to help human moderators dealing with abuse reports, and the approach yields sufficiently good results to achieve it. However, in our case the user data necessary to replicate this approach is not available.

In our own previous work [15], we propose to detect abusive messages from chat messages using a wide array of language features (bag-of-words, *tf-idf* scores, sentiment scores, etc.) as well as context features derived from the language models of other users. We also try advanced preprocessing approaches. This method allowed us to reach a performance of 72.1% in terms of $F$-measure on an abusive message detection task.

## 2.2    Network Extraction from Raw Data

Although a major part of the solutions focus on content features of exchanged messages to address the abuse problem, it appears that a user with previous exposure to automatic moderation techniques can easily circumvent them [11]. To avoid this problem, a solution would be to not focus only on the direct content exchanged but on the interactions between the users through these messages.

The number of respondents to a given message appears frequently in the literature, as a classification feature, e.g. [6]. However, there are not many works dealing with the extraction of conversational networks. This may be due to the fact that the task can be far from trivial, depending on the nature of the available raw data: the task is much harder for chat logs than for structured messages board or Web forums, for instance. These networks have the advantage of including the mentioned feature, but also much more information regarding the way users interact.

In [13], Mutton proposes a strategy to extract such a network from IRC chat logs. The goal is to build a tool to visualize user interactions in an IRC chat room over time. The author uses a simple set of rules based on *direct referencing* (when a user addresses another one by using his nickname), temporal proximity of messages, and temporal density of messages. In this paper, we will adapt and expend on those rules. Specifically, while in a regular IRC channel timestamps are indeed useful to determine intended recipients of a message, in our case they are basically irrelevant, so this approach cannot be adapted as is.

Travassoli *et al.* [20] explore different methods to extract representative networks from group psychotherapy chat logs. One method includes fuzzy referencing to mitigate effects of misspelled nicknames, and rules for representing one-to-all messages. The bulk of the methods uses static patterns of exchanges to predict a receiver. Their system shows a good agreement score with a human annotator.

Sinha and Rajasingh [18] use only direct referencing, but with the same fuzzy matching strategy, in order to extract a network representing the activity in the #ubuntu IRC support channel. This method manages to expose high level components of the Ubuntu social network, which in turn allows for the qualification of user behaviors into specific classes. This method of building user models can be very interesting when the data describing the users are scarce, as is the case on IRC where everyone can join and there is no requirement to register.

## 3    Methods

In this section, we describe our proposed original approach to detect abusive messages. It basically consists in training a classifier on features corresponding to topological measures processed on conversational networks. The classifier is standard, so we focus on the processing of the features. Thereby, Subsect. 3.1 presents how we extract conversational networks from conversation logs, while Subsect. 3.2 describes the topological measures computed for these conversational networks, and later used as classification features.

## 3.1  Network Extraction

We extract networks representing conversations between users, through a textual discussion channel. They take the form of weighted undirected graphs, in which the vertices and edges represent the users and the communication between them, respectively. The edge weights are a score which is an estimation of the intensity of the communication between the two connected users. Note that each network is defined relatively to a *targeted message*, since the goal of this operation is to provide features used to classify the said message.

The first step consists in determining which messages are used to extract the network. For this purpose, we define a *context period*, which is centered on the *targeted message*, and spans symmetrically before and after its occurrence. We arbitrarily use a width of 200 messages in our experiments. The graph extracted from this context period contains only the vertices representing the users which posted at least once on this channel, during this period.

The second step is to add the appropriate edges to the network, and to process their weight. We use a method based on a sliding window, a choice that is justified by two properties of the user interface of the considered discussion channel: (1) when a user joins a channel, the server sends him only the last 20 messages posted on the channel; and (2) it is impossible for a user to scroll back the history further than 20 lines. In our experiments, we arbitrarily use a window of 10 messages. We apply an iterative process, consisting in sliding the window over the whole context one message at a time. We call *current message* the *last* message of the window taken at a given time. Our assumption is that this message is destined to the authors of the other messages present in the window at this time. Furthermore, we suppose it is more likely that the message concerns the users who posted the most recently. These hypotheses can be justified by another property of the user interface: by default, users do not know who is in the channel at a given time, in particular the join/part events are not shown to them.

Based on these hypotheses, we update the edges and weights in the following way. First, we list the authors of the messages currently present in the window, order them by last message posted, and discard the author of the *current message* (since it is possible that several of his messages appear in this window): this results in what we call the *neighbor list*. However, the user interface allows to *explicitly* mention users in a message by their name, and the game prevents the users from changing their name: we need to take this property into account. For this purpose, we move the users directly referenced in the *current message* at the top of our list. If a user was not even in the window, it is simply inserted at the top of the list. Each user in the neighbor list is assigned by a score, which is a decreasing function of both his position in the list and of the length of the *neighbor list*. We can then update the graph: we create an edge between each user in the neighbor list and the author of the *current message*, with a weight corresponding to the user's score. If this edge already exists, we increase its current weight by the user's score.

Our choice to create or update edges towards all users in the window even in case of direct referencing is based on several considerations. First, directly referencing a user does not imply that he is part of the conversation or that the message is directed towards him: for instance, his name could just be mentioned as an object of the sentence. Second, there can be multiple direct references in a single message. Third, when in online public discourse, directly addressing someone does not mean he is the sole intended recipient of the message. For instance when discussing politics, a question directed towards someone can have as a secondary objective to have the target expose his stance on an issue to the other participants.

Once the iterative process has been applied for the whole context period, we get what we call the *Full* network. For testing matters, we also process 2 lesser networks based on the same context: the *Before* and *After* networks are extracted using only the 100 messages preceding and following the *targeted message*, respectively, as well as the *targeted message* itself. Figure 1 shows an example of the three networks associated with an abusive comment.

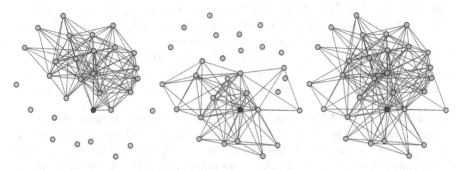

**Fig. 1.** Example of the 3 types of conversational networks extracted for a given context period: Before (left), After (center), and Full (right). The abusive user is represented in red. (Color figure online)

### 3.2 Features

The classification features we consider in this work are all topological measures, allowing to characterize graphs in various ways. We adopt an exploratory approach and consider a wide range of such measures, focusing on the most widespread in the literature. In the following, we describe them briefly, distinguishing between *local* ones, which characterize individual vertices, and *global* ones, which describe the whole graph at once. We process all the features for each of the 3 types of networks (Before, After, Full) described in the previous subsection.

**Local Topological Measures.** These measures are computed for the Vertex corresponding to the author of the targeted message.

The *Degree centrality* is a normalized version of the standard degree, which corresponds itself to the number of direct neighbors of the considered vertex. The

*Eigenvector Centrality* [2] can be considered as a generalization of the degree, in which instead of just counting the neighbors, one also takes into account their centrality: a central neighbor increases the centrality of the vertex of interest more than a peripheral one.

The *PageRank Centrality* [3] is also spectral (like the *Eigenvector Centrality*), but it is based on very different rationales. It models a random walk occurring on the network, and noticeably includes the possibility for the walker to teleport anywhere in the network at any step. The *Hub* and *Authority Scores* [12] are two complementary measures also based on random walks.

The *Betweenness Centrality* [9] is based on the number of shortest paths going through the considered vertex.

In communication networks, it is sometimes interpreted as the level of control the vertex of interest has over information transmission in the network. The *Closeness Centrality* [9] is the reciprocal of the total geodesic distance (i.e. the length of the shortest path) between the vertex of interest and the other vertices. It is generally considered it measures the efficiency of the vertex to spread a message over the graph, and its independence from the other vertices in terms of communication. The *Eccentricity* [10] is also distance-based, but to the contrary of the other selected measures, it quantifies how peripheral the vertex of interest is, by considering the distance to its farthest vertex.

Finally, the *Coreness Score* [17] is based on the notion of $k$-core, which is a maximal induced subgraph whose all vertices have a degree of at least $k$. The coreness of a vertex is the $k$ of the $k$-core of maximal degree to which it belongs.

**Global Topological Measures.** First, we use very classic statistics describing the graph size, the *Vertex* and *Edge Counts*. We also select the *Density*, which corresponds to the ratio of the number of existing edges to the number of edges in a complete graph containing the same number of vertices. In other words, the density corresponds to the proportion of existing edges, compared to the maximal possible number for the considered graph.

We also use two distance-related measures. The first is the *Diameter*, which corresponds to the highest distance found in the graph, i.e. the length of the longest shortest path. The second is the *Average Distance*, which is the average length of the shortest path processed over all pairs of vertices.

We process the total *Clique Count* in the network, where a clique is a complete induced subgraph. The *Degree Assortativity* [14] is also potentially interesting. It corresponds to the correlation processed between the series constituted of all connected vertices, and measures the statistical dependence between the degrees of two vertices and the presence of an edge connecting them. Finally, for each one of the 10 previously described local measures, we process the average over the whole graph.

## 4   Experiments

In this section, we first briefly present our corpus and our experimental setup (Subsect. 4.1), before describing and discussing our classification results (Subsect. 4.2).

## 4.1   Experimental Setup

We have access to a database of 4,029,343 messages that were exchanged by the users of a browser-based multi-player game. In the database, 779 messages have been flagged by one or more users as being abusive and subsequently confirmed as abusive by the game moderators. Each message belongs to a unique communication channel.

We further extract 2,000 messages at random from the messages not confirmed as abusive to constitute the non-abuse class. We previously experimented with this dataset in [15].

Because of the relatively small dataset, our experiment is set up for 10-Fold cross validation. We use a 70%-train/30%-test split.

We use Python-iGraph [7] to create the network and process the graph-based features for each message. As a classifier, we use an SVM, implemented in Sklearn under the name SVC (C-Support Vector Classification) [16] toolkit.

## 4.2   Results

Table 1 displays the results obtained for our random baseline, the content- and context-based classifier we previously presented in [15], and the graph-based classifier proposed in this article. The baseline uses the same classifier and architecture but the feature extraction step is replaced by a dummy function that yields two random values in $[0, 1]$. Our previous approach takes advantage of morphological, language and user behavior-based features, such as: message length, number of words, compressibility, bag of words with $tf$–$idf$ scores and probability of $n$-gram emission. In the present experiment, the training and testing sets were resampled, which explains why the values displayed for the content/context-based classifier are slightly different form the ones shown in [15].

With the graph-based approach, the performance is improved according to all 3 considered measures, compared to our previous effort. The overall performance is unexpectedly high for an approach that *completely ignores the content* of the messages. We suppose that this is mainly due to the fact that two thirds of the features include information regarding to the part of the conversations happening *after* the classified message, whereas this was the case for only *two* features (out of 67) in our content/context-based approach. Independently from this point,

**Table 1.** Classification results (in %) of the 3 abusive message classifiers: a random baseline, our previous approach [15], and the one presented in this article. All measures are computed for the *Abuse* class.

| Experiment | Precision | Recall | $F$-measure |
|---|---|---|---|
| Random baseline | 29.3 | 52.6 | 37.6 |
| Content/context-based classifier | 70.3 | 74.3 | 72.2 |
| Graph-based classifier | 76.8 | 77.2 | 77.0 |

the fact that both approaches reach relatively high performance levels is a very promising result: given that both classifiers are built on completely different features, combining them should even improve the overall performance.

Since our classifier is an SVM, we can use the Platt Scaling implementation of Sklearn to vary the decision threshold and therefore tune the system towards either high precision or high recall. The left plot of Fig. 2 shows the Precision-Recall curves of each of the 10 classifiers created for our experiment. One can see that by lowering the post probability threshold a little, it is possible to gain better coverage of the abuse class without losing too much precision. Therefore, we would argue that this system shows better promise as an alert system than as an automated moderation system.

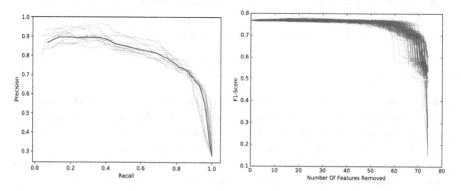

**Fig. 2.** Left: Precision-Recall curves of the 10 SVM classifiers. Right: Feature ablation curves of one classifier (200 runs). In both plots, the red curve represents the average. (Color figure online)

In order to estimate the importance of our features with regards to this classification task, we use the meta estimator *ExtraTreesClassifier* provided by the Sklearn toolkit. While the process is stochastic, it allows to give features a score indicating their contribution to the decisions of the classifier. We run further ablation runs, ordering the features by increasing impact of their removal on the classifier performance. This allows obtaining a smoother curve, with a performance drop on the right-side, corresponding to the removal of the most discriminant features (right plot of Fig. 2). Table 2 shows the 10 most discriminant features: using only these features, one can train a classifier obtaining a $F$-Measure score of 75.8%.

Overall, these features are quite heterogeneous, topologically speaking, in the sense they correspond to very different ways of characterizing graph structures. The *Degree Centrality*, *Edge Count*, and *Density* features are based on a microscopic view of the graph (vertices and edges are considered individually, or only with respect to their direct neighborhood). On the contrary, the *Betweenness Centrality*, *Hub Score* and *Eccentricity* are macroscopic, because they take advantage of paths spanning the whole graph. Finally, the *Coreness Score* is

**Table 2.** Most useful features of the graph-based classification approach.

| Network | Feature name | *F*-measure *before* ablation |
|---------|--------------|-------------------------------|
| Full | Average betweenness centrality | 75.8 |
| Before | Average coreness score | 75.4 |
| After | Edge count | 74.5 |
| After | Density | 73.1 |
| Full | Hub score | 72.9 |
| After | Degree centrality | 67.7 |
| Before | Edge count | 67.2 |
| Full | Average eccentricity | 58.4 |
| Before | Average eigenvector centrality | 56.6 |
| Full | Eccentricity | 35.0 |

mesoscopic, in the sense it is based on an intermediate view and considers subgraphs. This is consistent with the assumption that redundant features should not appear amongst the most discriminant ones.

At first sight, finding both *Edge Count* and *Density* can be surprising: given that the latter is a normalized version of the former, one could suppose they are redundant. However, this normalization is based on the number of vertices in the graph. Thus, in the present case, this simply means that the number of edges in our networks does not increase as a square function of number of vertices. On the contrary, certain features present in the table are part of some very correlated groups of features, which can be considered as almost inter-exchangeable. For instance, the *Average Eigenvector Centrality* and *Average Hub Score* for the *Before* graph have a 0.73 correlation.

All 3 considered types of graphs (*Before, After, Full*) are represented in these top features, which means they convey different information and are all of some help regarding the classification problem at hand. Moreover, it appears that certain related features appear together for several versions of the graph. This is the case for the *Edge Count* (*Before* vs. *After*), and of the Hub Score and Eigenvector Centrality (*Full* vs. *Before*). We assume that this reflects the fact abuses significantly modify the graph structure, according to these topological measures. In other words, they reflect strong changes in the conversation dynamics. When a measure appears only for the *Before* or *After* version of the graph, we conclude it allows characterizing only the pre- or post-state of the conversation, relatively to the abuse.

It is interesting that both the individual and average Eccentricity features are present in this table. A closer look reveals that their values are lower for graphs belonging to the *Abuse* class. This means that the maximal distance between the author of the targeted message and the rest of the graph decreases in case of abuse. More concretely, this user becomes less peripheral (or more central), and the same goes for the other users of the graph (in average). This fits in quite well with assumptions about how abuse impacts a discussion: an abuser would tend not to be peripheral in a conversation, while we can reasonably assume that the other participants will be piling on and therefore be less peripheral themselves. This may also explain why those features are, by far, the most discriminant ones.

## 5  Conclusion

In this article, we have presented an approach purely based on graph features to tackle the problem of automatically detecting online abuse. The method, while simple, yields reasonable results, besting the score obtained with our previous effort, which was content- and context-based.

However it is important to note a couple of important limitations. First, the amount of necessary computation is quite high if it is to be applied each time a new message is posted to the channel, compared to a pure content-based approach. Second, the method can only be applied after a delay when the necessary number of messages have been posted in response to the target message - this is not a method that can help *prevent* the $5^{th}$ message in a torrent of insults from reaching the channel. Rather, it could be used to perform some *a posteriori* moderation.

The next step in our study will be to assess the impact of different network construction strategies on the performance of the classifier. This will include experimenting with other weight distribution strategies, and different sizes for the context period and the sliding window. We will then aim to combine this system with our content-based classifier: in theory, they are both based on completely different types of information, so we can assume they are complementary and could lead to improved classification performance.

**Acknowledgments.** This work was financed by a grant from the *Provence Alpes Cte d'Azur* region (France) and the *Nectar de Code* company.

## References

1. Balci, K., Salah, A.A.: Automatic analysis and identification of verbal aggression and abusive behaviors for online social games. Comput. Hum. Behav. **53**, 517–526 (2015)
2. Bonacich, P.F.: Power and centrality: a family of measures. Am. J. Sociol. **92**, 1170–1182 (1987)
3. Brin, S., Page, L.E.: The anatomy of a large-scale hypertextual web search engine. Comput. Netw. ISDN Syst. **30**, 107–117 (1998)
4. Chavan, V.S., Shylaja, S.S.: Machine learning approach for detection of cyber-aggressive comments by peers on social media network. In: IEEE ICACCI, pp. 2354–2358 (2015)
5. Chen, Y., Zhou, Y., Zhu, S., Xu, H.: Detecting offensive language in social media to protect adolescent online safety. In: PASSAT/SocialCom, pp. 71–80 (2012)
6. Cheng, J., Danescu-Niculescu-Mizil, C., Leskovec, J.: Antisocial behavior in online discussion communities. Preprint arXiv:1504.00680 (2015)

7.  Csardi, G., Nepusz, T.: The igraph software package for complex network research. InterJournal Complex Syst. **1695**(5), 1–9 (2006)
8.  Dinakar, K., Reichart, R., Lieberman, H.: Modeling the detection of textual cyber-bullying. Soc. Mob. Web **11**, 02 (2011)
9.  Freeman, L.C.: Centrality in social networks i: conceptual clarification. Soc. Netw. **1**(3), 215–239 (1978)
10. Harary, F.: Graph Theory. Addison-Wesley, Reading (1969)
11. Hosseini, H., Kannan, S., Zhang, B., Poovendran, R.: Deceiving Google's perspective API built for detecting toxic comments. Preprint arXiv:1702.08138 (2017)
12. Kleinberg, J.: Authoritative sources in a hyperlinked environment. J. Assoc. Comput. Mach. **46**(5), 604–632 (1999)
13. Mutton, P.: Inferring and visualizing social networks on internet relay chat. In: 8th International Conference on Information Visualisation, pp. 35–43 (2004)
14. Newman, M.E.J.: Assortative mixing in networks. Phys. Rev. Lett. **89**(20), 208701 (2002)
15. Papegnies, E., Labatut, V., Dufour, R., Linares, G.: Impact of content features for automatic online abuse detection. In: International Conference on Computational Linguistics and Intelligent Text Processing (2017)
16. Pedregosa, F., Varoquaux, G., Gramfort, A., Michel, V., Thirion, B., Grisel, O., Blondel, M., Prettenhofer, P., Weiss, R., Dubourg, V., et al.: Scikit-learn: machine learning in Python. J. Mach. Learn. Res. **12**, 2825–2830 (2011)
17. Seidman, S.B.: Network structure and minimum degree. Soc. Netw. **5**(3), 269–287 (1983)
18. Sinha, T., Rajasingh, I.: Investigating substructures in goal oriented online communities: case study of Ubuntu IRC. In: IEEE International Advance Computing Conference, pp. 916–922 (2014)
19. Spertus, E.: Smokey: automatic recognition of hostile messages. In: 14th National Conference on Artificial Intelligence and 9th Conference on Innovative Applications of Artificial Intelligence, pp. 1058–1065 (1997)
20. Tavassoli, S., Moessner, M., Zweig, K.A.: Constructing social networks from semi-structured chat-log data. In: IEEE/ACM International Conference on Advances in Social Networks Analysis and Mining, pp. 146–149 (2014)
21. Yin, D., Xue, Z., Hong, L., Davison, B.D., Kontostathis, A., Edwards, L.: Detection of harassment on web 2.0. In: WWW Workshop: Content Analysis in the WEB 2.0 (2009)

# Exploring Temporal Analysis of Tweet Content from Cultural Events

Mathias Quillot[✉], Cassandre Ollivier, Richard Dufour,
and Vincent Labatut

LIA, University of Avignon, Avignon, France
{mathias.quillot,cassandre.ollivier,richard.dufour,
vincent.labatut}@univ-avignon.fr

**Abstract.** Online social networking platforms are an important communication medium for cultural events, as they allow exchanging opinions almost in real-time, by publishing messages during the event itself, but also outside of this period. Word embedding has become a popular way to represent and extract information from such messages. In this paper, we propose a preliminary work aiming at assessing the benefits of taking temporal information into account when modeling messages in the context of a cultural event. We perform statistical and visual analyses on two word different representations: one including temporal information (Temporal Embedding), the second ignoring it (Word2Vec approach). Our preliminary results show that the obtained models exhibit some similarities, but also differ significantly in the way they represent certain specific words. More interestingly, the temporal information conveyed by the Temporal Embedding model allows to identify more relevant word associations related to the domain at hand (cultural festivals).

**Keywords:** Word embedding · Temporal representation · Statistical analysis · Cultural events

## 1  Introduction

Social networks have become a new way of communicating and sharing information and views that can be accessed by billions of people. These social interactions may take the form of short text messages, such as the Twitter platform, where users have the possibility to instantly send a message (here a tweet) containing around 140 characters.

Thanks to its ease of use, Twitter is currently an essential platform for the exchange of messages. For particular events (news, concerts, festivals, presidential elections, etc.), users are increasingly inclined to express themselves through these short messages. Although this platform has become a formidable object of study for a variety of domains ranging from sociology [8,10,13] to automatic information extraction [3,11,15], the short format of the messages and the large size of the corpora often both make them difficult to analyze. In this article,

© Springer International Publishing AG 2017
N. Camelin et al. (Eds.): SLSP 2017, LNAI 10583, pp. 82–93, 2017.
DOI: 10.1007/978-3-319-68456-7_7

we analyze messages exchanged through the Twitter platform in the context of cultural events, with a particular focus on festivals. More precisely, we seek to account for shared content, through the words contained in tweets. The major difficulty of this type of analysis lies in the duration of the considered events: although a festival takes place over a defined period (ranging from a few days to several weeks), the activity of users on social networks can intervene at any time (before, during, or after the festival).

We assume that it is difficult to reveal all the information conveyed through the discussions (the tweets) in a global way without taking into account the temporal aspect of the messages (*i.e.* their emission date). A global model would have a tendency to reveal only the frequently shared information, ignoring the uncommon ones that could nonetheless be important over a particular period of time. This is problematic when these models are used as input for information retrieval tasks, such as automatic event extraction, automatic summarization, etc. Based on this observation we propose a preliminary work close to those initiated in [1,7] that seeks to compare two word embedding-based models: one ignoring the temporal aspect of the messages, using the state-of-the-art *Word2Vec* model [9], and one taking advantage of the emission date of tweets, using the *Temporal Embedding* approach.

The rest of this article is organized as follows. Section 2 presents the different methods used to analyze the impact of time in the events analysis. We present the experimental protocol as well as the results obtained in Sect. 3. Finally, we conclude and give some perspectives in Sect. 4.

## 2   Methods

We seek to highlight the interest of taking into account the temporal information conveyed by messages emitted through social networks, in a context of cultural events analysis. In Subsect. 2.1, the two compared word embedding representations are described: one considers the complete set of messages regardless of their emission date (the *Word2Vec* neural network method), while the second one takes into account the chronology of the documents (*Temporal embedding* approach). In Subsect. 2.2, we describe the methods used to compare the word embedding models, which include both subjective and objective tools. The goal here is to identify the points on which the models differ or converge. Figure 1 summarizes our overall framework, and is detailed in the rest of this section. Our different models are specified there in bold.

We note $N$ the total number of unique words in the corpus. When performing the analysis, one can focus on a list of $n$ words of interest corresponding to a subset of the corpus lexicon: this allows the end user to adopt either a *verification*-directed approach. If the user has some *a priori* knowledge and would like to check certain assumptions regarding the corpus, or a *exploratory* approach, consisting in using the whole lexicon as the word list (in this case, $n = N$).

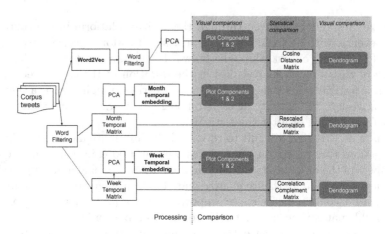

**Fig. 1.** Overview of the method proposed to process the models (left side) and compare them (right side). The blue boxes represent plots, which are compared visually as explained in Subsect. 2.2. The models are represented in bold. (Color figure online)

## 2.1   Word Embedding Representations

We briefly describe the classic *Word2Vec* model, before explaining how the Temporal Embedding model is extracted. Two distinct temporal resolutions are considered in this work (weeks vs. months).

*Word2Vec Neural Network.* *Word2Vec* models [9] are based on the hypothesis that semantically similar words tend to have similar contextual distributions. Concretely, this context is a window whose size is expressed in a number of words. This is centered on the word of interest. For our experiments we use the CBOW method, it seeks to predict the word reference given a context. For instance, a context of 2 words, the CBOW neural network model takes an input taking the form of a sequence of 4 words $w_{i-2}$, $w_{i-1}$, $w_{i+1}$ and $w_{i+2}$, and outputs a word $w_i$. We only use the hidden layer of the neural networks, which means each word is represented by a vector. The length of this vector is specified by the user as a parameter $d$ which is 200 by default in literature. We use the complete word vocabulary to train the CBOW model, the method outputs an $N \times d$ matrix. More information about *Word2Vec* models can be found in [9].

*Temporal Embedding.* Instead of globally taking all the corpus words into account for this method, as for the *Word2Vec* model, we focus on a predefined list of $n$ words of interest (represented by the *Word filtering* box in Fig. 1). For each word in this list we count its number of occurrences by time unit where the time unit is either one month or one week, depending on the considered model. In order to avoid zero values, which can be frequent when considering weekly occurrences, we smooth the numbers of occurrences through a moving average. This results in an $n \times m$ occurrence matrix called *Temporal matrix*, where $m$ is the period covered by the corpus expressed in time units. We then perform a *Principal*

*Component Analysis* (PCA) on this matrix. This provides us with the temporal embedding: another $n \times m$ matrix, whose columns are the obtained components ordered by increasing informativeness. The interest of the PCA is to get a more compact representation of the temporal embedding by focusing on the first few components.

## 2.2  Model Comparison

Comparing models is usually a difficult problem, the most frequent solution being to compare their performance on a targeted task (for example, in speech recognition, the best models are those that allow the lowest word error rate). In this paper, we investigate the interest of including temporal information in word representations from cultural events. To correctly evaluate the impact of each model, an objective ground truth would be necessary, i.e. knowing which words clearly represent a particular event and should be associated with it (and therefore which words have no interest). This would reveal the model that best represents a targeted cultural event. Since no ground truth is available, we first perform an objective comparison relying on statistical tests. These results are generally considered more reliable (and more easily comparable because experiments are reproducible). We seek to know with this first evaluation how close or different the models are. Then, we perform a more detailed analysis through a subjective comparison based on a visual human interpretation in order to investigate the contribution of the temporal information to cultural event representation.

*Objective Comparison.* Two statistical tests are used to compare the models globally: Wilcoxon-Mann-Whitney and Kendall's $\tau$. These non-parametric tests check the hypothesis whether two samples originate from the same distribution. They are complementary, in the sense the former can be considered as a median-based version of the $t$-test, whereas the latter focuses on rank correlation. We apply them to the Rescaled Correlation and Cosine Distance matrices which were previously computed for the models when extracting the plots and dendrograms. The tests allow us to compare the way the pairs of words are ordered in the different models based on their distances. In other words, they compare the models based on the *relative* positions of the words in the model spaces.

Besides this global comparison, we also perform a local one by focusing on each word separately. For a given word, we perform the same Kendall's and Wilcoxon-Mann-Whitney tests as before, on the distances between this word of interest and the other considered words. When comparing two models, we ultimately identify two groups of words: those whose relative positions are significantly different in these models (using a significance level of .05), and the others, whose relative positions are supposedly similar in both models.

*Subjective Comparison.* The second comparison is visually performed by humans, based on graphical representations of the models. We consider two of them: (1) the projection of the words in a 2D space, and (2) dendrograms. The former representation allows us to identify opposition between words, whereas the second one focuses on their associations.

The 2D space representation is obtained by considering the first 2 components (*i.e.* the most informative ones) of a PCA. Since the Temporal Embedding model includes a PCA, no additional process is required. For the *Word2Vec* approach, some additional processing is needed: we extract the $n$ rows corresponding to the word list from the $N \times d$ matrix (this step is represented as the *Word filtering* box in Fig. 1), resulting in the $n \times d$ so-called *Filtered Matrix*, on which a PCA is performed. The subjective comparison is conducted by checking how the words are spatially separated by the plot axes and how this differs from one model to the other. Put differently, if an axis separates two pairs of words and these words are away from this axis, we will consider that they are in opposition. If an opposition is present in both models, we consider this as a similarity between the models. On the contrary, if an opposition is found in one model only, we consider it as a difference between the models.

The dendrograms allow us to identify how the models gather words and organize them hierarchically. We obtain them using the standard hierarchical clustering algorithm available in the R Language. Note that this implementation requires a dissimilarity matrix as its input. Moreover, we use the complete linkage approach in order to favor compact clusters with small diameters. In the case of the temporal embedding, we first build an $n \times n$ correlation matrix based on the (week/month) temporal matrix (*i.e.* without the PCA). The correlation between two words is obtained by processing Pearson's coefficient for the two rows associated to these words in the matrix. When the correlation is negative, we set it to zero: this is a common practice when dealing with temporal series because in this context they are generally considered as noise. The dissimilarity is then obtained through the following rescaling: $1 - Cor$, which in our case produces values ranging from 0 (similar) to 1 (dissimilar). This gives us an $n \times n$ matrix, which is called *Rescaled Correlation Matrix* in Fig. 1. For the *Word2Vec* model, we build an $n \times n$ *Distance Matrix* based on the previously computed $n \times d$ filtered matrix. Each element of this distance matrix corresponds to the Cosine distance between two words of the list. Let us note $w_1$ and $w_2$ the respective word embeddings of these two words, then the distance is given by: $d(w_1, w_2) = 1 - Sim(w_1, w_2)$, where $Sim(w_1, w_2)$ is the classic Cosine similarity between the words. Here the Cosine approach seems more appropriate than the Correlation used on the temporal data because we know that the rows of the considered matrix are inter-dependent by construction. After having generated the dendrograms, we make the visual comparison by checking if two words which are contiguous in one dendrogram are also placed together in another dendrogram.

## 3   Experiments

In this section, we briefly present the data analyzed in this study (Subsect. 3.1), before describing and discussing the obtained results (Subsect. 3.2).

## 3.1 Corpus and List of Words

We use the corpus provided for the MC2 CLEF 2017 lab[1] which contains 70 million tweets [6]. These were automatically retrieved from Twitter using a predefined set of keywords related to cultural festivals in the world. They cover a period ranging from May 2015 to November 2016 and are composed by 134 different languages [4,5].

We focus on a manually curated list of words of interest. It was originally designed by cultural sociologists to focus on festivals. We extended the list in the following way. Firstly, we added certain cities of interest based on various general and specialized sources: Wikipedia's *List of the world's most liveable* (31 cities) [17], BFM Business's *List of the top 20 European cities* [2], and festival cities from Wikipedia's *List of theatre festivals* (30 cities) [18], Red Bulletin's *List of the top 15 music festivals* (12 cities) [12], Sky Scanner's *List of the top 10 music festivals* [14], and Temps de Vivre's *List of top cinema festivals* [16]. Second, we added other words related to the concept of festival in general: "theater", "music", "film". Third, we added some commercial brands also related to festivals, such as *apple* or *deezer* (33 words). In total, the list contains 119 different words.

## 3.2 Results

In this subsection we compare two word embedding representations under the form of 3 distinct models[2]. One *Word2Vec* model and two Temporal Embedding models based on two different time units (weeks and months respectively). We first discuss the outcome of the statistical tests before presenting the visual comparison of the plots and dendrograms (see Sect. 2).

*Statistical Methods.* We apply both the Wilcoxon-Mann-Whitney and Kendall's tests on all 3 models: *Month vs. Week Temporal Embeddings, Month Temporal Embedding vs. Word2Vec,* and *Week Temporal Embedding vs. Word2Vec.* All tests return a $p$-value smaller than $10^{-15}$: for these implementations of the tests, this means that they always reject model independence. Kendall's $\tau$, which is an association measure ranging from $-1$ to $+1$, is 0.68 when comparing the *Month vs. Week Temporal Embeddings*: this corresponds to a strong correlation between these models. For the *Word2Vec vs. Month and Week Temporal models,* we get $\tau = -0.06$ for both comparisons, which means that Word2Vec is almost independent from our temporal models. According to these tests, the information encoded in the temporal models is not the same as the one conveyed by Word2Vec. Thus, they can be considered as complementary and are likely to lead to different results depending on the task at hand.

After the global analysis we switch to individual words to compare the models in terms of which words have significantly different relative positions in the two considered models or a similar position in both models. Based on Kendall's test, we consider a similar representation when $p < 0.05$ and $\tau > 0.7$. For

---

[1] http://mc2.talne.eu/.

[2] http://tac.talne.eu.

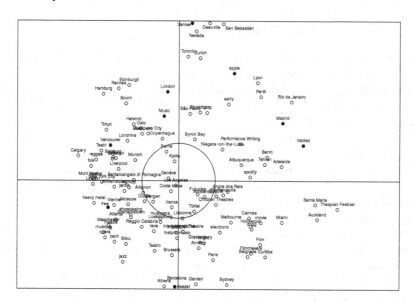

**Fig. 2.** Two first principal components of the Month Temporal Embedding model. (Color figure online)

the *Word2Vec vs. Month Temporal Embedding* models, out of the 119 words constituting our list, only 27 words are represented differently in the models. For the *Month Temporal Embedding vs. Week Temporal Embedding* and *Week Temporal Embedding vs. Word2Vec*, 74 and 25 words are represented differently. With Wilcoxon-Mann-Whitney, we test at $p < 0.05$ and get (in the same order): 64, 25 and 33 differing words. These results show that most words have the similar representation in the different tested model which in accordance to our global test results, means that the difference for the remaining words are very important.

*2D Representation.* We now switch to the visual comparison starting with the 2D plot based on the 2 main components obtained from the PCA. The semantics of these components are not available for the *Word2Vec* model so we do not discuss them. We only consider the positions of word pairs in the graphical representation, and stress the presence of oppositions.

We first compare the Month and Week Temporal Embedding models whose PCA are shown in Figs. 2 and 3, respectively. Globally, they seem to present the same oppositions. For instance, "Deezer" vs. "Venise" (in red in all the figures), and "Vancouver" vs. "Valdez" (in blue). This comparison is in line with our statistical results and seems to indicate that it is not necessary to use the week as a time unit because the month-based model requires less data and captures roughly the same information. We then compare both temporal models with the *Word2Vec* one, whose PCA is shown in Fig. 4. Visually, the oppositions seem to differ more than previously. For instance, "deezer" and "Venise" are not opposed

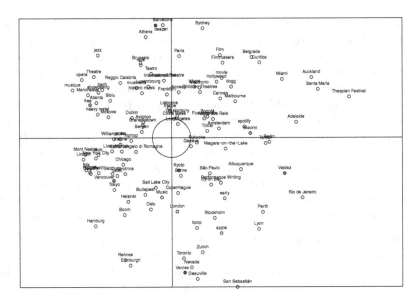

**Fig. 3.** Two first principal components of the Week Temporal Embedding model. (Color figure online)

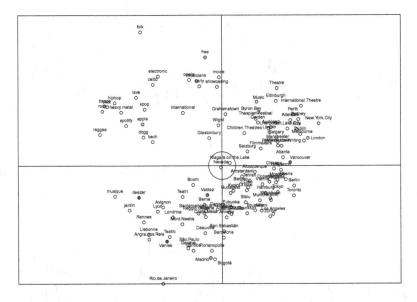

**Fig. 4.** Two first principal components of the Word2Vec model. (Color figure online)

anymore, while "Vancouver" and "Valdez" are less opposed. However we can also see some oppositions such as "Madrid" vs. "free" (in green), which are common to all 3 models.

Let us focus on an illustrative example: the *Apple Music Festival*, which takes place in London. Both Temporal Models tend to group the 3 concerned words "London", "Apple" and "music" (represented in yellow in the figures), whereas *Word2Vec* does not. This means that temporal models tend to gather words from a same cultural event. If we examine finely the *Word2Vec* representation, we can observe it groups words by semantic category: there are several clusters of cities, whereas "Film", "Filmmakers", "Hollywood" and "movie" are together (cinematic items), and so are "musique", "opera", "theatre", "jazz" and "Bach" (musical items). This is of course consistent of how *Word2Vec* is supposed to work. In conclusion of this visual comparison, we can state that not only are the temporal and *Word2Vec* models different as shown by the statistical tests, but the chronological information encoded by the former also allows to identify relevant groups of words relatively to what we know of the studied corpus.

*Dendograms.* The dendrograms of the Month and Week Temporal models, which are represented in Figs. 5 and 6, respectively, look strongly similar. In particular, we get the same direct connection between "Kyoto" and "Albuquerque" (in blue

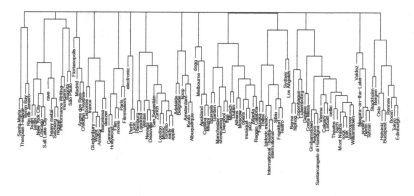

**Fig. 5.** Dendrogram of the Month Temporal Embedding model. (Color figure online)

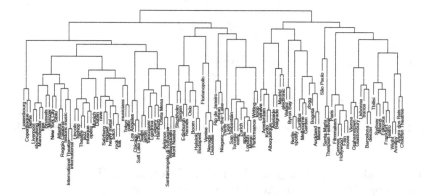

**Fig. 6.** Dendrogram of the Week Temporal Embedding model. (Color figure online)

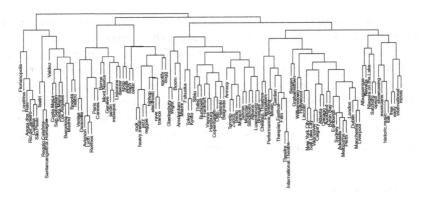

**Fig. 7.** Dendrogram of the Word2Vec model. (Color figure online)

in all the figures), or "Frankfurt" and "Teatro" (in purple). Note that they nevertheless differ on some pairs of words such as "Berne" and "Teatro" (in green), which are directly connected in the month model whereas "Berne" is connected to "hiphop" in the week model. Like before, there are visible differences between the representations of the temporal models and that of the *Word2Vec*, represented in Fig. 7. Focusing on the same pairs of words, we see that "Kyoto" is not connected with "Albuquerque" anymore, and neither are "Frankfurt" and "Teatro". Moreover, "Berne" is neither connected with "Teatro" or "hiphop", but rather with "MontNaeba" (a Japanese mountain).

In the case of *Word2Vec* the words are grouped by semantic similarity. The hierarchical nature of the groupings seem to connect them according to hyper/hyponymy relationships. For instance, "rock" is connected to "heavy.metal" (in orange in the figure), the latter is a subgenre of the former. Both are close to "Jazz" in the dendrogram, itself connected to "reggae" (in red) which can be considered as musical style strongly influenced by certain forms of jazz. Elsewhere, "Tokyo" is connected to "Kyoto" (both Japanese cities, Kyoto is in blue), "Santa Maria" to "Los Angeles" (both Californian cities, in Cyan) or "Paris" and "Cannes" (both French cities, in yellow).

Unlike *Word2Vec*, the temporal models lead to clusters of words which are consistent with the fact that the corpus is related to festivals. For instance in the dendrograms of both temporal models, we observe that "Cannes" (French city hosting a Cinema festival) and "Hollywood" (Californian center of the movie industry) are directly connected (both in orange in the figures), they are close to "Film" and "movie" (themselves directly connected aas well and represented in red). We also have "Avignon" (French city) connected to "Santarcangelo di Romagna" (Italian city): both cities (in cyan in the figures) host a renown theater festival. An other example is "London", directly connected to "Music" and close to "Apple" (all in yellow): this would represent the previously mentioned *Apple Music Festival* in London. Generally speaking in the dendrograms, festival cities are associated to words which are semantically related to the nature of their festival.

The visual comparison of the dendrograms confirm and complete our previous observations. The *Word2Vec* model is definitely different from the temporal ones which are relatively similar to each other. The context-based approach of *Word2Vec* does not capture the temporal information conveyed by the other models. In both cases, words are grouped according to their relative semantics. But in the case of *Word2Vec* these groups are built on semantic proximity. However with the Temporal Embeddings words from the same groups are indirectly related through a festival. In conclusion, the latter models seem more appropriate to a festival-oriented analysis of this corpus.

## 4   Conclusion

In this preliminary work, our objective was to study how considering temporal information affects the word embedding-based modeling of text corpora. We built two Temporal Embedding models and one *Word2Vec* model based on a corpus of tweets focusing on cultural events. We studied and compared them objectively through statistical tests and visually through PCA plots and dendrograms. It turns out that both temporal models appear to be highly similar according to their PCA plots and dendrograms whereas they seem different from the Word2Vec model. The statistical tests conclude that when taken globally, they are all significantly different from on an other. However, when considering each word separately we show that they differ only on a small proportion of words albeit with a large magnitude. These first results show that temporal information is not completely captured on *Word2Vec* model and *Temporal Embedding* is worth considering, at least when dealing with analyzing temporally messages related to cultural events.

However, a finer analysis is indispensable to correctly characterize the contribution of time information. To this end, we must carry on several steps. Firstly, we could adopt a more exploratory approach by expanding our analysis to the whole lexicon instead of focusing on a predefined list of words. Secondly, we could compare our models built on the considered corpus with out-of-the-box *Word2Vec* models in order to analyze the contribution of the corpus regarding how words are represented.

**Acknowledgments.** This work was funded by the GaFes project supported by the French National Research Agency (ANR) under contract ANR-14-CE24-0022.

## References

1. Basile, P., Caputo, A., Semeraro, G.: Analysing word meaning over time by exploiting temporal random indexing. In: First Italian Conference on Computational Linguistics CLiC-it (2014)
2. BFM Business: list of the top 20 European cities (2017). http://bfmbusiness.bfmtv.com/diaporama/le-top-20-des-villes-europeennes-ou-il-fait-bon-vivre-2832/20-dublin-1/

3. Endarnoto, S.K., Pradipta, S., Nugroho, A.S., Purnama, J.: Traffic condition infor-mation extraction & visualization from social media Twitter for Android mobile application. In: ICEEI, pp. 1–4 (2011)
4. Ermakova, L., Goeuriot, L., Mothe, J., Mulhem, P., Nie, J.Y., SanJuan, E.: Cul-tural micro-blog contextualization 2016 workshop overview: data and pilot tasks. In: CLEF Working Notes, pp. 1197–1200 (2016)
5. Goeuriot, L., Mothe, J., Mulhem, P., Murtagh, F., SanJuan, E.: Overview of the CLEF 2016 cultural micro-blog contextualization workshop. In: Fuhr, N., Quaresma, P., Gonçalves, T., Larsen, B., Balog, K., Macdonald, C., Cappellato, L., Ferro, N. (eds.) CLEF 2016. LNCS, vol. 9822, pp. 371–378. Springer, Cham (2016). doi:10.1007/978-3-319-44564-9_30
6. Goeuriot, L., Mulhem, P., SanJuan, E.: CLEF 2017 MC2 search and time line tasks overview. In: CLEF Working Notes, pp. 1197–1200 (2017)
7. Hamilton, W.L., Leskovec, J., Jurafsky, D.: Diachronic word embeddings reveal statistical laws of semantic change. arXiv preprint arXiv:1605.09096 (2016)
8. Java, A., Song, X., Finin, T., Tseng, B.: Why we Twitter: understanding microblog-ging usage and communities. In: WebKDD/SNA-KDD Workshop on Web Mining and Social Network Analysis, pp. 56–65 (2007)
9. Mikolov, T., Chen, K., Corrado, G., Dean, J.: Efficient estimation of word repre-sentations in vector space. arXiv:1301.3781 (2013)
10. Murthy, D.: Towards a sociological understanding of social media: theorizing Twit-ter. Sociology **46**(6), 1059–1073 (2012)
11. Nakov, P., Ritter, A., Rosenthal, S., Sebastiani, F., Stoyanov, V.: SemEval-2016 task 4: sentiment analysis in Twitter. In: SemEval, pp. 1–18 (2016)
12. Red Bulletin: list of the top 15 music festivals (2017). https://www.redbulletin.com/fr/fr/culture/les-15-meilleurs-festivals
13. Romero, D.M., Meeder, B., Kleinberg, J.: Differences in the mechanics of informa-tion diffusion across topics: idioms, political hashtags, and complex contagion on Twitter. In: WWW, pp. 695–704 (2011)
14. Sky Scanner: list of the top 10 music festivals (2017). https://www.skyscanner.fr/actualites/les-10-meilleurs-festivals-de-musique-du-monde
15. Sriram, B., Fuhry, D., Demir, E., Ferhatosmanoglu, H., Demirbas, M.: Short text classification in Twitter to improve information filtering. In: ACM SIGIR, pp. 841–842 (2010)
16. Temps de Vivre: list of the top cinema festivals (2017). http://www.tempsdevivre.org/cinebook/festivals.htm
17. Wikipedia: list of the world's most liveable cities (2017). https://fr.wikipedia.org/wiki/Classement_des_villes_les_plus_agreables_a_vivre
18. Wikipedia: list of theatre festivals (2017). https://en.wikipedia.org/wiki/List_of_theatre_festivals

# Towards a Relation-Based Argument Extraction Model for Argumentation Mining

Gil Rocha[✉] and Henrique Lopes Cardoso

LIACC/DEI, Faculdade de Engenharia, Universidade do Porto,
Rua Dr. Roberto Frias, 4200-465 Porto, Portugal
{gil.rocha,hlc}@fe.up.pt

**Abstract.** Argumentation mining aims to detect and identify the argumentative content expressed in text. In this paper we present a relation-based approach that aims to capture the relation of inference between the premise and conclusion. We follow a supervised machine learning approach and explore features at different levels of abstraction. Then, we apply this system for the task of argumentative sentence detection and compare the performance of the system with a competitive baseline approach. The corpus used in our experiments was annotated with arguments from textual resources written in Portuguese, namely opinion articles. The proposed system outperforms the baseline system, achieving 0.75 of f1-score on the test set.

**Keywords:** Information extraction · Argumentation mining · Machine learning · Natural language processing

## 1 Introduction

Argumentation is the process whereby arguments are constructed, presented and evaluated. An argument is composed by a set of propositions, where some of them (the premises) are pieces of evidence offered in support of a conclusion. The conclusion is a proposition that has truth-value (which is either true or false), put forward by somebody as true on the basis of the premises. As an example of an argument, consider the following two sentences: "All men are mortal and Socrates is a man. Therefore, Socrates is mortal.". In this simple example, the conclusion is "Socrates is mortal." and the premises are "All men are mortal" and "Socrates is a man". Each piece of text that constitutes an argument component (*i.e.* premise or conclusion) is known as an *Argumentative Discourse Unit* (ADU) [16]. The aim of *Argumentation Mining* (AM) from text, a sub-domain of text mining, is the automatic detection and identification of the argumentative structure contained within a piece of natural language text. As input, this process receives a piece of natural language text. If the text under analysis contains argumentative content, AM aims to detect all the arguments that are present in the text document, the relations between them and the internal structure of each individual argument. In the end, this process should

© Springer International Publishing AG 2017
N. Camelin et al. (Eds.): SLSP 2017, LNAI 10583, pp. 94–105, 2017.
DOI: 10.1007/978-3-319-68456-7_8

be able to output the corresponding argument diagram: the visual representation of the arguments presented in the text. The full task of AM can be decomposed into several subtasks [17], namely: text segmentation, identification of ADUs, ADU type classification (*i.e.* premise or conclusion), relation identification and relation type classification (*i.e.* support or conclusion).

In this paper we address the task of *Argumentative Sentence Detection* (ASD) following a supervised machine learning approach and employing different formulations to address this task. We explore several machine learning (ML) and natural language processing (NLP) techniques and features at different levels of abstraction: lexical, syntactic, structural and semantic. Some of these features were constructed using external resources, such as: a part-of-speech tagger, fuzzy wordnet and a model developed to recognize textual entailment and paraphrases.

*Recognizing Textual Entailment* (RTE) [4], a NLP task closely related to AM, aims to find entailment relations between text fragments. Given two text fragments, typically denoted as 'Text' (T) and 'Hypothesis' (H), RTE is the task of determining whether the meaning of the Hypothesis (H, *e.g.* "Joe Smith contributes to academia") is entailed (can be inferred) from the Text (T, *e.g.* "Joe Smith offers a generous gift to the university") [21]. In other words, a sentence T entails another sentence H if after reading and knowing that T is true, a human would infer that H must also be true. We may think of textual entailment and paraphrasing in terms of logical entailment ($\models$) (see [2] for more details).

This paper is structured as follows: Sect. 2 presents related work on argumentation mining. Section 3 introduces the corpus that was used in our experiments. Section 4 describe some of the external resources that were used to performed some of the NLP tasks and employ some of features described in this paper. Section 5 describes the methods that were used to address the task of ASD using supervised ML algorithms. Section 6 presents the results obtained by the system described in this paper. Finally, Sect. 7 concludes and points to directions of future work.

## 2   Related Work

Most argumentation mining approaches follow a machine learning paradigm, relying on heavily engineered NLP pipelines, extensive manual creation of features and making several simplifying assumptions for each subtask of the process.

Identifying arguments and their components are the first steps of an argumentation mining system. The former is typically formulated as a binary classification problem. Most existing systems make the simplifying assumption that ADUs are sentence level and employ wide variety of ML algorithms, including SVM [20,23], Logistic regression [18], Naïve Bayes [5], Maximum Entropy [14] and Decision Trees [5,23]. Employed features can be divided into lexical, syntactic, structural and semantic. Performance of state-of-the-art systems ranges from 0.55 to 0.77 of F1-score. Fine-grained approaches to determine the exact boundaries of ADUs usually apply state-of-the-art sequential models, such as

HMM and CRF [10, 11, 22], with performance ranging from 0.2 to 0.42 F1-score. An exception [22] reports F1-score of 0.867, though limited to a specific text genre (persuasive essays).

ADU classification aims to classify each ADU according to its argumentative role. Approaches vary mainly in the adopted argumentation theory, leading to different sets of labels. Typically, systems employ supervised ML algorithms and specialized features: lexical, syntactic, structural, topic, sentiment and semantic [14, 18, 22]. Performance varies from 0.17 to 0.83 F1-score, depending on the type of texts and assumptions made.

The last two steps of the process comprise the identification and classification of rhetorical relations between ADUs, aiming to obtain an argument diagram. Few state-of-the-art argumentation mining systems address these subtasks: [3] uses textual entailment; [14] uses a context-free grammar; [17] uses a minimum spanning tree algorithm; [12] combines methods from discourse analysis, topic modeling and supervised ML; [22] employs SVM using lexical, syntactic, discourse and structural features combined with a stance recognition model. Performance ranges from 0.51 to 0.83 F1-score, relying on simplifying assumptions regarding previous steps of the process and differing on the target argumentative relations and structure.

## 3  Corpus

The corpus used in the experiments reported in this paper, the *ArgMine* corpus[1], consists of a news articles collection, namely opinion articles, crawled from *SAPO* (a portal that aggregates news from several news providers in Portugal, amongst other services) and annotated with arguments by human annotators. An opinion article is an article published in a newspaper that reflects the author's opinion about a specific subject. One of the advantages of working with opinion articles is the richer argumentative content that is typically present, as compared to other types of news articles. On the other hand, authors tend to use refined vocabulary which can make the interpretation of the text more challenging. In addition, different authors tend to use different writing styles, which create some variability in the analyzed texts, and in turn complicate the task of machine learning algorithms. Another characteristic of opinion articles is their typical length: they are typically longer than other types of news articles.

Since longer text documents are more difficult and very time-consuming to annotate, each opinion article was divided into paragraphs. Consequently, for each annotation task a paragraph is presented to annotators instead of the complete opinion article. Providing paragraphs to annotators instead of the complete article can have some drawbacks, namely: when an argument is spread through several paragraphs, it is impossible to annotate it because each part of the argument will be presented in different annotation tasks; moreover, in some situations it can happen that some information in the remaining parts of the document could be useful and/or necessary to detect the arguments presented

---

[1] http://corpora.aifdb.org/ArgMine.

in one of the parts of the document. In the first case, we assume that this situation will not occur too often. A paragraph corresponds to a distinct section in a document, usually dealing with a single topic and terminated by a new line. Since arguments have to be about some topic and changes in topic can indicate that different arguments are being expressed, as explored in [12], then this assumption seems reasonable. In some situations where they are spread through several paragraphs, arguments require complex reasoning and knowledge about the world that are beyond the scope of the approaches presented in this paper.

In each annotation task, the annotators were asked to annotate all the arguments that are explicitly stated in the corresponding paragraph. These annotations consist of argument diagrams (*i.e.* a graph structure, where each node corresponds to an ADU and arrows indicate relations of support or conflict between ADUs) following the premise-conclusion argumentation model.

More details regarding the characteristics of the *ArgMine* corpus are presented in the following sections.

## 4   Resources

Here we introduce external resources used as auxiliary tools by the methods employed in this paper.

### 4.1   Data Preparation

To transform each sentence into the corresponding set of tokens and to obtain for each token the corresponding lemma and part-of-speech information (including syntactic function, person, number, tense, amongst others) we used the *CitiusTagger* [8] NLP tool. This tool includes a named entity recognizer trained in natural language text written in Portuguese.

Several experiments were made using different NLP techniques to process the sentences received as input: removing stop-words and auxiliary words (*i.e.* words relevant for the discourse structure but not domain specific, such as: prepositions, determiners, conjunctions, interjections, numbers and some adverbial groups) and lemmatization. Transforming each token in the corresponding lemma is a promising approach because it will make explicit that some of the words are repeated in both sentences, even if small variations of these words are used (*e.g.* different verb tenses). After this step, each sentence was represented in a structured format (set of tokens) and annotated with some additional information regarding the content of the text (*e.g.* part-of-speech tags).

### 4.2   Semantic Resources

Knowledge about the words of a language and their semantic relations with other words can be exploited with large-scale lexical databases. To enrich the feature set shown in Tables 1 and 2 with semantic knowledge, we explored external semantic resources. By exploiting these resources we aim to enable the system to deal better with the diversity and ambiguity of natural language text.

Similarly to WordNet [6] for the English language, CONTO.PT [9] is a fuzzy wordnet for Portuguese, which groups words into sets of cognitive synonyms (called *synsets*), each expressing a distinct concept. In addition, synsets are interlinked by means of conceptual and semantic relations (*e.g.* "hyperonym" and "part-of"). Synsets included in CONTO.PT were automatically extracted from several linguistic resources. All the relations represented in CONTO.PT (i.e. relations between words and synsets, as well as relations between synsets) include degrees of membership. Two tokens (obtained after tokenization and lemmatization) are considered synonyms if they occur in the same synset. One token $T_i$ is considered hyperonym of $T_j$ if there exists a hyperonym relation ("hyperonym_of") between the synset of $T_i$ and the synset of $T_j$. Similarly, $T_i$ is considered meronym of $T_j$ if there exists a meronym relation ("part_of" or "member_of") between the synset of $T_i$ and the synset of $T_j$.

Finally, we exploit a distributed representation of words (word embeddings). These distributions map a word from a dictionary to a feature vector in high-dimensional space in an unsupervised setting (without human intervention). This real-valued vector representation tries to arrange words with similar meanings close to each other based on the co-occurrences of these words in large-scale (non-annotated) corpora. Then, from these representations, interesting features can be explored, such as semantic and syntactic similarities. In our experiments, we used a pre-trained model provided by the *Polyglot*[2] tool [1], in which a neural network architecture was trained with Portuguese *Wikipedia* articles.

In order to obtain a score indicating the similarity between two text fragments $T_i$ and $T_j$, we compute the cosine similarity between the vectors representing each of the text fragments in the embedding space. Each text fragment is projected into the embedding space as $\vec{T_i} = \sum_{k=1}^{n} \vec{e}(w_k)n^{-1}$, where $\vec{e}(w_k)$ represents the embedding vector of the word $w_k$ and $n$ corresponds to the number of words contained in the text fragment $T_i$. Then, we compute the final value of the cosine similarity $\delta_{\vec{T_i},\vec{T_j}} = \cos(\vec{T_i}, \vec{T_j})$, $\delta_{\vec{T_i},\vec{T_j}} \in [-1, 1]$ followed by the following rescaling and normalization: $(1.0 - \delta_{\vec{T_i},\vec{T_j}})/2.0$. The entailment versor ($\hat{d}$) corresponds to the normalized direction vector obtained by subtracting the projection of $T$ in the embedding space, $\vec{e}(T)$, from the projection of $H$, $\vec{e}(H)$.

Additionally, we made use of an external system for recognizing textual entailment and paraphrases in text written in the Portuguese language [19]. This system receives as input a pair of sentences $\langle T, H \rangle$, where $T$ corresponds to the *Text* sentence and $H$ to the *Hypothesis* sentence. Given that the problem was formulated as a multi-class classification problem, the system classifies each $\langle T, H \rangle$ with one of the labels *Entailment* (if $T \models H$), *Paraphrase* (if $T \models H$ and $H \models T$, *i.e.*, if $T$ is paraphrase of $H$), or *None* (if $T$ and $H$ are not related with one of the previous labels). The system was trained in the ASSIN corpus [7], which corresponds, to the best of our knowledge, to the first corpus annotated with pairs of sentences written in Portuguese that is suitable for this task. It contains 5000 pairs of sentences extracted from news articles written

---

[2] http://polyglot.readthedocs.io/en/latest/index.html.

in European-Portuguese (EP) and 5000 pairs of sentences written in Brazilian-Portuguese (BP), obtained from *Google News* Portugal and Brazil, respectively. The model for recognizing textual entailment and paraphrases used in this paper was trained and evaluated in the EP partition of the corpus using a maximum entropy model. This model achieved an overall 0.83 of accuracy on the test set.

## 5   Methods

We here describe the approach we followed to address the task of argumentative sentence detection from natural language Portuguese text. We formulate the problem as a binary classification problem, following two distinct settings, as described in Sects. 5.1 and 5.2.

### 5.1   Sentence-Based Approach

In the first setting, each learning instance corresponds to a sentence and we aim to classify each sentence as *Argumentative* (Arg), if it contains one complete argument or at least one argumentative discourse unit (ADU), or *Non-argumentative* (NArg) otherwise. Following this setting, we make the simplifying assumption that an ADU or complete argument (*i.e.* containing at least two ADUs, the conclusion and one premise) corresponds to a single sentence. This is a strong assumption because some of the ADUs that can be found in the corpus have intra-sentence boundaries. However, learning intra-sentence boundaries to retrieve the exact boundaries of each ADU requires a corpus containing a considerable amount of intra-sentence annotations, something that the *ArgMine* corpus is lacking at this moment. We argue that making this assumption is the most adequate approach (given the corpus) to the problem.

This experimental setting can be seen as our baseline approach since it corresponds to the simplest way of formulating the problem.

**Data Preparation.** For each news article $a_i$, where $a_i \in C^{argmine}$, we divided $a_i$ into sentences using the *Citius Tagger* tool [8], which offers the functionality of dividing a given text in different sentences as part of the process of part-of-speech tagging. Concatenating all the sentences obtained from each article $a_i \in C^{argmine}$, we obtain dataset $X$, which will be used for the task of argumentative sentence detection. For each sentence $x_j \in X$, we determine the corresponding target value $y_j \in Y$, where $Y$ represents the set of target values, by performing the following procedure: consider news article $a_i$, where $x_j \in a_i$, and let $Z$ be the set of ADUs annotated for news article $a_i$. We consider that sentence $x_j$ has argumentative content ($y_j = 1$) if $\exists z_i \in Z : (z_i \subseteq x_j)$ *or* $(x_j \subseteq z_i)$. Otherwise, we consider that sentence $x_j$ has no argumentative content ($y_j = 0$).

**Features.** As listed in Table 1, we employ features at different levels of abstraction, namely: lexical, syntactic, structural and semantic-level.

**Table 1.** Feature set for Sentence-based approach

| Feature | Description |
|---|---|
| **Lexical** | |
| Bag-of-words | Contiguous sequence of 1 to $N$ tokens from a given sentence. We encode the presence of unigrams ($N = 1$), bigrams ($N = 2$), and trigrams ($N = 3$) in the sentence. Experiments were made with one-hot encoding and TF-IDF encoding; |
| Clue words | If contains words typically found in argumentative content; |
| Word couples | All possible combinations of word pairs within a sentence. Experiments were made constraining the pair of words to include one or two clue words. Experiments were made with one-hot encoding and TF-IDF encoding; |
| **Syntactic** | |
| Stats | Statistics regarding some of the part-of-speech tags occurring in the sentence, namely: adverbs, modal auxiliary, verbs and punctuation marks. Experiments were made with normalized counters and one-hot encoding; |
| Verb tense | Verb tense changes between sentence and surrounding sentences. |
| **Structural** | |
| Sentence_length | Number of tokens in the sentence; |
| Avg. word length | Averaged number of letters in each word in the sentence; |
| Relative position | Sentence relative position in the document. |
| **Semantic** | |
| Domain words overlap | Overlap of domain words (nouns, adjectives, verbs) between the sentence and the surrounding sentences. Each pair of words is considered an overlap if they have the same lemma or one of the following relations: synonym, hypernym and meronym. |
| RTE prediction | If RTE system predicts that the sentence entails or is entailed by any other sentence in the same document |
| Cosine_similarity | Cosine similarity between the embedding vector $\vec{e}(s_i)$ and the embedding vector $\vec{e}(s_j)$, with $j \in \{i-1; i+1\}$. |
| Entail_versor | Entailment versor ($\vec{d}$) in the word embeddings space. |

## 5.2 Relation-Based Approach

In this setting the problem is formulated in two steps: (a) a binary classifier is trained to distinguish whether a pair of sentences constitutes a simple argument or not (binary classifier). Here we assume that each sentence is an ADU and that one of the sentences plays the role of premise and the other plays the role of conclusion, composing a simple argument; (b) each sentence is classified as an argumentative sentence (Arg) if the classifier described in (a) predicts that the sentence is part of an argument (premise or conclusion) when paired with any other sentence within a given document. Otherwise, the sentence is classified as non-argumentative (NArg).

We hypothesize that the second formulation yields better predictions for ASD since it encapsulates and focuses on the notion that an argument is made of the least two components: one conclusion and at least one premise. In Sect. 6, experiments made to validate this hypothesis are presented.

**Data Preparation.** Similarly to the procedure performed with the method described in Sect. 5.1, we divided each news article $a_i$ into sentences using the *Citius Tagger* tool [8]. For each sentence $s_j \in a_i$, a pair of sentences is created with each of the remaining sentences $s_k$, with $k \in [1, |a_i|] \land k \neq j$. A positive (argumentative) pair is created with the first sentence (P) playing the role of premise and the second sentence (C) playing the role of conclusion in the corresponding annotated argument diagram. Otherwise, the pair of sentences is considered a negative (non-argumentative) pair. We followed this setup for the following reasons: (a) this approach follows the formulation used by the system for RTE and paraphrases. Consequently, predictions made by this system can be directly applied as a feature; (b) this is a consistent way of creating the learning instances (*i.e.* the premise is always the first sentence and the conclusion always the second sentence), which is an important requirement for the learning process when employing machine learning algorithms.

**Table 2.** Feature set for relation-based approach

| Feature | Description |
|---|---|
| *Lexical* | |
| Word couples | All possible combinations of word pairs between the sentences (one word in P and other word in C). Experiments were made constraining the pair of words to include one or two clue words. Experiments were made with one-hot and TF-IDF encoding |
| Clue Words | If exists premise keyword in P and conclusion keyword in C; |
| *Syntactic* | |
| Stats | Statistics regarding some of the part-of-speech tags occurring in P and C, namely: adverbs, modal auxiliary, verbs and punctuation marks. Experiments were made with normalized counters and one-hot encoding; |
| Verb tense | Changes in the verb tense between P and C. |
| *Structural* | |
| Sentence_length | Number of tokens in P and C; |
| Avg. word length | Averaged number of letters in each word in P and C; |
| Relative position | Absolute distance in number of sentences between P and C. |
| *Semantic* | |
| Domain words overlap | Overlap of domain words between P and C. An overlap occurs when two words have the same lemma or are synonyms |
| Hyperonym | % of tokens in $T$ hyperonyms of tokens in $H$. And vice-versa. |
| Meronym | % of tokens in $T$ meronyms of tokens in $H$. And vice-versa. |
| Antonym | % of tokens in $T$ antonyms of tokens in $H$. And vice-versa. |
| RTE prediction | RTE system predicts that P entails C |
| Cosine_similarity | Cosine similarity between the embedding vector $\vec{e}(P)$ and $\vec{e}(C)$ |
| Entail_versor | Entailment versor ($\hat{d}$) in the word embeddings space. |

Due the characteristics of the corpus, where the number of sentences containing ADUs is lower than the number of sentences that do not contain any ADU, the number of non-argumentative pairs generated with this approach is much larger than the number of argumentative sentence pairs. Consequently, we obtained a dataset that is extremely unbalanced. To overcome this problem, we performed *random undersampling* [13] to generate a balanced dataset, by randomly removing some of the learning instances (non-argumentative pairs).

**Features.** As listed in Table 2, we employ features at different levels of abstraction, namely: lexical, syntactic, structural and semantic-level.

**Resolution Step.** After training the model to classify each pair of sentences as argumentative or non-argumentative we have to translate these predictions to classify each sentence as argumentative or not (ASD), which corresponds to the task we aim to address in this paper.

First, for all possible pairs of sentences in a given document the model previously described predicts if the sentences constitute an argument or not. Then, for all sentences within a document we retrieve all the predictions where the target sentence was used (as P or C). If at least one of these predictions indicates that the target sentence forms an argument with any other sentence, then we indicate that the target sentence is an argumentative sentence. This procedure can be seen as a resolution step where we retrieve all pair-wise predictions and transform them into sentence-level predictions.

## 6   Experiments

The results presented in this section were obtained using the methods described in Sect. 5 and exploring the corpus described in Sect. 3.

For each classification task, we have run several experiments exploring some well known state-of-the-art algorithms, namely: *Support Vector Machine* (SVM) using linear and polynomial kernels, *Maximum Entropy model* (MaxEnt), *Adaptive Boosting* algorithm (AdaBoost) using *Decision Trees* as weak classifiers, *Random Forest Classifier* using *Decision Trees* as weak classifiers, and *Multilayer Perceptron Classifier* (Neural Net) with one hidden layer. All the ML algorithms previously mentioned were employed using the *scikit-learn* library [15] for the *Python* programming language. Since the MaxEnt model performed better for all the experiments presented in this paper, the results depicted in this section were all obtained using this model.

First, we report on 5-fold cross validation results over all the training examples available in the corpus described in Sect. 3 and using the model described in Sect. 5.1. The system obtained using this experimental setup is our baseline. Results are shown in Table 3.

In the second evaluation scenario we report results obtained using the method presented in Sect. 5.2. Since the number of non-argumentative (NArg) sentence pairs is substantially higher than the number of argumentative (Arg) sentence pairs, we employed methods to generate balanced datasets. To obtain the dataset

**Table 3.** Sentence-based approach scores

|        | Prec. | Rec. | F1 | # Sentences |
|--------|-------|------|------|-------------|
| **NArg** | 0.55 | 0.57 | 0.56 | 291 |
| **Arg** | 0.49 | 0.47 | 0.48 | 255 |

presented in Table 4, we used the random undersampling technique [13] by randomly removing some of the NArg examples until the number of NArg examples is the same as the number of Arg examples. The results shown in Table 4 were obtained in a 5-fold cross validation scenario.

**Table 4.** Relation-based approach scores

|        | Prec. | Rec. | F1 | # Sentence Pairs |
|--------|-------|------|------|------------------|
| **NArg** | 0.94 | 0.81 | 0.87 | 114 |
| **Arg** | 0.83 | 0.95 | 0.89 | 114 |

Finally, the results depicted in Table 5 were obtained using the test set partition from the corpus described in Sect. 3. The test set consists of 50 sentences: 37 non-argumentative sentences (NArg) and 13 argumentative sentences (Arg). From the analysis of the results, we conclude that the Relation-based approach yields the best overall results and, therefore, corresponds to the model that generalizes better to unseen data. This results confirm the hypothesis formulated in this paper: the Relation-based approach seems to provide a better formulation for the Argumentative Sentence Detection task.

**Table 5.** ASD test set scores

|        | Sentence-based approach | | | Relation-based approach | | |
|--------|-------|------|------|-------|------|------|
|        | Prec. | Rec. | F1 | Prec. | Rec. | F1 |
| **NArg** | 0.81 | 0.57 | 0.67 | 0.88 | 0.76 | 0.81 |
| **Arg** | 0.33 | 0.62 | 0.43 | 0.5 | 0.69 | 0.58 |

## 7   Conclusions

In this paper we address the task of argumentative sentence detection from text written in the Portuguese language. We aim to classify each sentence as containing argumentative content (*i.e.* containing a premise, conclusion or complete argument) or not. We formulate the task following two different approaches: sentence-based and relation-based approach. Validating our hypothesis, the relation based approach outperformed the sentence-based approach in the test set, demonstrating that the relation-based system generalizes better to unseen data for the task of ASD. In future work, we aim to replicate these experiments in a different corpus to validate the conclusions reported in this paper for texts written in other languages and with a corpora containing more annotated data. Furthermore, we aim to improve the quality of the semantic-based features. Even though semantic-based features were shown to have a positive impact in the predictions made by the system, we noticed some problems regarding coverage and propagation of errors caused by the external tools employed in this paper. Better computations (*e.g.* metrics to evaluate semantic similarity in the embeddings space and fuzzy wordnet), different sentence-level representations (*e.g.* exploring tree and dependency parsers) and approaches to deal with problems of coverage that were experienced when employing external resources are promising directions to improve the results presented in this paper that we aim to pursue.

**Acknowledgments.** The first author is partially supported by a doctoral grant from Doctoral Program in Informatics Engineering (ProDEI) from the Faculty of Engineering of the University of Porto (FEUP).

# References

1. Al-Rfou, R., Perozzi, B., Skiena, S.: Polyglot: distributed word representations for multilingual NLP. In: Proceedings of the 17th Conference on Computational Natural Language Learning, pp. 183–192. ACL, Sofia, August 2013
2. Androutsopoulos, I., Malakasiotis, P.: A survey of paraphrasing and textual entailment methods. J. Artif. Int. Res. **38**(1), 135–187 (2010)
3. Cabrio, E., Villata, S.: Natural language arguments: a combined approach. In: ECAI, vol. 242, pp. 205–210 (2012)
4. Dagan, I., Roth, D., Sammons, M., Zanzotto, F.M.: Recognizing Textual Entailment: Models and Applications. Synthesis Lectures on Human Language Technologies. Morgan & Claypool Publishers, San Rafael (2013)
5. Eckle-Kohler, J., Kluge, R., Gurevych, I.: On the role of discourse markers for discriminating claims and premises in argumentative discourse. In: Proceedings of the Conference on Empirical Methods in NLP, Lisbon, Portugal, 17–21 September 2015, pp. 2236–2242 (2015)
6. Fellbaum, C. (ed.): WordNet: An Electronic Lexical Database Language, Speech, and Communication. MIT Press, Cambridge (1998)
7. Fonseca, E., Santos, L., Criscuolo, M., Aluisio, S.: ASSIN: Avaliacao de similaridade semantica e inferencia textual. In: Computational Processing of the Portuguese Language - 12th International Conference, Tomar, Portugal, 13–15 July 2016 (2016)
8. Garcia, M., Gamallo, P.: Yet another suite of multilingual NLP tools. In: Sierra-Rodríguez, J.-L., Leal, J.P., Simões, A. (eds.) SLATE 2015. CCIS, vol. 563, pp. 65–75. Springer, Cham (2015). doi:10.1007/978-3-319-27653-3_7
9. Oliveira, H.G.: CONTO.PT: groundwork for the automatic creation of a fuzzy Portuguese wordnet. In: Silva, J., Ribeiro, R., Quaresma, P., Adami, A., Branco, A. (eds.) PROPOR 2016. LNCS, vol. 9727, pp. 283–295. Springer, Cham (2016). doi:10.1007/978-3-319-41552-9_29

10. Goudas, T., Louizos, C., Petasis, G., Karkaletsis, V.: Argument extraction from news, blogs, and social media. In: Likas, A., Blekas, K., Kalles, D. (eds.) SETN 2014. LNCS, vol. 8445, pp. 287–299. Springer, Cham (2014). doi:10.1007/978-3-319-07064-3_23

11. Habernal, I., Gurevych, I.: Exploiting debate portals for semi-supervised argumentation mining in user-generated web discourse. In: Proceedings of the Conference on Empirical Methods in NLP, pp. 2127–2137. Association for Computational Linguistics, Lisbon (2015)

12. Lawrence, J., Reed, C.: Combining argument mining techniques. In: Proceedings of the 2nd Workshop on Argumentation Mining, pp. 127–136. ACL (2015)

13. More, A.: Survey of resampling techniques for improving classification performance in unbalanced datasets. Computing Research Repository (CoRR) (2016)

14. Palau, R.M., Moens, M.F.: Argumentation mining: the detection, classification and structure of arguments in text. In: Proceedings of the 12th International Conference on Artificial Intelligence and Law, pp. 98–107. ACM, New York (2009)

15. Pedregosa, F., Varoquaux, G., Gramfort, A., Michel, V., Thirion, B., Grisel, O., Blondel, M., Prettenhofer, P., Weiss, R., Dubourg, V., Vanderplas, J., Passos, A., Cournapeau, D., Brucher, M., Perrot, M., Duchesnay, E.: Scikit-learn: machine learning in python. J. Mach. Learn. Res. **12**, 2825–2830 (2011)

16. Peldszus, A., Stede, M.: From argument diagrams to argumentation mining in texts: a survey. Int. J. Cogn. Inf. Nat. Intell. (IJCINI) **7**(1), 1–31 (2013)

17. Peldszus, A., Stede, M.: Joint prediction in MST-style discourse parsing for argumentation mining. In: Proceedings of the Conference on Empirical Methods in NLP, pp. 938–948. ACL, Lisbon, September 2015

18. Rinott, R., Dankin, L., Perez, C.A., Khapra, M.M., Aharoni, E., Slonim, N.: Show me your evidence - an automatic method for context dependent evidence detection. In: Proceedings of the Conference on Empirical Methods in NLP, Lisbon, Portugal, 17–21 September 2015, pp. 440–450 (2015)

19. Rocha, G., Lopes Cardoso, H.: Recognizing textual entailment and paraphrases in Portuguese. In: Oliveira, E., Gama, J., Vale, Z., Lopes Cardoso, H. (eds.) EPIA 2017. LNCS, vol. 10423, pp. 868–879. Springer, Cham (2017). doi:10.1007/978-3-319-65340-2_70

20. Rocha, G., Lopes Cardoso, H., Teixeira, J.: ArgMine: a framework for argumentation mining. In: Computational Processing of the Portuguese Language - 12th International Conference, PROPOR 2016, Student Research Workshop, Tomar, Portugal, 13–15 July 2016 (2016)

21. Sammons, M., Vydiswaran, V., Roth, D.: Recognizing textual entailment. In: Bikel, D.M., Zitouni, I. (eds.) Multilingual Natural Language Applications: From Theory to Practice, pp. 209–258. Prentice Hall, Upper Saddle River (2012)

22. Stab, C.: Argumentative writing support by means of natural language processing. Ph.D. thesis, Technische Universität Darmstadt, Darmstadt (2017)

23. Stab, C., Gurevych, I.: Identifying argumentative discourse structures in persuasive essays. In: Proceedings of the Conference on Empirical Methods in NLP, Doha, Qatar, 25–29 October 2014, pp. 46–56 (2014)

# Post-processing and Applications of Automatic Transcriptions

# Three Experiments on the Application of Automatic Speech Recognition in Industrial Environments

Ferdinand Fuhrmann[(✉)], Anna Maly, Christina Leitner, and Franz Graf

Joanneum Research,
Institute for Information and Communication Technology,
Steyrergasse 17, Graz, Austria
ferdinand.fuhrmann@joanneum.at

**Abstract.** In this work we examine the performance of automatic speech recognition (ASR) in industrial applications. We particularly present three experiments relating to the capturing device applied, the signal pre-processing employed, and the recognition engine used. Here, our aim was to create experimental conditions as close as possible to the envisioned application, i.e., an industrial adoption of ASR. Our results show the existence of evident dependencies between the recognition engine, the type of capturing device, and the noise type on the one side, and the complexity of the task, the present Signal-to-Noise-Ratio (SNR), and the minimum-acceptable SNR value on the other side. In summary, this work gives an overview of the capabilities and limitations of nowadays ASR systems for an application in an industrial context.

**Keywords:** Speech recognition · Industrial application · Noise

## 1 Introduction

Due to the spread of mobile devices, applications of ASR now range across almost all commercial and professional areas. Its employment in the producing industry is, however, still very limited, since companies often rely on traditional, thus reliable systems, which are hard to enhance with new technologies. Nevertheless, speech technologies have been in increased focus of industrial research effort for the last years.

Research on ASR goes back to the 1950s and has been continuously developed over the last decades [4]. Conventional systems based on Hidden-Markov-Models (HMMs) seemed to stagnate in performance around 20 years ago. Recently, advances in computing power and the availability of masses of internet-sourced acoustic and text data have provided a new performance boost. A good overview of the conventional approach and nowadays ASR systems can be found in [8] and [3], respectively.

The work reported in this article has been supported by the Austrian Ministry for Transport, Innovation and Technology (bmvit).

© Springer International Publishing AG 2017
N. Camelin et al. (Eds.): SLSP 2017, LNAI 10583, pp. 109–118, 2017.
DOI: 10.1007/978-3-319-68456-7_9

Industrial applications provide several difficulties for ASR applications. First of all, the acoustic environment exhibits non-stationary, often very loud noise sources. Here, noise reduction or speech enhancement algorithms must be applied to improve the recognition performance of the ASR system [1,5]. Moreover, users of such systems often speak with strong accents due to dialects or migration backgrounds. Modern ASR systems use speaker adaptation methods to adapt their acoustic models to the specific properties of the user(s) [9]. Finally, the used vocabulary depends heavily on the application domain context, which is often not covered by multi-purpose systems.

In this work we aim at comparing different speech recognition setups for industrial applications. These setups differ in the acoustic environment, the applied microphone type, the used pre-processing, the speech recognition engine, and the task. In particular, we compare different industrial noise types, distant vs. close microphone settings, different passive and active signal enhancement methods, scalable vs. cloud-based ASR, as well as various task settings (e.g., vocabulary complexity) in terms of the resulting recognition performance. All of the used resources (data, ASR signal chain, application tasks, etc.) were obtained and designed considering conditions as realistic as possible. The presented results give us clear insights into the potentials and limitations of the evaluated ASR setups in the context of industrial applications.

The remainder is organized as follows: Sect. 2.1 provides information about the data applied, while Sect. 2.2 describes the method used to generate the different ASR setups as well as the ASR systems employed. Section 3 outlines the three experiments while Sect. 4 presents and discusses the results obtained. Finally, Sect. 5 concludes this article.

## 2    Materials and Methods

In this section, we describe the main methodology behind the experiments. The aim was to compare the performance of different ASR systems in industrial applications under the most realistic conditions. We therefore first recorded real speech and environmental audio (noise) signals from production sites. Second, we generated evaluation data by recording the simulataneous-emitted speech and noise signals at various SNR values with different signal capturing and pre-processing devices. We then performed speech recognition from the generated evaluation data and compared the resulting performance of 2 different ASR systems. In all experiments, we used a sampling frequency of 44.1 kHz and 16-bit amplitude quantization, the employed language was German.

### 2.1    Data

The data used for the evaluation experiments were taken solely from recordings of real target signals. We recorded speech utterances from subjects working in the industrial production-site area and gathered environmental noise signals from different production sites.

We visited 6 different production sites (PS) where we were able to record indoor environmental audio signals. We placed several omnidirectional microphones at different positions at the production site and captured up to 6 h of audio data at each site. The visited production sites range from heavy machinery production (injection molding with dozens of machines in the hall), over semi-automatic production lines with humans and robots working side-by-side, to the set-up of automatic production lines where humans install and teach robots. We recorded during the production process, hence the full range of manufacturing sounds was captured. Measured sound pressure levels ranged from 60 to 82 dBA, Fig. 1 additionally shows the average Power Spectral Densities (PSD) of the six noise types recorded in PS1 to PS6.

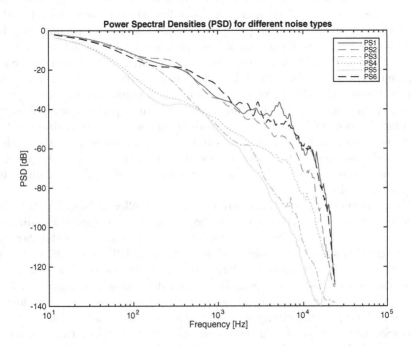

**Fig. 1.** Power Spectral Densities (PSD) for noise types PS1 to PS6. Data was obtained by averaging the power spectra of 30 s of audio followed by a gaussian smoothing. For better comparability, resulting spectra were aligned to start at 0 dB

Speech signals were acquired by recording 52 (39 males and 13 females) subjects working in the industrial production-site domain. Many of them had strong accents due to a migration background or a specific regional dialect. We prepared 70 short utterances in German which included short commands, typical menu navigation utterances, search requests to a machine, and some out-of-context sentences. Subjects were asked to read from a printed sheet of paper and advised to speak as natural as possible. The speech signals were captured via an AKG HSC271 headset microphone, we used the headphones to playback noise samples previously recorded (PS1 with a SPL of 75 dBA measured with a reference microphone placed at the outer ear channel entrance) to account for the Lombard effect [7].

## 2.2   Method

Next, we describe the process of generating evaluation data from the previously recorded speech and noise signals. The aim was to create the most comparable and realistic settings among different ASR setups for a given acoustic environment. The experimental setup was arranged in an acoustically treated room with semi-aneochic reflective behavior. A NTi TalkBox reproduced the target speech signal which was captured by different sound receiving devices at different distances. We particularly used a high-quality headset microphone H (AKG HSC271) at the mouth position (i.e. very close), a shotgun microphone S (AKG C300B with CK98 capsule), a cardioide microphone C (AKG C300B with CK91 capsule), and a 14 cm 1-dimensional microphone array with 8 equally spaced measurement microphones (PCB 130D20), each of the 3 latter in a 1 m distance to the sound emitter. Noise was input via 4 studio loudspeakers, placed all around the emitter-receiver setup, with membranes pointing away from the recording devices. This enabled a quasi-diffuse acoustic noise environment. A reference microphone was placed at 0.5 m distance, used to measure the levels of speech and noise signals. Levels were measured using the A-weighted equivalent continuous sound level ($LA_{eq}$) over an integration time of 1 s. Figure 2 illustrates the experimental setup.

The microphone array signals were processed by different beamforming algorithms to enhance the received speech audio signal. Here, we applied two different beamforming algorithms, Delay & Sum (D&S) beamforming and the Parsimonious Excitation-based Generalized sidelobe canceller (PEG) [6]. D&S compensates the phase differences of the target direction to enforce signals from the target direction and average signals from other directions by summation of the delayed signals [2]. The D&S introduces no artifacts but shows weak interference suppression and poor directivity behavior for low frequencies. The PEG localizes and tracks the target speaker in every frequency bin and utilizes the information achieved to build the blocking matrix and the adaptive noise canceller of the subsequent Generalized Sidelobe Canceller (GSC). We implemented this beamformer with a shrink threshold, defining the opening angle of the main lobe, of $\pm 9°$. The PEG shows high suppression performance but introduces artifacts by adaptively subtracting noise.

ASR was then applied by either a local, scalable system (LSS) or an online, cloud-based system (OCS). The main difference between these two ASR approaches is that the former is fully configurable while the latter can be viewed as a black-box where the speech signal is input and a text string is output. We particularly provided a restricted phrase-based dictionary, i.e. finite-state grammar (FSG), in the LSS for each application scenario (see Sect. 3), while the OCS is only updated with the respective context-domain words. Moreover, the

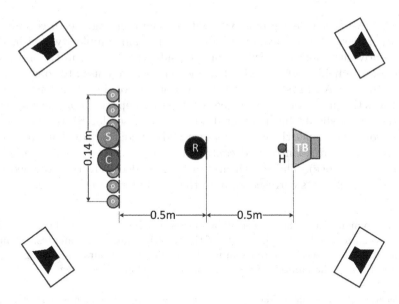

**Fig. 2.** Schematic illustration of the experimental setup for data set generation. Speech signals are emitted by the TalkBox speaker (TB) and received by the headset (H), shotgun (S), cardioid (C), and an array of omnidirectional microphones (O). Noise is generated by the 4 loudspeakers positioned around the emitter-receiver setup. The reference microphone (R) is used to measure the levels of speech and noise

SDK applied for the LSS (see below) offered full adaptation functionality for the acoustic models. We therefore adapted the underlying acoustic models for both the speaker and the acoustic background[1].

## 3   Experiments

As a first step, we generated evaluation data at different SNRs, i.e., $-5$, 0, 5, and 10 dB, using all speech data for all noise types. In a second step, the recorded signals are sent to the ASR systems and the output is evaluated for different task scenarios to infer the performance of the ASR setups examined. Here, we defined the following three scenarios to represent different possible applications of ASR systems for industrial tasks:

- **S1 - C&C.** Typical command and control utterances, consisting of at most 2 words, e.g., "greater", "engine on".
- **S2 - Menu.** Menu navigation via speech utterances for a machine operation, e.g., "open the setup mask".
- **S3 - Search.** Parameter search in a complex machine user interface, including many domain words, e.g., "where can I find the heating parameters".

---

[1] We note that detailed information related to the acoustic models applied by the ASR systems cannot be provided here, since these parts of the systems are meant to be used as black-boxes. We nevertheless assume that both ASR systems adopt state-of-the-art acoustic models following a DNN-HMM structure.

From these scenarios we built vocabularies representing the three different application tasks. These vocabularies differ in size and complexity, and include the respective utterances recorded from the subjects working in the industrial production-site field (see Sect. 2.1). The vocabularies were used to construct the FSG for the LSS ASR system, as well as the domain-word context for the OCS. Table 1 lists the properties of the three vocabularies depending on the respective ASR system. It should be noted that we explicitly set for S1 the number of domain-words to 0, since the used utterances did not contain any domain-specific semantics, rather frequently-used control sequences such as *"start engine"* or *"greater"* (see above). The speech utterances recorded were then divided into three different datasets corresponding to the vocabularies outlined above.

**Table 1.** Comparison of the defined vocabularies, resulting from the three scenarios. For each scenario, the total audio length of the resulting dataset in minutes, the maximum length of the speech utterance in words, the number of phrases in the grammar (for the LSS), and the number of domain words (for the OCS) is shown.

| Scenario | Length [min] | Max. length | # phrases (LSS) | # domain words (OCS) |
|---|---|---|---|---|
| S1 - C&C | 12.0 | 2 | 27 | 0 |
| S2 - menu | 12.4 | 8 | 112 | 20 |
| S3 - search | 20 | 9 | 957 | 33 |

We used a commercial available ASR SDK, typically used for low SNR applications, as the LSS, and Google's Cloud Speech API[2] as the OCS system. Language settings were set to German for both systems.

In order to evaluate the performance of the different ASR setups we compared the Word Accuracy (WA) of the resulting output. For each scenario (S1 to S3) applied in the subsequent experiments, we record all relevant utterances from all 52 subjects with all capturing devices at each considered SNR.

From these data, we derived the 3 experiments. They will be examined in detail in the remainder of this work.

- **E1 - Capturing.** We compare the capturing devices applied for a given acoustic environment at various SNRs and for the different scenarios. Following Sect. 2.2, these devices are H, S, C, D&S, and PEG. Hence, we can compare the ASR performance depending on the devices for a determined SNR and a specific task. As described above, the given scenario defines the language model setting applied. We use the LSS ASR system in this experiment.
- **E2 - Noise.** We compare the different noise types recorded at the production sites for various SNRs for a given capturing device. Thus, we can identify the influence of the acoustic environment on the ASR performance. Again, we use the LSS in this experiment.

---

[2] https://cloud.google.com/speech/.

- **E3 - ASR method.** We compare the two ASR methods described in Sect. 2.2 for a given acoustic environment at various SNRs and for different tasks. This is done for selected capturing devices. Here, we can assess the applicability of each method for a determined SNR in each task.

## 4 Results

Table 2 shows the results for E1. Here, a comparison of the WA for all capturing devices dependent on the SNR and the scenario is shown. In the following, we will particularly refer to three aspects covered, (a) the difference between close vs. a distant microphones, (b) the performance of the active and passive pre-processing devices, and (c) the influence of the task.

**Table 2.** Results for E1. WA is shown for different capturing devices at different SNR values and different scenarios, for a given noise type (PS1). The devices are headset (H), shotgun (S), cardioid (C), delay-&-sum (D&S), and parsimonious excitation-based generalized sidelobe canceller (PEG). The LSS system is used in this experiment for recognition

| SNR | Scenario | H | S | C | D&S | PEG |
|---|---|---|---|---|---|---|
| 10 dB | C&C | 98.2 | 95.9 | 93.6 | 92.7 | 90.4 |
| | Menu | 93.6 | 91.5 | 86.0 | 86.7 | 88.3 |
| | Search | 85.7 | 79.4 | 72.4 | 69.8 | 70.0 |
| 5 dB | C&C | 97.3 | 93.1 | 88.0 | 86.6 | 89.1 |
| | Menu | 92.8 | 86.2 | 81.3 | 81.0 | 85.7 |
| | Search | 83.7 | 72.1 | 62.2 | 58.1 | 64.5 |
| 0 dB | C&C | 96.6 | 86.4 | 77.9 | 73.3 | 82.1 |
| | Menu | 92.5 | 79.6 | 73.4 | 72.9 | 78.9 |
| | Search | 81.1 | 54.3 | 40.1 | 36.1 | 48.5 |
| −5 dB | C&C | 95.2 | 66.7 | 53.8 | 52.4 | 70.9 |
| | Menu | 90.4 | 69.2 | 62.9 | 64.4 | 70.1 |
| | Search | 75.9 | 24.4 | 13.2 | 12.3 | 23.3 |

(a) A first comparison of close vs. distant microphones obviously reveals differences between H and the others for any SNR setting. Only for small vocabularies and high SNR values (10 dB and more) these differences become small and a comparable ASR performance can be expected. For SNR values of 5 dB and less the close microphone settings clearly outperform all other microphones in terms of speech recognition accuracy.
(b) We show that the best performing active pre-processing algorithm (PEG) can achieve a recognition rate as good as S, but only for low SNR values. Since the PEG algorithm heavily distorts the enhanced microphone signal, recognition performance is degraded for high SNR values and improved for

lower ones when compared to S. Moreover, D&S does not perform better than C for any of the evaluated configurations. We note that array geometry is an important factor influencing the performance of the beamformer. Here we applied a very simple geometry with a very limited number of microphones. We expect performance to improve for all array setups with enhanced geometries (nested, spirals, etc.) and more microphones.

(c) Evaluating the simulated scenarios we can see that C&C (S1) combined with H always achieve acceptable performance. Contrary, a single cardioid microphone setup (C) only works for high SNR values (5 dB and greater), and passive signal pre-processing via S may be used for SNRs greater than 0 dB. Menu navigation (S2) via speech input may be useful for SNRs of 10 dB and greater with a single microphone, while a complex search functionality (S3) is only applicable with H and SNRs of 5 dB and greater.

**Table 3.** Results for E2. WA is shown for different noise types at various SNR values and three different scenarios, for a given capturing device (i.e. shotgun). Again, the LSS system is used in this experiment for recognition

| SNR | Scenario | PS1 | PS2 | PS3 | PS4 | PS5 | PS6 |
|---|---|---|---|---|---|---|---|
| 10 dB | C&C | 95.9 | 96.1 | 97.3 | 97.1 | 97.8 | 96.2 |
|  | Menu | 91.5 | 91.8 | 92.2 | 92.5 | 93.7 | 92.3 |
|  | Search | 79.4 | 79.8 | 83.3 | 82.3 | 85.1 | 79.1 |
| 5 dB | C&C | 93.1 | 92.6 | 95.9 | 96.0 | 96.8 | 93.3 |
|  | Menu | 86.2 | 86.7 | 91.7 | 91.0 | 93.3 | 88.6 |
|  | Search | 72.1 | 71.5 | 80.3 | 77.4 | 83.4 | 73.2 |
| 0 dB | C&C | 86.4 | 85.2 | 92.5 | 92.8 | 95.2 | 83.3 |
|  | Menu | 79.6 | 77.3 | 88.3 | 88.0 | 91.3 | 79.7 |
|  | Search | 54.3 | 52.5 | 72.1 | 69.7 | 79.5 | 54.7 |
| −5 dB | C&C | 66.7 | 61.0 | 83.2 | 81.5 | 92.7 | 61.1 |
|  | Menu | 69.2 | 66.5 | 79.8 | 79.1 | 85.1 | 68.1 |
|  | Search | 24.4 | 19.0 | 52.0 | 50.3 | 69.0 | 26.7 |

Table 3 shows the results for E2. Here, we compare for a given ASR setup – shotgun microphone S with LSS system – the WA for the different recorded noise types. As can be seen, the performance greatly depends on the noise type. For instance, the scenario *menu* is applicable for all depicted SNRs for noise type PS5, while for most other noise types a SNR of 5 dB is the absolute minimum. However, looking at the figures alone is not clearly conclusive. Therefore, inspection of Fig. 1 together with a perceptual evaluation revealed the differences in the respective noise types; the more energy in the frequency band 500 Hz to 5 kHz, the worse the performance of the ASR setup. Perceptually speaking, the brighter the noise sounds, the more problems should be expected for speech interfaces (e.g. the acoustic environment of PS1 and PS2 consisted of the noise of dozens heavy machines, while the PS3 settings only included a few machines making far less mid and high frequency noise). This observation makes sense since those frequencies which are relevant for human perception to recognize speech fall inside the above-mentioned frequency interval. In conclusion, SNR values only approximately give hints about the applicability of speech interfaces. It seems that the properties of the noise type should be considered as important as the SNR value.

**Table 4.** Results for E3. WA is shown for the two different ASR systems and two selected capturing devices (headset and cardioid) for a given noise type (PS1)

| SNR | Scenario | Headset (H) | | Cardioid (C) | |
|-----|----------|-----|-----|-----|-----|
| | | LSS | OCS | LSS | OCS |
| 10 dB | C&C | 98.2 | 72.8 | 93.6 | 65.4 |
| | Menu | 93.6 | 94.6 | 86.0 | 92.5 |
| | Search | 85.7 | 80.0 | 72.4 | 74.0 |
| 5 dB | C&C | 97.3 | 73.1 | 88.0 | 55.8 |
| | Menu | 92.8 | 94.5 | 81.3 | 83.9 |
| | Search | 83.7 | 79.1 | 62.2 | 67.6 |
| 0 dB | C&C | 96.6 | 68.7 | 77.9 | 26.9 |
| | Menu | 92.5 | 93.4 | 73.4 | 59.9 |
| | Search | 81.1 | 79.2 | 40.1 | 47.4 |
| −5 dB | C&C | 95.2 | 63.8 | 53.8 | 7.0 |
| | Menu | 90.4 | 91.6 | 62.9 | 18.4 |
| | Search | 75.9 | 78.2 | 13.2 | 12.2 |

Finally, Table 4 shows the results for E3. Here, we compare, for the microphones headset (H) and cardioid (C), the performance between the two different recognition engines described in Sect. 3. Due to their difference in language modeling, we can observe great recognition performance differences dependent on the SNR value and the task (S1-S3). First, OCS is not an option for the C&C task, as the performance of this system is worse for all SNRs and evaluated microphones. This result seems obvious since the language context, which is heavily utilized by OCS systems, is not present in this task. For syntactically more complex utterances as represented in S2 and S3, OCS exhibits similar performance figures as the LSS, even though we can observe a performance drop for lower SNR values when compared to the LSS. In summary, for the majority of tasks employed in industrial applications, regardless of SNR and used signal capturing device, FSG-based systems (LSS) are the best choice. However, if more language context is provided in the utterances, OCS systems offer the potentials of a natural language recognition and an easy integration via cloud APIs, with a similar performance compared to the LSS.

At last it should be noted that the here-adopted experimental method including the separation of data collection and laboratory experimentation allows for a wide range of parameter adjustments and thus experimental evaluation. Nevertheless, this method may introduce some minor misalignment in the resulting figures of the tables above. In particular, the used playback level and noise type applied in the recording of the speech utterances do not always align with the corresponding settings of those parameters in the laboratory experiments. Nevertheless, we think that the bias introduced by this misalignment can safely be neglected when looking at the broad scope of the obtained results.

## 5    Conclusions

In this work we present 3 experiments related to ASR for industrial applications. In a thorough experimental methodology we evaluated the different components in a typical ASR system signal chain. We only applied realistic audio data from industrial production sites in our evaluation experiments. Our results give a clear picture regarding the main influencing factors determining the applicability of ASR in industrial context. Depending on the task, the SNR and the frequency characteristics of the acoustic environment, one has to choose the proper signal capturing, pre-processing and recognition system. Here, our tabular comparison provides clear suggestions towards the right setup given the determining factors. In conclusion, speech recognition is ready for the industry but must be carefully implemented with respect to the identified conditions.

## References

1. Acero, A.: Acoustical and Environmental Robustness in Automatic Speech Recognition, vol. 201. Springer Science+Business Media, Heidelberg (2012)
2. Brandstein, M., Ward, D.: Microphone Arrays: Signal Processing Techniques and Applications. Springer Science+Business Media, Heidelberg (2013)
3. Deng, L., Li, X.: Machine learning paradigms for speech recognition: an overview. IEEE Trans. Audio Speech Lang. Process. **21**(5), 1060–1089 (2013)
4. Huang, X., Baker, J., Reddy, R.: A historical perspective of speech recognition. Commun. ACM **57**(1), 94–103 (2014)
5. Li, J., Deng, L., Gong, Y., Haeb-Umbach, R.: An overview of noise-robust automatic speech recognition. IEEE/ACM Trans. Audio Speech Lang. Process. **22**(4), 745–777 (2014)
6. Madhu, N., Martin, R.: A versatile framework for speaker separation using a model-based speaker localization approach. IEEE Trans. Audio Speech Lang. Process. **19**(7), 1900–1912 (2011)
7. Pick, H.L., Siegel, G.M., Fox, P.W., Garber, S.R., Kearney, J.K.: Inhibiting the Lombard effect. J. Acoust. Soc. Am. **85**(2), 894–900 (1989)
8. Rabiner, L., Juang, B.H.: Fundamentals of speech recognition (1993)
9. Shinoda, K.: Speaker adaptation techniques for automatic speech recognition. In: Proceedings of the APSIPA ASC (2011)

# Enriching Confusion Networks for Post-processing

Sahar Ghannay$^{(\boxtimes)}$, Yannick Estève, and Nathalie Camelin

LIUM - Le Mans University, Le Mans, France
{sahar.ghannay,yannick.esteve,nathalie.camelin}@univ-lemans.fr

**Abstract.** The paper proposes a new approach for *a posteriori* enrichment of automatic speech recognition (ASR) confusion networks (CNs). CNs are usually needed to decrease word error rate and to compute confidence measures, but they are also used in many ways in order to improve post-processing of ASR outputs. For instance, they can be helpfully used to propose alternative word hypotheses when ASR outputs are corrected by a human on post-edition. However, CNs bins do not have a fixed length, and sometimes contain only one or two word hypotheses: in this case the number of alternatives to correct a misrecognized word is very low, reducing the chance of helping the human annotator.

Our approach for CN enrichment is based on a new similarity measure presented in this paper, computed from acoustic and linguistic word embeddings, that allows us to take into consideration both acoustic and linguistic similarities between two words.

Experimental results show that our approach is relevant: enriched CNs (for a bin size equals to 6) increase the potential correction of erroneous words by 23% than initial CNs produced by an ASR system. In our experiments, a spoken language understanding task is also targeted.

**Keywords:** Confusion networks · Post processing · Acoustic and linguistic word embeddings · Similarity measurey measure

## 1 Introduction

Despite of the recent advances in the field of speech processing, automatic speech recognition errors are still unavoidable. This reflects the sensitivity of this technology to variability, *e.g.* to acoustic conditions, speaker, language style, *etc.*

These errors can have a considerable impact on applications based on the use of automatic transcriptions, like subtitling, computer assisted transcription, speech to speech translation, spoken language understanding, information retrieval, *etc.* Error detection and correction aim to improve the exploitation of ASR outputs by downstream applications.

Many studies have focused on ASR error detection and correction, some of them [1–3] have attempted to improve recognition accuracy for many tasks such as keyword search, spoken language understanding and other tasks by using discriminative post-processing on ASR outputs.

Other studies consider the use of automatic speech recognition confusion networks (CNs) to decrease word error rate and to compute confidence measure.

© Springer International Publishing AG 2017
N. Camelin et al. (Eds.): SLSP 2017, LNAI 10583, pp. 119–130, 2017.
DOI: 10.1007/978-3-319-68456-7_10

These networks were introduced in [4]. They rely on *posterior* probabilities and were used to represent a set of alternative sentences. Authors in [5] propose an approach to automatically correct erroneous words in the CNs. It depends on the use of the n-grams and the semantic score between words that are located far from each other based on Normalized Relevance Distance. Confusion networks can be used as well in many ways to improve post-processing of ASR outputs. For instance, they can be used to propose alternative word hypotheses when ASR outputs are corrected by a human on post-edition [6]. However, CN bins, sets of competing hypothesis between two nodes in the CN, do not have a fixed length, and sometimes contain only one or two word hypotheses: in this case the number of alternatives to correct a misrecognized word is very low, reducing the chance of helping the human annotator.

In this study, we propose an approach for CN enrichment, which is based on a similarity measure computed from acoustic and linguistic word embeddings. This measure allows us to take into consideration both acoustic and linguistic similarities between two words. Since our assumption is that word hypotheses in a same bin should be close in term of acoustics and/or linguistics, we propose to use this particularity to enrich confusion networks by applying this new similarity measure. This enrichment will be evaluated in the context of a human post-edition of automatic transcripts. Moreover, the proposed similarity measure can be used in a spoken language understanding (SLU) system in order to propose semantically relevant alternative words to ASR outputs. Last, this similarity measure is applied as well for prediction of potential ASR errors for rare words.

## 2    Word Embeddings

Many neural approaches have been proposed to build word embeddings, they can be based on continuous bag of words, syntactic dependency, co-occurrences matrix, and even audio signal. Hence, they can capture different types of information: semantic, syntactic, and acoustic.

### 2.1    Linguistic Word Embeddings

Word embeddings were initially introduced through the construction of neural language models [7,8]. They are defined as projections in a continuous space of words in a manner that preserve semantic and syntactic similarities.

Following the results published in [9] in which different kinds of word embeddings are evaluated and different word embeddings combinations are compared, we use a combination of word embeddings to get better results. It has been shown in [9] that the combination through PCA (Principal Component Analysis) achieves the best performance in the analogical and similarity tasks. Since the approach we propose is based on the cosine similarity too, we suggest to use PCA to combine *word2vecf* on dependency trees [10], *skip-gram* provided by *word2vec* [11], and *GloVe* [12]. The description of these embeddings and the combination approaches is presented in our previous study [9].

We considered word embeddings presented here as linguistic representations of words, since they are built based on lexical, contextual, and syntactic information.

## 2.2   Acoustic Word Embeddings

Recent studies have started to reconsider the use of whole words as the basic modeling unit in speech recognition and query applications, instead of phonetic units. These systems are based on the use of a function that embeds an arbitrary or fixed dimensional speech segments into a continuous space, named acoustic embeddings, in a such way that speech segments of words that sound similarly will have similar embeddings. These representations were successfully used in a query-by-example search system [13,14], in a segmental ASR lattice re-scoring system [15] and recently for ASR error detection [16].

In [15], the authors proposed an approach to build acoustic word embeddings of words observed in an audio corpus, and also of words never observed in this corpus, by exploiting their orthographic representation. Moreover, a such acoustic word embedding derived from an orthographic representation can be perceived as a canonical acoustic representation for a word, since different pronunciations imply different acoustic embeddings. This approach relies on the use of two neural architectures: a convolutional neural network classifier over words trained independently to build acoustic embeddings, and a deep neural network (DNN) trained by using a triplet ranking loss function [15,17,18] in order to project the orthographic word representation to the acoustic embeddings space, that results the acoustic word embeddings $\mathbf{w}^+$. The orthographic word representation consists on a bag of n-grams of letters ($n \leq 3$), in which we reduce its dimension using an auto-encoder, that results the orthographic embeddings $\mathbf{o}^+$.

In another previous study [19], we have investigated the evaluation of the intrinsic performances of acoustic word embeddings, and compare them to their orthographic embeddings, on orthographic, phonetic similarities and homophone detection tasks. As a reminder, we report in Table 1 some results obtained in that study.

**Table 1.** Evaluation results of similarity ($\rho$) and homophone detection tasks (*precision*). $\rho$ corresponds to the Spearman's rank correlation coefficient

| Tasks | 160K Vocab. | |
|---|---|---|
| | $\mathbf{o}^+$ | $\mathbf{w}^+$ |
| Orthographic | **0.569** | 0.510 |
| Phonetic | 0.414 | **0.468** |
| Homophone | 0.528 | **0.593** |

As shown in this table, the acoustic word embeddings are better than orthographic ones to measure the phonetic proximity between two words. Moreover, they are better too to detect homophone words. These results confirm that acoustic word embeddings have captured additional information about word pronunciation in addition to the information carried by their spelling. In this study, the acoustic word embeddings are used as acoustic representations of the words.

## 3    Similarity Measure to Enrich Confusion Networks

In this study, we propose to use both linguistic and acoustic word embeddings to *a posteriori* enrich confusion networks, in order to improve post-processing of ASR outputs. Due to the nature of acoustic and language models involved in an ASR system, our assumption is that word hypotheses in a same bin should be close from acoustic and/or linguistic points of view.

Since we aim to enrich confusion networks by adding nearest neighbors of the recognized word hypotheses, this neighborhood has to be characterized: it is done through the cosine similarities of acoustic and linguistic word embeddings.

With the purpose to take benefit from both linguistic and acoustic similarities, we propose to use a linear interpolation to combine them. This results to a similarity called $LA_{SimInter}$, defined as:

$$LA_{SimInter}(\lambda, x, y) = (1 - \lambda) \times L_{Sim}(x, y) + \lambda \times A_{Sim}(x, y) \tag{1}$$

where $x$ and $y$ are two words, $\lambda$ is the interpolation coefficient, while $L_{Sim}$ and $A_{Sim}$ are respectively the linguistic and acoustic similarities computed with the cosine similarity applied to respectively the linguistic and acoustic word embeddings of $x$ and $y$.

Since our goal is to enrich confusion networks and use them to propose alternative word hypotheses to correct ASR outputs, we aim to optimize the $\lambda$ value for this purpose. To estimate $\lambda$, a list of known substitution errors made by an ASR system is used. The construction details of this list is presented in the Sect. 4.1.

Let define $h$ an erroneous word hypothesis and $\bar{r}$ the reference word that is substituted with $h$. For each word pairs $(h, \bar{r})$ in the list, we compute the probability of using $h$ when the reference word $\bar{r}$ is wrong, *i.e.* the probability of substituting the reference word with the hypothesis one, which is defined as:

$$P(h|\bar{r}) = \frac{\#(h, \bar{r})}{\#\bar{r}} \tag{2}$$

where $\#(h, \bar{r})$ refers to the number of occurrences of the word pair and $\#\bar{r}$ is the number of errors (deletion + substitution) on the reference word.

Based on the similarity score $LA_{SimInter}(\lambda, h, \bar{r})$ and the probability $P(h|\bar{r})$, we choose the interpolation coefficient $\hat{\lambda}$ that minimizes the mean squared error (MSE) as:

$$\hat{\lambda} = \underset{\lambda}{\operatorname{argmin}} MSE(\forall (h, \bar{r}) : P(h|\bar{r}), LA_{SimInter}(\lambda, h, \bar{r})) \tag{3}$$

This choice is not optimal since similarities are not normalized in function of the number of errors related to one word in the vocabulary whereas probabilities are, but we assume this approach provides an acceptable approximation in the search of the $\lambda$ value that aims to combine $L_{Sim}(x,y)$ and $A_{Sim}(x,y)$ in order to predict the confusability between $x$ and $y$.

By using $LA_{SimInter}$ with $\hat{\lambda}$, it is now possible to propose for a given word its linguistically and acoustically nearest neighbors.

Table 2 shows an example of hypothesis word and its nearest neighbors. As expected, the neighbors of any given word seem linguistically similar when induced by linguistic word embeddings, and sound like it when they are induced by the acoustic ones. By combining acoustic and linguistic word similarities ($LA_{SimInter}$), it is also possible to restrict the neighborhood to words close to any given word both linguistically and acoustically.

**Table 2.** Nearest neighbors of the hypothesis word 'portables', with some translations in English and their pronunciation in French. 'portables' is a French word pronounced pɔʁtabl that can be translated to the same word 'portables' in English

| Nearest neighbors of the French word 'portables', pronounced /pɔʁtabl/ | | |
|---|---|---|
| $L_{Sim}$ | $A_{Sim}$ | $LA_{SimInter}$ |
| téléphones, ordinateurs, portable, portatif | portable, portant, portants, portait | portable, portant, portatif, portait |
| *telephones, computers, portable, portable* | *portable, carrying, racks, carried* | *portable, carrying, portative, carried* |
| / telefɔn/ / ɔʁdinatœʁ/ / pɔʁtabl/ / pɔʁtatif/ | / pɔʁtabl/ / pɔʁtã/ / pɔʁtã/ / pɔʁtɛ/ | / pɔʁtabl/ / pɔʁtã/ / pɔʁtã/ /pɔʁtɛ/ |

# 4 Experimental Setup

We present in this section the performance of the similarity measure $LA_{SimInter}(\lambda, h, \bar{r})$ on two tasks: prediction of ASR potential errors for rare words and enrichment of confusion networks.

## 4.1 Computation of Linguistic and Acoustic Embeddings

The 100-dimensional linguistic word embeddings results from the combination of word2vecf, skip-gram, and GloVe, using PCA. The word embeddings were computed from a large textual corpus composed of about 2 billions of words. This corpus was built from articles of the French newspaper "Le Monde", the French Gigaword corpus, articles provided by Google News, and manual transcriptions of about 400 h of French broadcast news. It contains dependency parses used to train word2vecf embedding, while the unlabeled version is used to train skip-gram and GloVe [20].

The training set for the convolution neural network consists of 488 h of French Broadcast News with manual transcriptions. This dataset is composed of data coming from the ESTER1 [21], ESTER2 [22] and EPAC [23] corpora. It contains 52K unique words that are seen at least twice each in the corpus. All of them corresponds to a total of 5.75 millions occurrences.

Acoustic features provided to the convolution neural network are log-filterbanks, computed every 10 ms over a 25 ms window yielding a 23-dimension vector for each frame. Each word is represented by 100 frames, thus, by a vector of 2300 dimensions. When words are shorter they are padded with zero equally on both ends, while longer words are cut equally on both ends. Once the acoustic embeddings are built, we build orthographic embeddings from the vocabulary compose of 52K words, and train the DNN architecture to build the acoustic word embeddings.

The resulting model, is used to build 100-dimensional acoustic word embeddings from the same vocabulary as the linguistic ones. A detailed description of the training data of the architectures used to build these acoustic word embeddings is presented in [16].

### 4.2   Experimental Data

Experimental data is based on the entire official ETAPE corpus [24], composed by audio recordings of French broadcast news shows, with manual transcriptions. This corpus is enriched with automatic transcriptions generated by the LIUM ASR system, detailed in [25], which won the ETAPE evaluation campaign. Its vocabulary contains 160K words.

The automatic transcriptions have been aligned with reference transcriptions using the *sclite*[1] tool. From this alignment, one can derive the lists of errors produced by our ASR system. The experimental data is divided into two sets: Train and Test, which are composed respectively of 399K and 58K words. Their word error rates are 25.2% and 21.9% respectively. More, they have respectively 10.3% and 8.3% of substitution errors.

For this task, we will use the list of substitution errors of Train to compute the interpolation coefficient $\hat{\lambda}$, while the list of Test will be used to evaluate the performance of our approach to enrich confusion networks and to correct erroneous word hypotheses. This list is composed of 4678 occurrences of substitution error pairs, named $Sub_{Test}$ further in the paper. For these substitution error pairs we use their corresponding confusion bins.

Figure 1 illustrates the percentage of the confusion bins according to the number of their alternative words (*i.e* words in concurrence with the 1-best

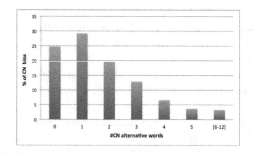

**Fig. 1.** Percentage of confusion network bins according to their size

---

[1] http://www.icsi.berkeley.edu/Speech/docs/sctk-1.2/sclite.htm.

hypothesis). The CN bins do not have a fixed length and 55% of them contain none or only one alternative word, that justify our aim about CN enrichment. The CN bins that have a size between 6 and 12 are grouped into a single class [6-12].

## 4.3    Tasks and Evaluation Score

We propose in this study two evaluation tasks: the prediction of errors for rare words (task1) and the correction of ASR errors (task2).

Given a word pair (a, b) in a list $L$ of $m$ substitution errors, the evaluation tasks consist on looking for the word $b$ in the list $N(a, \Gamma, n)$ of the $n$ nearest neighbors of $a$, computed through the similarity $\Gamma$. In our experiments, the similarity can be $L_{Sim}$, $A_{Sim}$ or $LA_{SimInter}$.

The evaluation score is calculated by varying the size $n$ and computing the precision at $n$ of finding the word $b$. The precision at $n$ computed for all the word pairs in the list $L$, taking into account their occurrence frequencies in the evaluation corpus, is called $S(\Gamma, n)$ and computed as follows:

$$S(\Gamma, n) = \frac{\sum_{i=1}^{m} f(i, \Gamma, n) \times \#(a_i, b_i)}{\sum_{i=1}^{m} \#(a_i, b_i)} \tag{4}$$

where $f$ is defined as:

$$f(i, \Gamma, n) = \begin{cases} 1 & \text{if } b_i \subset N(a_i, \Gamma, n) \\ 0 & \text{otherwise} \end{cases}$$

where $i$ corresponds to the $i^{th}$ word pair $(a_i, b_i)$ of $L$, $a_i$ and $b_i$ are defined according to the evaluation tasks:

- task1: $b_i$ corresponds to the hypothesis word $h$ and $a_i$ is the reference word $\bar{r}$,
- task2: $b_i$ corresponds to reference word $\bar{r}$ and $a_i$ is the hypothesis word $h$.

# 5    Experimental Results

## 5.1    Prediction of Potential ASR Errors for Rare Words

To compare the performance of the combined similarity to the linguistic and acoustic ones, we evaluate them on ASR errors prediction task for rare words. These latter are defined as the reference words not seen in the training corpus of the ASR system. This is why the $Sub_{Test}$ list was filtered to keep only the errors (misrecognized reference words) not seen in Train. It is composed of 538 pairs of substitution errors, named $Sub_{TestRarewords}$. For each reference word $\bar{r}$ in the $Sub_{TestRarewords}$ we derive their 30 nearest neighbors from the ASR vocabulary, based on linguistic, acoustic or combined similarities. That results to three similarity lists named respectively $List_{SimL}$, $List_{SimA}$, and $List_{SimInter}$.

Figure 2 illustrates the results of predicting errors for rare words using the lists described above, by varying their sizes from 1 to 30. We observe that the results are in favor of $List_{SimInter}$ followed by $List_{SimA}$: this shows that our proposition to optimize the interpolation weight to combine $List_{SimL}$ and $List_{SimA}$ is relevant. The interesting area in this figure is the left part, which shows the results of the prediction when the list of errors is short. When this list is composed of only one word, the prediction is correct 11% of the time. This must be analyzed in light of the vocabulary size of the ASR system, which contains 160K words: each word of the vocabulary can be selected in a list of error prediction. The prediction is correct 20% of the time when the size of the $List_{SimInter}$ list is 12. It seems that this list reaches then a plateau. The combined similarity will be used for the remaining experiments.

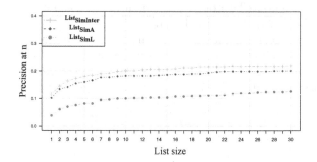

**Fig. 2.** Performance of predicting ASR errors for rare words by varying the size of the lists

## 5.2    Enrichment of Confusion Networks

The enrichment of confusion networks can be used for post-processing of automatic transcriptions, or to enrich the automatic transcriptions provided for a spoken language understanding system.

**Post-processing of Automatic Transcriptions.** For each hypothesis word ($h$) in $Sub_{Test}$ we derive their 6 nearest neighbors from the ASR vocabulary, based on the combined linguistic and acoustic similarity $LA_{SimInter}$. The resulting list is named $List^h_{SimInter}$. Then, for each word pair $(h, \bar{r})$ in $Sub_{Test}$ we enrich their corresponding confusion bins with the nearest neighbors of the hypothesis word ($h$) from $LA_{SimInter}$, to have a bin size equals to 6 (this size seems relevant to visualize alternative words in a graphical user interface in a computer-assisted transcription software [26]). The list of competing words in the confusion bin is named $List_{CN}$, and the one in the enriched confusion bin is named $List_{EnrichCN}$.

We evaluate the performance of the resulting lists on erroneous word hypotheses correction task. In this task, we try to see, when there is a recognition error, whether the correct word ($\bar{r}$) was in the nearest neighbors (or confusion bin) of the recognized word ($h$). As shown in Table 3, experimental results show that

**Table 3.** Performance of CN enrichment: evaluation of $List_{CN}$ and $List_{EnrichCN}$ on erroneous word hypotheses correction task in terms of precision at 6

|      | $List_{CN}$ | $List_{EnrichCN}$ |
|------|------|------|
| P@6  | 0.17 | **0.21 (+23.5%)** |

our approach is relevant: enriched confusion networks permit to increase the precision at 6 of more than 23% in comparison to CN produced by our ASR system. Notice that the P@6 value for the alternative words proposed by the $List_{SimInter}$ alone is 0.11.

**Spoken Language Understanding Task.** The approach we propose can be useful for a spoken language understanding task, to correct the semantically relevant erroneous word hypotheses. However, in the case of only the 1-best ASR hypotheses was provided to the dialogue manager, one can use the proposed similarity metric to expand this 1-best hypotheses and build confusion networks. This scenario in which getting access to only the 1-best ASR hypotheses is frequent in industry, especially when the semantic interpretation module is fed by outputs generated by an ASR system from a third party in the cloud.

For this experiment, we use the automatic transcriptions of the French MEDIA corpus [27,28] generated by a variant of the ASR system developed by LIUM that won the last evaluation campaign on French language [29]. This variant system contains 2.5K words and its language model is adapted to the MEDIA data. The purpose of the MEDIA corpus is to evaluate spoken language understanding systems. Often the SLU task derived from the MEDIA corpus consists on labeling recognized words with semantic concepts [30]. For a such task, a misrecognized word implies usually an error of labeling, that can be prevented by using confusion networks or word-lattices [31], when available. We expect to propose relevant alternative words in the scenario where only the 1-best hypothesis is available.

The automatic transcriptions were aligned with the reference ones in order to extract the list of substitution errors produced by the ASR system. This list is divided into two sets: Train to compute the interpolation coefficient $\hat{\lambda}$, which is enriched with Train Etape used for the previous experiments. Test is used for the evaluation, and has been filtered to keep only 1204 occurrences of semantically relevant erroneous words, based on the semantic labels. Since the size of MEDIA vocabulary is limited to 2.5K words, it is enriched with the vocabulary composed with 160K words.

For each hypothesis word ($h$) in Test list, we derive their 6 nearest neighbors from the ASR MEDIA vocabulary, based on the combined linguistic and acoustic similarity $LA^h_{SimInter}$.

By using the resulting list, one can enrich the one-best hypotheses produced by the ASR system and compute how many times we propose the correct word to recognize as an alternative in this list. Experimental results show that, thanks to our proposition, it is possible to potentially retrieve 20.6% of semantically relevant words that were initially misrecognized.

# 6  Conclusions

Assuming that word hypotheses in a same confusion network bin should be close in term of acoustics and/or linguistics, we propose to take benefit from linguistic and acoustic word embeddings to enrich confusion networks, in order to improve post-processing of ASR outputs.

We propose an approach to compute a similarity function, called $LA_{SimInter}$, which is optimized to ASR error correction. We show that this function allows us to compute relevant lists of nearest neighbors linguistically and acoustically. This list is used successfully to enrich the confusion networks and to increase the potential correction of erroneous words by 23% in comparison to initial confusion networks provided by the ASR system. Moreover, when used in a spoken language understanding task, this approach permits to propose 6 alternative words to 1-best hypotheses carrying on semantics to be exploited by the SLU module. When the ASR hypothesis is wrong on a word supporting a semantic concept, these alternatives contain the correct word in 20.6% of the cases.

Through our contribution and experimental results, we show that it is possible to relevantly enrich confusion networks by applying a similarity computed from linguistic and acoustic word embeddings. In addition, once we have the linguistic and the acoustic word embeddings, one can derive easily the lists of nearest neighbors linguistically and acoustically and use them as additional information to improve downstream applications.

# References

1. Stoyanchev, S., Salletmayr, P., Yang, J., Hirschberg, J.: Localized detection of speech recognition errors. In: 2012 IEEE Spoken Language Technology Workshop (SLT), pp. 25–30. IEEE (2012)
2. Pincus, E., Stoyanchev, S., Hirschberg, J.: Exploring features for localized detection of speech recognition errors. In: Proceedings of the SIGDIAL Conference, pp. 132–136. ACL (2013)
3. Soto, V., Cooper, E., Mangu, L., Rosenberg, A., Hirschberg, J.: Rescoring confusion networks for keyword search. In: 2014 IEEE International Conference on Acoustics, Speech and Signal Processing (ICASSP), pp. 7088–7092. IEEE (2014)
4. Mangu, L., Brill, E., Stolcke, A.: Finding consensus in speech recognition: word error minimization and other applications of confusion networks. Comput. Speech Lang. **14**(4), 373–400 (2000)
5. Fusayasu, Y., Tanaka, K., Takiguchi, T., Ariki, Y.: Word-error correction of continuous speech recognition based on normalized relevance distance. In: IJCAI, pp. 1257–1262 (2015)
6. Laurent, A., Meignier, S., Merlin, T., Deléglise, P.: Computer-assisted transcription of speech based on confusion network reordering. In: 2011 IEEE International Conference on Acoustics, Speech and Signal Processing (ICASSP), pp. 4884–4887. IEEE (2011)
7. Bengio, Y., Ducharme, R., Vincent, P., Janvin, C.: A neural probabilistic language model. JMLR **3**, 1137–1155 (2003). JMLR.org
8. Schwenk, H.: CSLM-a modular open-source continuous space language modeling toolkit. In: INTERSPEECH, pp. 1198–1202 (2013)

9. Ghannay, S., Favre, B., Estève, Y., Camelin, N.: Word embedding evaluation and combination. In: Language Resources and Evaluation Conference (LREC 2016), Portorož, Slovenia, 10th edn., pp. 23–28, May 2016

10. Levy, O., Goldberg, Y.: Dependency based word embeddings. In: Proceedings of the 52nd Annual Meeting of the Association for Computational Linguistics, vol. 2, pp. 302–308 (2014)

11. Mikolov, T., Chen, K., Corrado, G., Dean, J.: Efficient estimation of word representations in vector space. In: Proceedings of Workshop at ICLR (2013)

12. Pennington, J., Socher, R., Manning, C.D.: Glove: global vectors for word representation. In: Proceedings of the Empirical Methods in Natural Language Processing (EMNLP 2014), vol. 12 (2014)

13. Kamper, H., Wang, W., Livescu, K.: Deep convolutional acoustic word embeddings using word-pair side information. arXiv preprint arXiv:1510.01032 (2015)

14. Levin, K., Henry, K., Jansen, A., Livescu, K.: Fixed-dimensional acoustic embeddings of variable-length segments in low-resource settings. In: 2013 IEEE Workshop on Automatic Speech Recognition and Understanding (ASRU), pp. 410–415. IEEE (2013)

15. Bengio, S., Heigold, G.: Word embeddings for speech recognition. In: INTERSPEECH, pp. 1053–1057 (2014)

16. Ghannay, S., Estève, Y., Camelin, N., Deleglise, P.: Acoustic word embeddings for ASR error detection. In: INTERSPEECH 2016, San Francisco, CA, USA, 9–12 September 2016

17. Wang, J., Song, Y., Leung, T., Rosenberg, C., Wang, J., Philbin, J., Chen, B., Wu, Y.: Learning fine-grained image similarity with deep ranking. In: Proceedings of the IEEE Conference on Computer Vision and Pattern Recognition, pp. 1386–1393 (2014)

18. Weston, J., Bengio, S., Usunier, N.: Wsabie: scaling up to large vocabulary image annotation. In: IJCAI, vol. 11, pp. 2764–2770 (2011)

19. Ghannay, S., Estève, Y., Camelin, N., et al.: Evaluation of acoustic word embeddings. In: ACL 2016, p. 62 (2016)

20. Ghannay, S., Estève, Y., Camelin, N., Dutrey, C., Santiago, F., Adda-Decker, M.: Combining continuous word representation and prosodic features for ASR error prediction. In: Dediu, A.-H., Martín-Vide, C., Vicsi, K. (eds.) SLSP 2015. LNCS, vol. 9449, pp. 84–95. Springer, Cham (2015). doi:10.1007/978-3-319-25789-1_9

21. Galliano, S., Geoffrois, E., Mostefa, D., Choukri, K., Bonastre, J.-F., Gravier, G.: The ESTER phase II evaluation campaign for the rich transcription of French Broadcast News. In: INTERSPEECH 2005, pp. 1149–1152 (2005)

22. Galliano, S., Gravier, G., Chaubard, L.: The ESTER 2 evaluation campaign for the rich transcription of French radio broadcasts. In: INTERSPEECH, vol. 9, pp. 2583–2586 (2009)

23. Estève, Y., Bazillon, T., Antoine, J.-Y., Béchet, F., Farinas, J.: The EPAC corpus: manual and automatic annotations of conversational speech in French broadcast news. In: LREC. Citeseer (2010)

24. Gravier, G., Adda, G., Paulsson, N., Carr, M., Giraudel, A., Galibert, O.: The ETAPE corpus for the evaluation of speech-based TV content processing in the French language. In: Proceedings of the Eight International Conference on Language Resources and Evaluation (LREC 2012) (2012)

25. Deléglise, P., Estève, Y., Meignier, S., Merlin, T.: Improvements to the LIUM French ASR system based on CMU Sphinx: what helps to significantly reduce the word error rate? In: INTERSPEECH, Brighton, UK, September 2009

26. Cardinal, P., Boulianne, G., Comeau, M., Boisvert, M.: Real-time correction of closed-captions. In: Proceedings of the 45th Annual Meeting of the ACL on Interactive Poster and Demonstration Sessions, pp. 113–116. Association for Computational Linguistics (2007)

27. Bonneau-Maynard, H., Quignard, M., Denis, A.: MEDIA: a semantically annotated corpus of task oriented dialogs in French. Lang. Resour. Eval. **43**(4), 329 (2009)

28. Devillers, L., Maynard, H., Rosset, S., Paroubek, P., McTait, K., Mostefa, D., Choukri, K., Charnay, L., Bousquet, C., Vigouroux, N., et al.: The French MEDIA/EVALDA project: the evaluation of the understanding capability of spoken language dialogue systems. In: LREC. Citeseer (2004)

29. Rousseau, A., Boulianne, G., Deléglise, P., Estève, Y., Gupta, V., Meignier, S.: LIUM and CRIM ASR system combination for the REPERE evaluation campaign. In: Sojka, P., Horák, A., Kopeček, I., Pala, K. (eds.) TSD 2014. LNCS, vol. 8655, pp. 441–448. Springer, Cham (2014). doi:10.1007/978-3-319-10816-2_53

30. Raymond, C., Riccardi, G.: Generative and discriminative algorithms for spoken language understanding. In: INTERSPEECH, pp. 1605–1608 (2007)

31. Servan, C., Raymond, C., Béchet, F., Nocéra, P.: Conceptual decoding from word lattices: application to the spoken dialogue corpus media. In: The Ninth International Conference on Spoken Language Processing (INTERSPEECH 2006-ICSLP) (2006)

# Attentional Parallel RNNs for Generating Punctuation in Transcribed Speech

Alp Öktem[1(✉)], Mireia Farrús[1], and Leo Wanner[1,2]

[1] Universitat Pompeu Fabra, Barcelona, Spain
{alp.oktem,mireia.farrus,leo.wanner}@upf.edu
[2] Catalan Institute for Research and Advanced Studies (ICREA), Barcelona, Spain

**Abstract.** Until very recently, the generation of punctuation marks for automatic speech recognition (ASR) output has been mostly done by looking at the syntactic structure of the recognized utterances. Prosodic cues such as breaks, speech rate, pitch intonation that influence placing of punctuation marks on speech transcripts have been seldom used. We propose a method that uses recurrent neural networks, taking prosodic and lexical information into account in order to predict punctuation marks for raw ASR output. Our experiments show that an attention mechanism over parallel sequences of prosodic cues aligned with transcribed speech improves accuracy of punctuation generation.

**Keywords:** Speech transcription · Recurrent neural networks · Prosody · Punctuation generation · Automatic speech recognition

## 1 Introduction

The introduction of punctuation marks into the automatic speech recognition (ASR) output is an important issue in applications such as automatic transcription/subtitling, speech-to-speech translation, language analysis, etc. Punctuation is essential for grammaticality, readability, and (in the case of a number of different tasks), subsequent processing. Thus, correct sentence segmentation and punctuation of recognized speech improves the quality of machine translation [6,7,24,26], and missing periods and commas in machine generated text results in suboptimal information extraction from speech [13,15]. Also, most of the data-driven parsing models use punctuation as features.

In spoken language, punctuation is influenced by two intertwined linguistic phenomena: (1) syntax and (2) prosody. Syntax determines the distribution of punctuation marks in accordance with the grammar of a language. Prosody realization in speech (such as, e.g., word grouping, pausing, emphasis, rising-falling intonation, etc.) tends also to signal the position and type of the punctuation marks. For instance, a pause after consecutive words might signal an enumeration, which requires comma, and rising intonation at the end of a sentence is a likely indicator of a question.

© Springer International Publishing AG 2017
N. Camelin et al. (Eds.): SLSP 2017, LNAI 10583, pp. 131–142, 2017.
DOI: 10.1007/978-3-319-68456-7_11

However, state-of-the-art approaches to punctuation generation are mainly driven by only syntactic (and lexical) criteria. In particular, recent data-driven approaches that use recurrent neural networks (RNN) proved to be competitive due to RNN's ability to capture long and short term syntactic dependencies. Models that account for prosodic features [30, 31] rely merely on pause duration between words; other prosodic features such as fundamental frequency (f0) and intensity information are ignored. Another shortcoming of the state-of-the-art is that the models are trained on either only written data [2] or on a combination of written and spoken data (with, again, a dominance of written material) [31]. This makes the trained models biased towards written data.

In what follows, we present a neural network setup that is able to process lexical and prosodic information in parallel for punctuation generation in raw speech data. This is different to, e.g., [31], which processes syntactic and prosodic information in sequence (and thus loses the linguistic evidence that both are intertwined). The proposed model makes it possible to integrate any desired feature (be it lexical, syntactic or prosodic) and allows us to test which prosodic features influence punctuation placement to what extent. Unlike previous works, we furthermore use in our experiments only spoken data and exploit various prosodic features that influence the usage of punctuation marks in automated transcriptions. The source code of our model is made publicly available together with a link to the dataset we used in our experiments in https://github.com/TalnUPF/punkProse.

The remainder of the paper is structured as follows. In Sect. 2, we describe the main architecture of our model. The experimental setup and the results of the experiments are outlined in Sect. 3 and discussed in Sect. 4. Section 5 summarizes briefly recent related work, and, finally, Sect. 6 concludes the paper and sketches some of the main lines proposed for future work.

## 2   Our Model

Our model is inspired by Tilk et al.'s work [31]. Tilk et al. use a bidirectional recurrent network [27] for keeping track of the word context in two directions. Their model is a two-stage model. In the first stage, syntactic and lexical features are processed. In the second stage, pauses between words (as prosodic features) are also taken into account.

As Tilk et al., we use *gated recurrent units* (GRU) [8] for the RNN layers. Introduced as a simpler variate of *long short-term memory* (LSTM) units [11], GRUs make computation simpler by having fewer parameters. Number of gates in hidden units are reduced to two: (a) the *reset* gate determines whether the previous memory will be ignored, and (b) the *update* gate determines how much of the previous memory will be carried on.

Our modification to their proposal is that instead of passing continuous prosodic feature values to the second stage, we discretize the feature values and input them to the model through separate parallel GRU layers that are tuned in one single stage. Figure 1 illustrates our model.

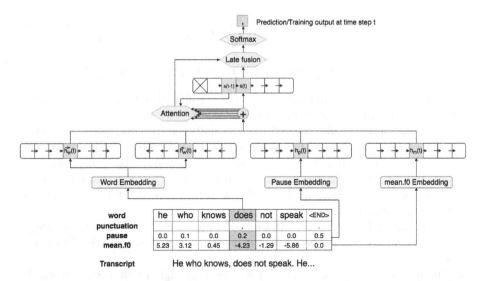

**Fig. 1.** Our neural network architecture depicting processing of a speech data sample with pause and mean f0 features aligned at the word level

For the sake of simplicity, we assume that the model is trained only with sequences of words ($w$), pause durations ($p$) and mean fundamental frequency ($m$). In this setting, the model has 4 GRU units: bidirectional layers for words, a unidirectional layer for pauses coming before the words, and a unidirectional layer for mean f0 values of words. GRU layers are preceded by embedding layers for words ($W_e$), pauses ($W_p$) and mean f0 ($W_m$). Inputs to the embedding layers are one-hot encoded vectors of sizes respective to their vocabulary sizes. The hidden states of the GRU layers at time step $t$ are:

$$\overrightarrow{h_w}(t) = GRU(x(t)W_e, \overrightarrow{h_w}(t-1))$$
$$\overleftarrow{h_w}(t) = GRU(x(t)W_e, \overleftarrow{h_w}(t+1))$$
$$h_p(t) = GRU(p(t)W_p, h_p(t-1))$$
$$h_m(t) = GRU(m(t)W_m, h_m(t-1))$$

where $x(t)$, $p(t)$ and $m(t)$ are the word index, pause level and mean f0 level respectively at time step $t$. The parallel GRU states are concatenated to form the context vector $h(t)$ before being passed over as input to another unidirectional GRU layer:

$$h(t) = \left[\overrightarrow{h_w}(t), \overleftarrow{h_w}(t), h_p(t), h_m(t)\right]$$
$$s(t) = GRU(h(t), s(t-1))$$

The attention mechanism combines all input states into a weighted context vector $a(t)$ which is then late-fused with the state $s(t)$ of the output GRU:

$$a(t) \; = \sum_{i=1}^{N} h(t)\alpha_{t,i}$$

$$f(t) \; = \; a(t)W_{fa} \bigodot \sigma(a(t)W_{fa}W_{ff} \; + \; s(t)W_{fs} \; + \; b_f) \; + \; s(t)$$

where $\alpha_{t,i}$ is the weight that determines the amount of influence of each input state to the current output and N is sequence size.

The attention mechanism is useful for the neural network to identify positions in a sequence where important information is concentrated [1]. For words, it helps to focus on positions of words and word combinations that signal the introduction of a punctuation mark. For prosodic features, it either remembers a salient point in the sequence or detects a certain movement that could help determining a punctuation mark at a certain position.

The output GRU layer uses a late-fusion approach, which lets the context gradient carry on easily by preventing it passing through many activation functions [33].

Finally, the late-fused context $f(t)$ is passed through a *Softmax* layer, which outputs the probability of the punctuation mark to be placed between the current and the previous word (starting from the second word in sequence):

$$y(t) \; = \; Softmax(f(t)W_y \; + \; b_y)$$

## 3   Experiments

### 3.1   Data

The experiments presented in this paper were performed on a corpus consisting of TED (Technology, Entertainment, Design) talks[1]. TED talks are a set of conference talks lasting approximately 15 min each that have been held worldwide in more than 100 languages. They include a large variety of topics, from technology and design to science, culture and academia. The corpus consists of 1046 talks by 884 English speakers, uttering a total amount of 156034 sentences. The corresponding transcripts, as well as audio and video files, are available on TED's website; they were created by volunteers and include punctuation and paragraph breaks [12]. The subtitle timings of TED transcripts do not always correspond to sentences in the transcript. To overcome this limitation, precise word timings were first obtained through Viterbi forced alignment using an automatic speech recognition system. The word timings were then further used to automatically obtain sentence boundaries and thus sentence timings [12].

As for the prosodic features, three main prosodic elements were extracted following the methodology in [12] in order to analyze their influence on punctuation generation: pauses, fundamental frequency (f0), and intensity. Pause durations were extracted from the provided word timings, while f0 and intensity contours were extracted at 10 ms precision using Praat software [5] with linear interpolation and octave jump removal for fundamental frequency provided by Praat.

---

[1] http://www.ted.com.

f0 measurements were converted to semitones relative to speaker mean f0 value for normalization, while the speaker mean intensity over a talk was subtracted from the intensity values for the same purpose, so that speaker mean values were represented by zero values in both cases.

## 3.2   Features Extraction and Preprocessing

The prosodic TED corpus is processed in order to be fed into the neural network. Firstly, the following aligned sequences are extracted for each talk:

*word* stands for the words that are uttered by the speaker. Abbreviations are decomposed into the letters they consist of (e.g., 'DIY' to three separate words 'D' 'I' 'Y'). Numbers are converted into text and separated (e.g., '93' to 'ninety three').

*punctuation* marks the symbol coming before the corresponding word. We limited the symbol vocabulary to period ('.'), comma (','), question mark ('?'), exclamation mark ('!'), colon (':'), semicolon (';'), dash ('-') and 'no punctuation'. In the cases when more than one punctuation mark occur before a word (e.g., in a quotation), the most important punctuation mark is chosen as the symbol at that position.

*reduced-punctuation* is the reduced version of *punctuation*. Exclamation mark, dash, colon, and semicolon are mapped to a period.

*pause* holds the silence duration in milliseconds coming before the corresponding word. It is calculated from the word timings information obtained from speech alignment.

*mean.f0* and *mean.i0* are the mean fundamental frequency and intensity values (in semitones) for the corresponding word.

*range.f0* and *range.i0* are calculated by subtracting the minimum f0/intensity value from the maximum f0/intensity value for the corresponding word.

Secondly, taking into account that the number of words per sentence in our corpus is 15–20 in average, the data is sampled into sequences of size 50, each sample starting with a new sentence and ending with an END token. With this setting, more than one sentence fits into a sample. Sentences are placed in samples in the same order in the speech data. If the sample end is reached before the end of a sentence, the sentence portion that fits is kept in that sample and the next sample starts from the beginning of that sentence. We avoided putting together data from different talks in the same sample by discarding the last unfinished sample from a talk. Also, sentences with more than 50 words are discarded.

59811 samples were extracted this way. 70% (41867 samples) of this data were allocated for training, 15% for testing and 15% for validation (8971 samples each).

The word vocabulary is created with the tokens that occur more than 7 times in the corpus and two extra tokens: *out-of-vocabulary* and *end-of-sequence*. This totaled up to 13830 tokens. The output punctuation vocabulary in our experiments is of size 4 (from the reduced punctuation set).

In order to input prosodic features to the RNNs, they had to be vocabularized as well. This is achieved by assigning a vocabulary index for certain ranges of the continuous feature values. The ranges were determined by dividing the feature value distribution according to the number of occurrences within that range. Via manual inspection, we divided the pause durations into 66 and semitone values distribution into 81 levels.

### 3.3 Experimental Setup

We used Theano [29] for implementing our models. In our experimental setup, the embedding vector sizes for words and prosodic features are set to 100 and 10 respectively. This is because prosodic feature vocabulary is significantly smaller than the word vocabulary. The hidden layer dimension of all GRU layers is also set to 100, except for pause durations, where a smaller dimension of 10 performed better in terms of validation scores, such that we set it to 10.

The model is trained in batches of size 128. The weight matrices are updated using the AdaGrad algorithm [10] with a learning rate of 0.05 for minimizing the negative log-likelihood of the predicted punctuation sequence.

### 3.4 Punctuation Generation Results

As the majority of the punctuation marks in our dataset consisted of the punctuation marks in the reduced set (comma, period and question mark), experiments were performed only with this set.

The two-stage method by Tilk et al. is used as a baseline by training over our data twice: first, only with text, and then together with the pause durations.

**Table 1.** Punctuation generation results for two stages [31] and our single-stage approach

| Model | Features | Comma | | | Period | | | Question | | | All | | |
|---|---|---|---|---|---|---|---|---|---|---|---|---|---|
| | | P | R | $F_1$ | P | R | $F_1$ | P | R | $F_1$ | P | R | $F_1$ |
| Two stages | word (w) | 56.9 | 36.6 | 44.5 | 67.6 | 62.5 | 64.9 | 68.5 | 46.9 | 55.7 | 63.2 | 49.0 | 55.2 |
| | w+pause(p) | 51.0 | 51.6 | 51.3 | 68.6 | 57.8 | 62.8 | 66.8 | 48.9 | 56.5 | 58.9 | 54.4 | 56.6 |
| Single stage | w+p | 61.6 | 44.5 | **65.6** | 71.7 | 72.5 | 72.1 | 66.5 | 64.7 | 65.6 | 67.3 | 58.2 | 62.4 |
| | w+p+range.f0 | 58.7 | 52.0 | 55.1 | 72.4 | 76.1 | 74.2 | 67.9 | 64.7 | 66.3 | 65.9 | 63.6 | 64.8 |
| | w+p+mean.f0 | 59.3 | 53.3 | 56.1 | 74.9 | 75.9 | 75.4 | 65.2 | **67.4** | 66.3 | 67.2 | **64.3** | **65.7** |
| | w+p+range.i0 | 55.0 | **54.3** | 54.6 | **75.0** | 70.3 | 72.5 | 70.0 | 58.7 | 63.9 | 64.5 | 61.9 | 63.2 |
| | w+p+mean.i0 | 58.4 | 53.4 | 55.8 | 74.5 | 74.3 | 74.4 | 68.8 | 63.9 | 66.3 | 66.6 | 63.5 | 65.0 |
| | w+p+range.f0+range.i0 | 60.9 | 45.5 | 52.1 | 71.9 | 76.0 | 73.9 | 71.5 | 61.0 | 65.9 | 67.3 | 60.2 | 63.6 |
| | w+p+range.f0+mean.i0 | 61.2 | 46.6 | 53.0 | 72.9 | 77.6 | 75.2 | **74.2** | 63.1 | **68.2** | 68.0 | 61.6 | 64.7 |
| | w+p+mean.f0+range.i0 | 61.6 | 47.9 | 53.9 | 73.1 | **79.6** | **76.2** | 74.1 | 62.0 | 67.5 | 68.2 | 63.1 | 65.6 |
| | w+p+mean.f0+mean.i0 | 56.9 | 52.2 | 54.4 | 77.1 | 70.4 | 73.6 | 71.3 | 61.6 | 66.1 | 66.7 | 60.9 | 63.7 |
| | w+p+mean.f0 +range.f0+range.i0 | **63.4** | 44.5 | 52.3 | 73.6 | 77.4 | 75.5 | 65.7 | 66.4 | 66.1 | **69.2** | 60.5 | 64.6 |

Tilk et al.'s models are based on BRNN with an attention mechanism, which provided the best results when compared to other models [31].

In our single stage approach, the use of only lexical information (words) provided the same scores as the use of only words in the two-stages approach, since only one step is involved in both approaches. Then, in order to assess the contribution of new prosodic information to our model, the extracted prosodic features were added one by one. The pause duration feature was always kept while trying combinations of new features, i.e., means and ranges of both f0 and intensity. The outcomes of our experiments in generating periods, commas and question marks are presented in Table 1 in terms of precision (P), recall (R), and $F_1$ scores.

## 4    Discussion

A significant improvement is achieved with the proposed parallel RNNs approach compared to the two-stage model when trained with the same dataset. We observe an overall improvement in $F_1$ score of 5.8% when same features (word and pause durations) are used with our model. The model opens the way for a further improvement of 3.3% with the addition of mean f0 feature into the model, resulting in an overall $F_1$ score of 65.7%.

We also see from the results that the inclusion of f0- and intensity-related prosodic features—apart from pauses—into the neural network improves the generation score for period and question marks. An improvement of 4.1% in $F_1$ score is observed for periods with the inclusion of mean f0 and intensity range features on top of pause features. For question marks, the best $F_1$ score is achieved with f0 range and mean intensity features on top of pause durations (improvement of 2.6%). For commas, we observe that precision and recall improve with different settings but when looked at the $F_1$ score best feature combination stays to be words and pause durations.

The best performing set of features seems to be the combination of pause and mean f0 when looked at the overall $F_1$ score. However, we see that each punctuation mark has a different set of features that improve their generation results the most. Combination of f0 range and mean of intensity gives best results for generating question marks (68.2% in $F_1$ score). For period, using mean f0 and intensity range features together yields the best result (76.2% in $F_1$ score). Recall that colons, semicolons, dashes and exclamation marks in our dataset are also mapped to periods.

It has to be stated that our evaluation method for the baseline does not corroborate the design decision of Tilk et al. Their two stage training helps building a more solid lexical model by training on a larger text corpus. In [31], they report an overall $F_1$ score of 72.2% trained only on written textual data, which further improves to an $F_1$ score of 77% with additional training on pause-annotated corpus. However for our purely spoken data, their model performs with an $F_1$ score of 56.6% which shows only an 1.4% improvement after the addition of pause features.

Our initial guess was that training with four prosodic features at once would oversaturate the model; however, the results for the feature set consisting of mean f0, f0 range and intensity range combined gives promising results. The best overall precision score (69.2%) and precision for generating commas (63.4%) are achieved with this feature set.

## 5  Related Work

The problem of punctuation determination has been addressed in several works in the literature—as has been the closely-related issue of boundary detection. Both problems have been tackled from diverse perspectives, and many of them only take into consideration the recognized ASR output text, ignoring the speech related information contained in the original speech, or they simply tackle the problem for textual data in which the correct punctuation is missing, e.g., in a sentence generation or a grammatical correction scenario. In [16], for instance, the punctuation detection is addressed from a syntax-based perspective by using the output of an adapted chart parser, which provides information on the expected punctuation placement. Also in [32] and in [23] the punctuation generation task is carried out without taking prosodic cues into account. In the former, several textual features including language model scores, token n-grams, sentence length and syntactic information extracted from parse trees are combined using conditional random fields (CRF). In the latter, the task is based on dynamic conditional random fields and applied to a conversational speech domain. A more recent work [2] introduces a language-independent model with a transition-based algorithm using LSTMs [11], without any additional syntactic features.

Overall, it has been shown that prosodic features are highly indicative of sentence boundaries as well as of punctuation placement. Therefore, a great deal of effort has been put in several works into the use of prosodic features when original speech is available. In [3], sentence boundaries are characterized by prosodic features and modeled by decision tree classifiers. In [20], the authors successfully detect automatically full stops by using a neural network to estimate the weights assigned to pauses, f0 changes and amplitude range, which are later used by a punctuation mark classifier; commas are shown to be more difficult to detect.

Other studies, such as [17], combine prosodic, word and grammatical features by using SVM and CRF classifiers, and test the prediction experiments on different speech styles, validating the hypothesis that the punctuation problem is much more difficult to address in ASR output than in manual transcripts. Prosodic and textual cues are also combined in [22] and implemented in a decision tree classifier with the goal to detect sentence boundaries. A combination of lexical-, prosodic-, and speaker-based features is also found in [4] for the detection of full stops, commas, and question marks in a bilingual English-Portuguese broadcast news data, while [19] focuses on Czech broadcast news speech to detect commas and sentence boundaries by using a prosodic model in a decision trees and a multi-layer perceptron and N-gram models for language modeling. Similar works deal with the punctuation generation problem by using statistical models of prosodic features [9], the combination of both textual and prosodic features based on adaptive boosting [18], and a cross-linguistic study of prosodic features through two different approaches for feature selection: a forward search wrapper and feature filtering [14]. Although not using prosodic features strictly speaking, [25] takes advantage of the transcriptions of multiple parallel speech streams in four different languages in order to increase punctuation generation accuracy.

More recently, the already mentioned work by Tilk et al. addresses the use of textual features and pauses (as sole prosodic feature) in an LSTM recurrent neural network [30] and in a bidirectional recurrent neural network [31] in order to detect full stops and commas in the former, and also question marks in the latter. As already discussed above, Tilk et al. 's methodology combines syntactic and prosodic features in a two-stage model. Only textual features are learned from a large non-spoken text corpus in a first stage. Then, in a second stage, the model is retrained with pause durations on a smaller corpus. This approach follows the work from [28], in which the language model can be trained on large amounts of textual data—lacking of the corresponding spoken data—, while the acoustic model—also based only on pause duration—is trained on a smaller corpus.

## 6   Conclusions and Future Work

In this work, we have presented a recurrent neural network architecture that processes lexical and prosodic information in parallel for the generation of punctuation, avoiding the dominance of written data, and thus the bias of trained models towards written material. Our proposed model allows the integration of any desired feature (lexical, syntactic or prosodic) and thus a further analysis of the impact of every feature used on the punctuation generation. In addition, the current model achieves a significant improvement over previous works that used two stages and were biased to written data.

The results are significantly better also when prosodic features are added to the lexical information. Solely pauses—when trained with a separate RNN—improve considerably the lexical-based scores. Moreover, f0- and intensity-based prosodic features help to achieve a better period and question mark detection in terms of $F_1$ measure, and comma detection is improved in terms of precision and recall in some specific settings. All in all, the best combination of prosodic features is when our model is trained on words together with the preceding pause durations and their normalized mean f0 values.

As future work, we plan to experiment with more prosodic features (such as speech rate) and their combinations and also see whether other RNN types such as LSTM help solve the problem better. Also, a model that gives attention to different prosodic features for different punctuation marks is a field to explore.

Our model trains word embeddings together with the whole architecture. We believe that pre-trained word embeddings extracted from a larger speech corpus would improve the scores. Also it has been recently shown that character-based encodings improve results in neural network based applications by largely decreasing the word vocabulary size [21].

**Acknowledgements.** We would like to thank Francesco Barbieri for offering his technical insights throughout this work. This work is part of the KRISTINA project, which has received funding from the *European Union's Horizon 2020 Research and Innovation Programme* under the Grant Agreement number H2020-RIA-645012. The second author is partially funded by the Spanish Ministry of Economy, Industry and Competitiveness through the *Ramón y Cajal* program.

# References

1. Bahdanau, D., Cho, K., Bengio, Y.: Neural machine translation by jointly learning to align and translate. CoRR abs/1409.0473 (2014). http://arxiv.org/abs/1409.0473

2. Ballesteros, M., Wanner, L.: A neural network architecture for multilingual punctuation generation. In: Proceedings of the 2016 Conference on Empirical Methods in Natural Language Processing (2016)

3. Baron, D., Shriberg, E., Stolcke, A.: Automatic punctuation and disfluency detection in multi-party meetings using prosodic and lexical cues. Channels **20**(61), 41 (2002)

4. Batista, F., Moniz, H., Trancoso, I., Mamede, N.: Bilingual experiments on automatic recovery of capitalization and punctuation of automatic speech transcripts. IEEE Trans. Audio Speech Lang. Process. **20**(2), 474–485 (2012)

5. Boersma, P., Weenink, D.: Praat: doing phonetics by computer [computer program] (2016). http://www.praat.org/

6. Cho, E., Niehues, J., Kilgour, K., Waibel, A.: Punctuation insertion for real-time spoken language translation. In: Proceedings of the Eleventh International Workshop on Spoken Language Translation (2015)

7. Cho, E., Niehues, J., Waibel, A.: Segmentation and punctuation prediction in speech language translation using a monolingual translation system. In: International Workshop on Spoken Language Translation (IWSLT) 2012 (2012)

8. Cho, K., van Merrienboer, B., Gülçehre, Ç., Bougares, F., Schwenk, H., Bengio, Y.: Learning phrase representations using RNN encoder-decoder for statistical machine translation. CoRR abs/1406.1078 (2014). http://arxiv.org/abs/1406.1078

9. Christensen, H., Gotoh, Y., Renals, S.: Punctuation annotation using statistical prosody models. In: Proceedings of the ISCA Workshop on Prosody in Speech Recognition and Understanding, pp. 35–40 (2001)

10. Duchi, J., Hazan, E., Singer, Y.: Adaptive subgradient methods for online learning and stochastic optimization. J. Mach. Learn. Res. **12**, 2121–2159 (2011). http://dl.acm.org/citation.cfm?id=1953048.2021068

11. Dyer, C., Ballesteros, M., Ling, W., Matthews, A., Smith, N.A.: Transition-based dependency parsing with stack long short-term memory. CoRR abs/1505.08075 (2015). http://arxiv.org/abs/1505.08075

12. Farrús, M., Lai, C., Moore, J.D.: Paragraph-based prosodic cues for speech synthesis applications. In: Proceedings of the 8th International Conference on Speech Prosody (2016)

13. Favre, B., Grishman, R., Hillard, D., Ji, H., Hakkani-Tur, D., Ostendorf, M.: Punctuating speech for information extraction. In: IEEE International Conference on Acoustics, Speech and Signal Processing, ICASSP 2008, pp. 5013–5016. IEEE (2008)
14. Fung, J.G., Hakkani-Tür, D., Magimai-Doss, M., Shriberg, E., Cuendet, S., Mirghafori, N.: Cross-linguistic analysis of prosodic features for sentence segmentation. In: Eighth Annual Conference of the International Speech Communication Association (2007)
15. Hillard, D., Huang, Z., Ji, H., Grishman, R., Hakkani-Tur, D., Harper, M., Ostendorf, M., Wang, W.: Impact of automatic comma prediction on POS/name tagging of speech. In: Spoken Language Technology Workshop, pp. 58–61. IEEE (2006)
16. Jakubíček, M., Horák, A.: Punctuation detection with full syntactic parsing. Spec. Issue: Nat. Lang. Process. Appl. **46**, 335–346 (2010)
17. Khomitsevich, O., Chistikov, P., Krivosheeva, T., Epimakhova, N., Chernykh, I.: Combining prosodic and lexical classifiers for two-pass punctuation detection in a Russian ASR system. In: Ronzhin, A., Potapova, R., Fakotakis, N. (eds.) SPECOM 2015. LNCS, vol. 9319, pp. 161–169. Springer, Cham (2015). doi:10.1007/978-3-319-23132-7_20
18. Kolář, J., Lamel, L.: Development and evaluation of automatic punctuation for French and English speech-to-text. In: Proceedings of INTERSPEECH, pp. 1376–1379 (2012)
19. Kolář, J., Švec, J., Psutka, J.: Automatic punctuation annotation in Czech broadcast news speech. In: in Proceedings of the SPECOM (2004)
20. Levy, T., Silber-Varod, V., Moyal, A.: The effect of pitch, intensity and pause duration in punctuation detection. In: 2012 IEEE 27th Convention of Electrical and Electronics Engineers in Israel (IEEEI), pp. 1–4. IEEE (2012)
21. Ling, W., Trancoso, I., Dyer, C., Black, A.W.: Character-based neural machine translation. CoRR abs/1511.04586 (2015)
22. Liu, Y., Chawla, N.V., Harper, M.P., Shriberg, E., Stolcke, A.: A study in machine learning from imbalanced data for sentence boundary detection in speech. Comput. Speech Lang. **20**(4), 468–494 (2006)
23. Lu, W., Ng, H.T.: Better punctuation prediction with dynamic conditional random fields. In: Proceedings of the 2010 Conference on Empirical Methods in Natural Language Processing, pp. 177–186. Association for Computational Linguistics (2010)
24. Matusov, E., Mauser, A., Ney, H.: Automatic sentence segmentation and punctuation prediction for spoken language translation. In: International Workshop on Spoken Language Translation (IWSLT) 2006 (2006)
25. Miranda, J., Neto, J.P., Black, A.W.: Improved punctuation recovery through combination of multiple speech streams. In: 2013 IEEE Workshop on Automatic Speech Recognition and Understanding (ASRU). pp. 132–137. IEEE (2013)
26. Peitz, S., Freitag, M., Mauser, A., Ney, H.: Modeling punctuation prediction as machine translation. In: International Workshop on Spoken Language Translation (IWSLT) 2011 (2011)
27. Schuster, M., Paliwal, K.K., General, A.: Bidirectional recurrent neural networks. IEEE Trans. Signal Process. **45**(11), 2673–2681 (1997)
28. Shen, W., Yu, R.P., Seide, F., Wu, J.: Automatic punctuation generation for speech. In: IEEE Workshop on Automatic Speech Recognition and Understanding, ASRU 2009, pp. 586–589. IEEE (2009)

29. Theano Development Team: Theano: a Python framework for fast computation of mathematical expressions. arXiv e-prints abs/1605.02688, May 2016. http://arxiv.org/abs/1605.02688
30. Tilk, O., Alumäe, T.: LSTM for punctuation restoration in speech transcripts. In: Proceedings of INTERSPEECH, pp. 683–687 (2015)
31. Tilk, O., Alumäe, T.: Bidirectional recurrent neural network with attention mechanism for punctuation restoration. In: Proceedings of INTERSPEECH, pp. 3047–3051 (2016)
32. Ueffing, N., Bisani, M., Vozila, P.: Improved models for automatic punctuation prediction for spoken and written text. In: INTERSPEECH, pp. 3097–3101 (2013)
33. Wang, T., Cho, K.: Larger-context language modelling. CoRR abs/1511.03729 (2015). http://arxiv.org/abs/1511.03729

# Lightweight Spoken Utterance Classification with CFG, tf-idf and Dynamic Programming

Manny Rayner[(✉)], Nikos Tsourakis, and Johanna Gerlach

TIM/FTI, University of Geneva, Geneva, Switzerland
{Emmanuel.Rayner,Nikolaos.Tsourakis,Johanna.Gerlach}@unige.ch

**Abstract.** We describe a simple spoken utterance classification method suitable for data-sparse domains which can be approximately described by CFG grammars. The central idea is to perform robust matching of CFG rules against output from a large-vocabulary recogniser, using a dynamic programming method which optimises the tf-idf score of the matched grammar string. We present results of experiments carried out on a substantial CFG-based medical speech translator and the publicly available Spoken CALL Shared Task. Robust utterance classification using the tf-idf method strongly outperforms plain CFG-based recognition for both domains. When comparing with Naive Bayes classifiers trained on data sampled from the CFG grammars, the tf-idf/dynamic programming method is much better on the complex speech translation domain, but worse on the simple Spoken CALL Shared Task domain.

**Keywords:** Speech recognition · Spoken utterance classification · Robustness · Context-free grammar · tf-idf · Medical applications

## 1 Overview

Spoken utterance classification is generally agreed to be an important problem, but published work to date has concentrated on a small number of scenarios, the most common of which are call routing and slot-filling applications like ATIS. It is in most cases assumed that there will be substantial amounts of training data available [5, 7, 8]. There are, however, many practically interesting types of application requiring spoken utterance classification which do not fit well into this picture. Our primary focus of interest here is fixed-phrase medical speech translators ("medical phraselators"). A medical phraselator contains on the order of thousands to tens of thousands of source-language utterances relevant to medical situations, each one paired with predefined translations in the target languages. The doctor speaks, and the app attempts to find the stored utterance closest

The work described here was funded by the Fondation privée des Hôpitaux universitaires de Genève and Unitec. We would like to thank Pierrette Bouillon for semantic annotation of the test data and many helpful suggestions, and Nuance Inc for generously making their software available to us for research purposes.

© Springer International Publishing AG 2017
N. Camelin et al. (Eds.): SLSP 2017, LNAI 10583, pp. 143–154, 2017.
DOI: 10.1007/978-3-319-68456-7_12

to what they have said, showing it to the doctor to confirm that it has under-stood correctly; if the doctor approves the app's choice, it speaks a translation in the target language. The challenge is to make the matching process flexible and accurate, so that the users can express themselves reasonably freely and be correctly recognised most of the time. Since there are many semantic classes, and doctor time is scarce and hard to obtain, it is optimistic to expect more than small amounts of training data to be available until an advanced point in the project.

In the approach we describe here, we manually construct a CFG grammar which defines plausible variants for the questions, after which we robustly match spoken input to that CFG grammar. We have been surprised to find that a very simple matching method based on tf-idf indexing and dynamic programming gives quite good results. Although it seems plausible that a sophisticated modern deep learning method could achieve a lower error rate, the tf-idf method has definite advantages. It requires essentially no training data, is easy to implement, and is fast both at compile-time and at runtime. As noted, our main interest is in medical speech translation, but we also present results for the Spoken CALL Shared Task, an open dataset we recently have been involved in popularising.

The rest of the paper is organised as follows. Section 2 describes the two domains used. Section 3 describes the speech recognisers. Section 4 sketches experiments using Weka classifiers; these work well for the simple CALL domain, but much less well for the complex medical speech translation domain. The next two sections contain the main results of the paper: Sect. 5 describes the tf-idf/DP matching method, and Sect. 6 an evaluation on the two domains used. The final section concludes.

## 2   Domains Used

### 2.1   Medical Phraselators and the BabelDr Project

In the preceding section, we briefly outlined what we mean by a "medical phrase-lator". We have since 2015 been involved in a collaboration between the Geneva University Faculty of Translation and Interpreting and the Hôpitaux Universi-taires de Genève (HUG), Geneva's largest hospital, whose goal is to produce a system of this general type. It is worth pointing out that medical phraselators have not been rendered obsolete by Google Translate (GT). A 2014 study [9] suggests that GT may mistranslate typical medical questions as much as 30% of the time; recent experiments carried out by our own group produce broadly similar results [3]. The problem is not so much the high error rate in itself as the fact that the only feedback given to the user, the recognition result, is very unreliable; GT often produces an incorrect translation after correct recognition. A phraselator, in contrast is explicitly designed to give dependable feedback.

The system we have developed, BabelDr (http://babeldr.unige.ch/; [4]), sup-ports translation of medical examination questions from French into several lan-guages, prioritising coverage relevant to Arabic- and Tigrinya-speaking migrants presenting at HUG's Accident & Emergency and migrant health departments.

The grammar has been written manually in a simple formalism based on Synchronous CFG [1]. The structure is "flat" and consists of a large set of top-level rules defining the various question patterns, together with more rules that define various kinds of phrase. The size of the generated coverage is of the order of tens or hundreds of millions of possible source-language sentences, mapping into of the order of thousands of semantic concepts. An example of a BabelDr rule is shown in Fig. 1. The Source lines define the actual CFG rule; the line marked Target/french is the backtranslation shown to the user at runtime. The backtranslations can also be accessed through a searchable help pane in the GUI.

```
Utterance
Source depuis combien d'heures \
       ($avez_vous | $ça_fait | $ressentez_vous) $mal_au_ventre
Source depuis combien d'heures $c_est_douloureux
Source ?(est-ce que) (ça | cela) fait combien d'heures que vous \
       (avez | ressentez | souffrez de) $mal_au_ventre
Source combien d'heures (cela | ça) (fait | fait-il) que vous \
       (avez | ressentez | souffrez de) $mal_au_ventre
Source (il y a | ça fait) combien d'heures que vous \
       (avez | ressentez | souffrez de) $mal_au_ventre
Source combien d'heures (il y a | ça fait) que vous \
       (avez | ressentez | souffrez de) $mal_au_ventre
Target/french depuis combien d'heures avez-vous mal au ventre ?
EndUtterance
```

**Fig. 1.** BabelDr rule for the question *"Depuis combien d'heures avez-vous mal au ventre?"* ("For how many hours have you experienced stomach pain?"). We only show the source-language (French) side. Items starting with a dollar sign ($) are non-terminals.

In the initial version of the system, the grammar was compiled into a CFG-based language model and then into a recognition package that could be run on the Nuance Toolkit 10.2 engine [11]. This yielded a system which provided practically useful performance, but suffered from the usual problems associated with rule-based applications: performance was reasonably good for utterances inside grammar coverage but very poor on out-of-coverage ones, and it was too often difficult for the user to know where the dividing line went.

## 2.2 Data and CFG Grammars Used for Current Experiments

For the experiments carried out here, we had 965 utterances of recorded training data available. Test data was collected from medical students and doctors during December 2016 and January 2017, using a scenario in which the subjects used the earlier rule-based version of the system to communicate with simulated patients [3]. Data was logged and then transcribed and semantically annotated by the project member responsible for grammar development (not one of the authors). This produced a total of 827 utterances, of which 110 were annotated as being out of domain with respect to the grammar version used, i.e. not sufficiently closely associated with any of the semantic categories defined by the grammar. This left 717 in-domain utterances, containing 3794 words, which were used for the present experiments. Of these 717 in-domain utterances, 503 (70.2%) were inside grammar coverage.

The experiments described in this paper were performed using a version of the grammar chosen so that it predated the data collection exercise. The version used has a vocabulary of 2046 words, expands to about 45M possible strings, and defines 2187 possible semantic categories. Each semantic category has an associated backtranslation. We extracted the set of 2187 backtranslations, and used them as additional training data in ways described in more detail below.

## 2.3   The Spoken CALL Shared Task

The methods we describe here were motivated by the requirements of the BabelDr project, but in order to get some idea of their general applicability we also evaluated them on a second domain where we had suitable data readily available. The Spoken CALL Shared Task ([2]; https://regulus.unige.ch/spokencallsharedtask/) is a joint initiative by Geneva University, the University of Birmingham and Radboud University, whose goal has been to create an open challenge dataset in the area of prompt-response systems for speech-enabled Computer Assisted Language Learning ("spoken CALL"). Training data was released in July 2016, and test data in January 2017; the task received twenty submissions from nine different groups. Results were presented at the SLaTE workshop in August 2017 (http://www.slate2017.org/challenge.html).

The Shared Task dataset was collected using an online CALL app designed for Swiss German teens in their second or third year of learning English. Content was structured as a series of interactive dialogues, each one parametrized so that it could appear in many different variants, which allowed students to practice fluency and generative language skills. Like BabelDr, the CALL app used a Nuance recogniser with a language model derived from a CFG grammar, which associated each response with one or more prompts. This CFG grammar was made available as part of the Shared Task training data released. The grammar was not intended to be complete, and was only meant to be taken as providing a baseline.

A Shared Task item is a tuple consisting of the following elements: (a) a prompt; (b) a recorded audio file with the student's response; (c) a transcription; (d) a binary annotation (correct/incorrect) noting whether the audio file is a fully correct response to the prompt; (e) a binary annotation (correct/incorrect) noting whether the audio file is a semantically (but possibly not grammatically) correct response to the prompt. The last three fields are kept secret in the test data, and the task is to reproduce the (d) column. Shared Task data can easily be transformed into an utterance classification task by extracting the items where the response is marked as semantically correct. The semantic classification task is then to reconstruct the prompt given the audio file.

## 2.4    Data and CFG Grammars Used for Current Experiments

The training data used for the experiments was the 5222 utterance set released with the Spoken CALL Shared Task. This was available in two versions: as transcriptions, and as recognition results produced by the recogniser (cf. Sect. 3).

As test data, we used the portion of the Shared Task test data which was marked as semantically correct, transforming it as described above into data for an utterance classification task. The resulting dataset has 875 items containing 4630 words. 568 items (64.9%) were inside grammar coverage.

The grammar used was the one included in the Shared Task release. This has a vocabulary of 419 words, expands to about 45K possible strings, and defines 501 possible semantic categories.

# 3    Recognisers

In both domains, the baseline was thus defined by an annotated CFG grammar which also acted as a language model for a recogniser. The challenge was to make this baseline system robust to out-of-coverage utterances. We adopted an obvious strategy: use the available data to create a broad-coverage recogniser tuned to the domain and a robust classifier which associated recogniser output with the semantic classes defined by the CFG grammar. We start by describing the large-vocabulary recognisers, which were produced differently in the two domains:

**BabelDr.** We used the large vocabulary Nuance Transcription Engine, with an interpolated language model that combined the default language model with a model derived from the BabelDr training data.

**Shared Task.** We used the Kaldi recogniser developed by Mengjie Qian and colleagues at the University of Birmingham, the ASR data for which was publicly posted on the Shared Task site[1] under entry JJJ. The JJJ entry achieved the second best score on the Shared Task and is described in [10].

Table 1 presents basic performance results for the different recognisers when run on the test data, giving Word Error Rate (WER) and Sentence Error Rate (SER) for in-coverage, out-of-coverage and all data. For the grammar-based recogniser, we also present results for the portion of the test data where the confidence score is over the threshold. The threshold value of 0.65 was tuned on the Shared Task training data, also available from the Shared Task site. Performance was not sensitive to the exact setting, and threshold values between 0.60 and 0.70 gave similar results. Note that although the large-vocabulary recogniser strongly outperforms the grammar-based recogniser on the whole set, the converse relationship obtains on the "high confidence" subset of the data. As we will see later, this is why a hybrid system is able to outperform the plain robust system for both domains.

We now proceed to issues concerning semantic classification, which are the main subject of the paper.

---

[1] https://regulus.unige.ch/spokencallsharedtask, "Results" tab.

**Table 1.** Recogniser performance for BabelDr and Spoken CALL Shared Task domains on in-coverage, out-of-coverage and all data. The "%Data" column shows the proportion of the data for which the grammar-based recogniser is over the confidence threshold.

| Recogniser | %Data | IC | | OOC | | All | |
|---|---|---|---|---|---|---|---|
| | | WER | SER | WER | SER | WER | SER |
| *BabelDr* | | | | | | | |
| Grammar-based | (All) | 16.1 | 29.0 | 64.4 | 100.0 | 31.7 | 50.2 |
| Grammar-based (high confidence) | 38.9 | 3.4 | 13.5 | 39.5 | 100.0 | 6.8 | 19.7 |
| Large-vocabulary | (All) | 10.4 | 29.0 | 22.3 | 63.1 | 13.3 | 39.2 |
| *Spoken CALL Shared Task* | | | | | | | |
| Grammar-based | (All) | 17.2 | 28.2 | 53.4 | 99.3 | 30.0 | 53.1 |
| Grammar-based (high confidence) | 27.7 | 1.7 | 3.8 | 36.5 | 100.0 | 6.0 | 16.0 |
| Large-vocabulary | (All) | 7.9 | 21.1 | 18.6 | 52.8 | 11.7 | 32.2 |

## 4  Utterance Classification Using Weka

We began by testing performance, for the two domains used, of several popular classifiers supported by the Weka toolkit [6]. We report results for J48 decision trees, naive Bayes and SVM; other methods we tried gave clearly worse results. Our basic approach in all cases was to take labelled text data—sets of text strings representing utterances, each one paired with an associated semantic class—and extract unigram features, one for each word in the vocabulary.

For both domains we had a bit less than a thousand items of test data, in the form of labelled recognition results produced by recognisers. This data could reasonably be regarded as unseen for the purposes of the present experiments. The labelled training data we had available was fairly dissimilar for the two domains. For the Spoken CALL Shared Task, we had a substantial number (more than 5K) training examples, which were available both as transcriptions and as recognition results; for BabelDr, we had less than a thousand such examples. We did however have 2187 backtranslations, one for each semantic class. Since the backtranslations are both shown to the user after each turn and also available through the help system, users often imitated them, so we expected them to be a useful knowledge source.

Another important difference between the domains was in the grammars. The Spoken CALL Shared Task grammar was quite small; it was possible to expand it fully, giving about 45 thousand utterances, and use the whole grammar for training. The BabelDr grammar was much bigger, expanding to about 45 *million* utterances, and using the whole set was not feasible. Instead, we sampled the grammar randomly, creating 100 possible utterances from each rule. Table 2 summarises the domains and the available resources.

Table 3 presents the results. For the Spoken CALL Shared Task, the classification error was quite good even when training only on the transcriptions and recognition results, and improved further when the grammar data was added,

**Table 2.** Summary of available resources for the two domains

|  | SharedTask | BabelDr |
|---|---|---|
| *Grammar* | | |
| #semantic categories | 501 | 2187 |
| #words of vocabulary | 419 | 2046 |
| #utterances in coverage | ∼45K | ∼45M |
| #utterances used for training | ∼45K | ∼220K |
| *Recorded training data* | | |
| #utterances training data | 5222 | 965 |
| *Backtranslations* | | |
| #backtranslations | – | 2187 |
| *Recorded test data* | | |
| #utterances test data | 875 | 717 |

reaching 11.8% for the best method. The figures for BabelDr, the domain we were actually interested in, were much less satisfactory, with a best error rate of 28.8%. On examining the results more closely, we thought one problem might be the fact that we were only using a small portion of the grammar. We consequently searched for a method which would let us use the whole grammar in some suitable form.

**Table 3.** Classification error rates using Weka methods on unseen spoken test data for the two domains. "J48" = J48 decision tree method. "NBayes" = Naive Bayes method. SVM training exceeded resource bounds for the BabelDr data.

| Training data | J48 | NBayes | SVM |
|---|---|---|---|
| *BabelDr* | | | |
| Backtranslations | 87.7 | 54.3 | – |
| Backtranslations + Transcriptions + Rec results | 55.2 | 38.4 | – |
| Backtranslations + Transcriptions + Rec results + Grammar | 34.1 | 28.8 | – |
| *Spoken CALL Shared Task* | | | |
| Transcriptions + Rec results | 16.2 | 15.9 | 13.9 |
| Transcriptions + Rec results + Grammar | 14.1 | 13.0 | 11.8 |

# 5    Utterance Classification Using tf-idf and Dynamic Programming

Attempting to find a way to use the whole grammar, rather than only a small part of it, two possible ideas suggested themselves to us. One was simply to try to find some kind of closest match between the string returned by the recogniser and a grammar rule. The other was to recast the problem as a type of document-indexing task, where the "documents" are the grammar rules. Specifically, we could use some version of the well-known tf-idf method [12] to find the rules which had high tf-idf scores with respect to the recogniser string; the tf-idf score basically measures the extent to which a word is a useful "keyword", i.e. occurs only in a small number of rules. In fact, it turned out to be easy to combine both ideas and split the problem into three parts. First, use tf-idf to find a small number of rules whose associated keywords match words in the recognition hypothesis produced by the recogniser; second, find the closest match between the recognition hypothesis and each rule in the shortlist produced by the first step; third, use information obtained from the matches to reorder the shortlist.

We approximate by treating the recogniser hypothesis as a bag of words rather than as an ordered string. This is an acceptable approximation for grammars like those considered here, where word-order is rarely important. It makes it possible to implement the matching process as a simple dynamic programming algorithm which recursively expands out the chosen grammar rule, chooses the best match for each piece, and combines the pieces. Since each grammar constituent only needs to be considered once, the process is very fast. In a little more detail, the currently implemented method is as follows:

1. At compile time, index words to associate them with the top-level rules in which they occur. Assign a word an idf score which is high if it occurs in few rules, low if it occurs in many rules. The simplest way to do this is to define the idf score for a word $W$ to be $1/f_W$, where $f_W$ is the number of top-level rules in which $W$ can occur; we may also smooth, use a logarithmic scale, etc. Call this mapping of words to rules and idf scores the *word to rules table*.
2. At compile time, associate each top-level rule with the closure of the set of non-top-level rules it may link to. Order these rules by the maximum depth at which they can occur. Call this mapping of rules to ordered lists of non-top-level rules the *rule to rule closure table*.
3. At runtime, the matcher is presented with a recognition hypothesis from the large-vocabulary recogniser. Use tf-idf to find the $n$ top-level rules with the best scores according to a naive scoring method which totals the tf-idf scores for all the words that are both used by the rule and also occur in the recognition hypothesis. This gives us a preliminary ordering of the rules.
4. For each rule in the $n$-best list created by the preceding step, perform a dynamic programming (DP) match against the recognition hypothesis, treating the hypothesis as a bag of words weighted by tf-idf scores. This DP match can be performed efficiently, since it is linear in the size of the grammar closure for the rule and logarithmic in the length of the input string. In more detail, the match proceeds as follows:
(a) Begin by matching each phrasal rule in the rule closure list from (2), starting with the deepest ones, which are ordered to occur earliest in the list. The idea is that each rule will only be matched when all the non-terminals that can occur in it have already been matched. Associate each non-top-level rule with its best matching score and call the mapping of non-top-level rules to scores the *phrase score table*.

   (b) To match a word in a CFG rule, check to see if it is in the input bag of words. If it is, add the tf-idf score from (1). If it isn't, add a fixed no-match penalty.

   (c) To match a sequence $\langle P, Q \rangle$ in a CFG rule, match $P$ and $Q$ separately and assign a score which is the sum of the scores for $P$ and $Q$.

   (d) To match an alternation $(P \mid Q)$ in a CFG rule, match $P$ and $Q$ separately and assign a score which is the larger of the scores for $P$ and $Q$.

   (e) To match a nonterminal in a CFG rule, look up its best score in the phrase score table.

   (f) At the end, add the fixed no-match penalty for each word in the input that has not been matched.

5. When all items in the $n$-best list have been matched, reorder them using the scores obtained in the previous step.

## 5.1   Refinements to the Basic Method

We tried a variety of tweaks to the basic method described above, including replacing the plain tf-idf scores with logarithmic scores and rescoring using the edit distance to the best grammar match measured in terms of the number of characters, the number of words, or the number of words weighted by the td-idf scores of the words affected. The only modification which had a positive effect on development set performance was one designed to address the problem of very unspecific rules, for example the rule associated with questions semantically equivalent to *Avez-vous mal?* ("Does it hurt?"). The problem with rules like these is that utterances matching them may fail to contain any word with a high tf-idf score, meaning that they cannot rise to the top of the $n$-best list. After some experimentation, the best solution found was to order the rules by minimum possible score at compile time, and at runtime always to add the $m$ potentially lowest-scoring rules. Based on the development set, we put $m = 3$.

# 6   Evaluation of the tf-idf/DP Method

We carried out a series of experiments to evaluate the tf-idf/DP method using the BabelDr and Spoken CALL Shared Task domains. For each domain, we compared four different versions of the system:

**Rule-based.** The pure rule-based version. Recognition is performed by the grammar-based language model, and semantic interpretation by the CFG.

**tf-idf.** A minimal robust version using the large-vocabulary recogniser together with semantic interpretation using only tf-idf. For this to be possible, we expanded the CFG rules to remove all the non-terminals and leave a flat grammar where each rule gave a single semantic result, and only used steps (1) and (3) from the sequence in Sect. 5.

**tf-idf/DP.** The full robust version, which combines the large-vocabulary recogniser and the complete semantic interpretation method from Sect. 5, including both tf-idf and DP matching.

**Hybrid.** A version which uses a simple method to combine the **Rule-based** and **tf-idf/DP** versions. For 1-best, the hypothesis is chosen from the rule-based system if the recogniser's confidence score is over a threshold, otherwise it is chosen from the robust system. For $n$-best ($n > 1$), the hypotheses chosen are the 1-best result from the rule-based system and enough results from the robust system to make $n$ different hypotheses.

**Table 4.** 1-best and 2-best semantic classification error on unseen text and speech data for four different versions of the two systems, distinguishing between in-coverage, out-of-coverage and all input. Text input is transcribed speech input. "Rule-based" = pure rule-based system; "tf-idf" = robust system with only tf-idf; "tf-idf/DP" = full robust system; "hybrid" = hybrid system combining "rule-based" and "tf-idf/DP".

| Version | IC | | | | OOC | | | | All data | | | |
|---------|------|------|------|------|------|------|------|------|------|------|------|------|
| | Text | | Speech | | Text | | Speech | | Text | | Speech | |
| | 1-bst | 2-bst | 1-bst | 2-bst | 1-bst | 2-bst | 1-bst | 2-bst | 1-bst | 2-bst | 1-bst | 2-bst |
| *BabelDr* | | | | | | | | | | | | |
| Rule-based | (0) | (0) | 13.9 | 11.7 | (100) | (100) | 72.0 | 70.6 | 29.8 | 29.8 | 31.2 | 29.3 |
| tf-idf | 11.9 | 10.7 | 19.7 | 17.5 | 47.7 | 34.6 | 52.8 | 43.5 | 22.3 | 17.9 | 29.6 | 25.2 |
| tf-idf/DP | 1.2 | 0.0 | 8.5 | 6.2 | 43.5 | 28.5 | 48.1 | 39.3 | 13.8 | 8.6 | 20.4 | 16.0 |
| Hybrid | (0) | (0) | 6.4 | 1.6 | 43.5 | 28.5 | 48.1 | 38.8 | 13.8 | 8.6 | 18.8 | 12.7 |
| *Spoken CALL Shared Task* | | | | | | | | | | | | |
| Rule-based | (0) | (0) | 22.5 | 19.7 | (100) | (100) | 63.5 | 60.9 | 35.1 | 35.1 | 36.9 | 34.2 |
| tf-idf | 15.3 | 9.3 | 25.0 | 13.0 | 23.5 | 14.7 | 30.6 | 22.1 | 18.2 | 9.3 | 27.3 | 15.9 |
| tf-idf/DP | 1.8 | 0.5 | 11.8 | 7.2 | 20.2 | 14.0 | 30.9 | 21.2 | 8.2 | 5.3 | 18.5 | 12.2 |
| Hybrid | (0) | (0) | 9.5 | 6.0 | 20.2 | 14.0 | 30.6 | 22.5 | 8.2 | 5.3 | 16.9 | 11.8 |

Summary results for classification error on the test sets are presented in Table 4, which shows 1-best and 2-best error rates for text and speech input, and Table 5, which breaks down results for the robust versions as a function of the number of word errors in the large-vocabulary recogniser's output. Rather surprisingly, the first impression is that performance on the two domains is reasonably similar. Looking first at Table 3, we see that WER over the whole test set is 12–13% for the large-vocabulary recogniser. For the grammar-based recogniser it is about 30% for the whole set and about 6–7% for the subset where the confidence score is over the threshold.

Turning next to Table 4, we see that 1-best semantic classification error on the whole set using the pure rule-based system is about 30–40% for spoken input. This is reduced to 17–19% for the hybrid version. 2-best error reduces from 30–35% to about 12–13%. The relative improvement in 1-best error is 40% for BabelDr and 54% for Shared Task; for 2-best error, it is 57% for BabelDr and 65% for Shared Task. The larger improvement in the Shared Task system is consistent with the fact that its CFG grammar represents a much smaller development effort and is less carefully constructed. Comparing the lines for

**Table 5.** 1-best and 2-best semantic classification error as a function of number of word errors. #Errs = number of word errors in 1-best speech recognition hypothesis; #Sents = number of examples with given number of word errors

| #Errs | #Sents | tf-idf | | tf-idf/DP | | Hybrid | |
|---|---|---|---|---|---|---|---|
| | | 1-bst | 2-bst | 1-bst | 2-bst | 1-bst | 2-bst |
| *BabelDr* | | | | | | | |
| 0 | 453 | 18.3 | 15.0 | 7.5 | 4.0 | 7.3 | 5.5 |
| 1 | 121 | 41.3 | 37.2 | 30.6 | 28.2 | 27.3 | 19.8 |
| 2 | 75 | 54.7 | 44.0 | 54.7 | 42.7 | 49.3 | 30.7 |
| >2 | 68 | 55.9 | 51.5 | 50.0 | 45.6 | 47.1 | 27.9 |
| *Spoken CALL Shared Task* | | | | | | | |
| 0 | 593 | 15.3 | 7.1 | 5.6 | 2.0 | 5.9 | 3.5 |
| 1 | 117 | 39.3 | 30.8 | 36.8 | 28.2 | 32.5 | 21.4 |
| 2 | 110 | 56.4 | 26.4 | 49.1 | 29.1 | 40.0 | 30.9 |
| >2 | 55 | 72.7 | 58.2 | 58.2 | 52.7 | 56.4 | 41.8 |

plain tf-idf, tf-idf/DP and hybrid, we see that inclusion of the DP matching step makes a large difference, particularly on in-coverage data, and hybrid improves non-trivially on tf-idf/DP.

Finally, Table 5 measures robustness to recognition errors. The hybrid system achieves a 1-best classification error of 6–7% on utterances which are correctly recognised, falling to about 30% on utterances with one recognition error, 40–45% on utterances with two recognition errors, and 50–55% on utterances with more than two recognition errors. The contribution of DP matching is most important on correctly recognised utterances. The largest differences occur on text input, which we included to give a baseline approximating perfect recognition. The higher error rate on BabelDr data (13.8% versus 8.2%) probably reflects the more challenging nature of the domain.

The dynamic programming matching method is fast both at compile-time and at runtime. Running on a 2.5 GHz Intel laptop, compilation of the tables required by the tf-idf/DP method requires less than a minute for each domain. Average processing time at runtime is about 40 ms/utterance.

# 7  Conclusions and Further Directions

We have presented a simple spoken utterance classification method suitable for domains which have little training data and can be approximately described by CFG grammars, and evaluated it on two such domains. Compared to plain CFG-based classification, the method reduces 1-best error on spoken input by over a third on the well-tuned BabelDr domain and over a half on the poorly-tuned Shared Task domain. We find these results encouraging, not least because the methods so far implemented can very likely be improved. Two obvious things to try next are introducing a better treatment of OOV words, which at the moment are uniformly counted as skipped, and simply tuning the recogniser more.

Our practical goal in this project has been to improve the BabelDr system. From a theoretical point of view, however, the most interesting finding has been the contrast between the mainstream Weka methods and tf-idf/DP. On the small Shared Task domain, the Weka methods strongly outperform tf-idf/DP, with the Naive Bayes method achieving a classification error of 13.0% as compared to the "hybrid" method's 16.9%. On the much more challenging BabelDr domain, however, the pattern is reversed. Naive Bayes scores 28.8%—only slightly better than the baseline CFG—while "hybrid" reduces the error to 18.8%. As noted, we think the poor performance of the Weka methods may reflect the inadequacy of creating training data by random sampling from the grammar, and it is possible that some more intelligent sampling method may allow us to address the problem. We are currently investigating this.

# References

1. Aho, A.V., Ullman, J.D.: Properties of syntax directed translations. J. Comput. Syst. Sci. **3**(3), 319–334 (1969)
2. Baur, C., Chua, C., Gerlach, J., Rayner, E., Russell, M., Strik, H., Wei, X.: Overview of the 2017 spoken CALL shared task. In: Proceedings of the Seventh SLaTE Workshop, Stockholm, Sweden (2017)
3. Bouillon, P., Gerlach, J., Spechbach, H., Tsourakis, N., Halimi, S.: BabelDr vs Google Translate: a user study at Geneva University Hospitals (HUG). In: Proceedings of the 20th Conference of the European Association for Machine Translation (EAMT), Prague, Czech Republic (2017)
4. Bouillon, P., Spechbach, H.: BabelDr: a web platform for rapid construction of phrasebook-style medical speech translation applications. In: Proceedings of EAMT 2016, Vilnius, Latvia (2016)
5. Hakkani-Tür, D., Béchet, F., Riccardi, G., Tur, G.: Beyond ASR 1-best: using word confusion networks in spoken language understanding. Comput. Speech Lang. **20**(4), 495–514 (2006)
6. Holmes, G., Donkin, A., Witten, I.H.: Weka: A machine learning workbench. In: Proceedings of the Second Australian and New Zealand Conference on Intelligent Information Systems, pp. 357–361. IEEE (1994)
7. Kuo, H.K.J., Lee, C.H., Zitouni, I., Fosler-Lussier, E., Ammicht, E.: Discriminative training for call classification and routing. Training **8**, 9 (2002)
8. Mesnil, G., He, X., Deng, L., Bengio, Y.: Investigation of recurrent-neural-network architectures and learning methods for spoken language understanding. In: Interspeech, pp. 3771–3775 (2013)
9. Patil, S., Davies, P.: Use of Google Translate in medical communication: evaluation of accuracy. BMJ **349**, g7392 (2014)
10. Qian, M., Wei, X., Jancovic, P., Russell, M.: The University of Birmingham 2017 SLaTE CALL shared task systems. In: Proceedings of the Seventh SLaTE Workshop, Stockholm, Sweden (2017)
11. Rayner, M., Bouillon, P., Ebling, S., Strasly, I., Tsourakis, N.: A framework for rapid development of limited-domain speech-to-sign phrasal translators. In: Proceedings of the workshop on Future and Emerging Trends in Language Technology, Sevilla, Spain (2015)
12. Sparck Jones, K.: A statistical interpretation of term specificity and its application in retrieval. J. Doc. **28**(1), 11–21 (1972)

# Low Latency MaxEnt- and RNN-Based Word Sequence Models for Punctuation Restoration of Closed Caption Data

Máté Ákos Tündik[(✉)], Balázs Tarján, and György Szaszák

Department of Telecommunications and Media Informatics,
Budapest University of Technology and Economics,
Magyar Tudósok körútja 2, Budapest 1117-H, Hungary
{tundik,tarjanb,szaszak}@tmit.bme.hu

**Abstract.** Automatic Speech Recognition (ASR) rarely addresses the punctuation of the obtained transcriptions. Recently, Recurrent Neural Network (RNN) based models were proposed in automatic punctuation exploiting wide word contexts. In real-time ASR tasks such as closed captioning of live TV streams, text based punctuation poses two particular challenges: a requirement for low latency (limiting the future context), and the propagation of ASR errors, seen more often for informal or spontaneous speech. This paper investigates Maximum Entropy (MaxEnt) and RNN punctuation models in such real-time conditions, but also compares the models to off-line setups. As expected, the RNN outperforms the MaxEnt baseline system. Limiting future context results only in a slighter performance drop, whereas ASR errors influence punctuation performance considerably. A genre analysis is also carried out w.r.t. the punctuation performance. Our approach is also evaluated on TED talks within the IWSLT English dataset providing comparable results to the state-of-the-art systems.

**Keywords:** Punctuation recovery · Recurrent Neural Network · LSTM · Maximum Entropy · Low latency real-time modeling

## 1 Introduction

Punctuation insertion into the output of Automatic Speech Recognition (ASR) is a known problem in speech technology. The importance of having punctuations in automatically generated text – transcripts, indexing, closed captions, for metadata extraction etc. – has been outlined several times [1,16], as punctuation helps both human readability, and also eventual subsequent processing with text based tools, which usually require the punctuation marks at the very first step of their operation: the tokenization. In dictation systems, punctuation marks can be explicitly dictated; however, in several other domains where ASR is used, this is not possible.

© Springer International Publishing AG 2017
N. Camelin et al. (Eds.): SLSP 2017, LNAI 10583, pp. 155–166, 2017.
DOI: 10.1007/978-3-319-68456-7_13

Two basic approaches can be distinguished for automatic punctuation, although they are often used in combination: prosody and text based approaches. In general prosody based approaches require less computation, less training data and hence can result in lightweight punctuation models. They are also more robust to ASR errors; recently proposed text based approaches on the other hand provide mostly more accurate punctuation, but are more sensitive to ASR errors and may introduce high latency due to the processing of a wide context, requiring extensive computations and also future context which directly results in high latency.

In this paper we focus on reducing this latency by still maintaining the accuracy provided by text based models. We demonstrate systems intended to be used for punctuation of closed-captioned data. ASR technology is widely used by television companies to produce closed captions especially for live programs [21], which require almost real-time processing with little latency.

Much effort has been devoted to develop reliable punctuation restoration algorithms, early approaches proposed to add punctuation marks to the N-gram language model of the ASR as hidden events [8,23]. These models have to be trained on huge corpora to reduce data sparsity [8]. More sophisticated sequence modeling approaches were also inspired by this idea: a transducer alike approach getting a non-punctuated text as input is capable of predicting punctuation as was presented in numerous works [1,3,11], with frameworks built on top of Hidden Markov Models (HMM), Maximum Entropy (MaxEnt) models or conditional random fields, etc. MaxEnt models allow for any easy combination of textual and prosodic features into a common punctuation model [10]. In a comprehensive study [2], many features were compared in terms of their effect on punctuation accuracy of a MaxEnt model. It was found that the most powerful textual features were the word forms and part-of-speech (POS) tags, whereas the best prosodic feature was the duration of inter-word pauses.

Applying a monolingual translation paradigm for punctuation regarded as a sequence modeling task was also proposed in [5], which also allowed for considerably reducing time latency. Recently, sequence-to-sequence modeling deep neural network based solutions have been also presented: taking a large word-context and projecting the words via an embedding layer into a bidirectional Recurrent Neural Network (RNN) [22], high quality punctuation could be achieved. RNNs are successfully used in many sequence labeling tasks as they are able to model large contexts and to learn distributed features of words to overcome data sparsity issues. The first attempt to use RNN for punctuation restoration was presented in [24], where a one-directional LSTM [9] was trained on Estonian broadcast transcripts. Shortly after, Tilk and Alumäe introduced a bidirectional RNN model using GRU [7] together with attention mechanism, which outperformed previous state-of-the-art on Estonian and English IWSLT datasets [25]. In a recent study [15], capitalization and punctuation recovery are treated as correlated multiple sequence labeling tasks and modeled with bidirectional RNN. In [14], a prosody based punctuation approach was proposed using an RNN on top of phonological phrase sequence modeling.

In this paper, we introduce a lightweight RNN-based punctuation restoration model using bidirectional LSTM units on top of word embeddings, and compare its performance to a MaxEnt model. We pay a special attention to low latency solutions. Both approaches are evaluated on automatic and manual transcripts and in various setups including on-line and off-line operation. We present results on Hungarian broadcast speech transcripts and the IWSLT English dataset [4] to make the performance of our approach comparable to state-of-the-art systems. Apart form the purely prosody based approach outlined in [14], we are not aware of any prior work for punctuation restoration for Hungarian speech transcripts.

Our paper is structured in the following way: first we present the used datasets in Sect. 2, then we move on to presenting the experimental systems in Sect. 3. The results of Hungarian and English Punctuation Restoration tasks are presented and discussed in Sect. 4. Our conclusions and future ideas are drawn in Sect. 5.

## 2    Data

### 2.1    The Hungarian Broadcast Dataset

The Hungarian dataset consists of manually transcribed closed captions made available by the Media Service Support and Asset Management Fund (MTVA), Hungary's public service broadcaster. The dataset contains captions for various TV genres enabling us to evaluate the punctuation models on different speech types, such as weather forecasts, broadcast news and conversations, magazines, sport news and sport magazines. We focus on the restoration of those punctuations, which have a high importance for understandability in Hungarian: commas, periods, question marks and exclamation marks. The colons and semicolons were mapped to comma. All other punctuation symbols are removed from the corpora. We reserve a disjunct 20% of the corpus for validation and use a representative test set, not overlapping with training and validation subsets. For further statistics about training and test data we refer the reader to Table 1.

**Table 1.** Statistics of the Hungarian dataset

| Genres | Training & Validation | | | | | Test | | | | | |
|---|---|---|---|---|---|---|---|---|---|---|---|
| | #Words | #Com | #Per | #Que | #Excl | #Words | #Com | #Per | #Ques | #Excl | WER |
| Weather | 478 K | 40 K | 31.5 K | 30 | 730 | 2.4 K | 250 | 200 | 0 | 20 | 6.8 |
| Brc.-News | 3493 K | 279 K | 223 K | 3.5 K | 4.6 K | 17 K | 1.5 K | 1 K | 20 | 50 | 10.1 |
| Sport News | 671 K | 55 K | 39.5 K | 280 | 2 K | 6 K | 500 | 400 | 2 | 30 | 21.4 |
| Brc.-Conv. | 4161 K | 533 K | 225 K | 26.5 K | 4 K | 46.8 K | 6.3 K | 2.6 K | 250 | 130 | 24.7 |
| Sport mag. | - | - | - | - | - | 22.7 K | 2 K | 1.4 K | 100 | 50 | 30.3 |
| Magazine | 4909 K | 732 K | 376 K | 72 K | 36 K | 10.4 K | 1.5 K | 700 | 150 | 70 | 38.7 |
| Mixed | 1526 K | 187 K | 102 K | 11 K | 11.4 K | 30.7 | 4 K | 1.7 K | 280 | 150 | - |
| ALL | 15238 K | 1826 K | 997 K | 113 K | 58.8 K | 136 K | 16 K | 8 K | 800 | 500 | 24.2 |

The automatic transcription of the test set is carried out with an ASR system optimized for the task (close captioning of live audio) [27]. The language model for the ASR was trained on the same corpus as the punctuation model and was coupled with a Deep Neural Network based acoustic model trained on roughly 500 hours of speech using the Kaldi ASR toolkit [18]. The average word error rate (WER) of the automatic transcripts was around 24%, however showed a large variation depending on genre (see later Table 1). Note, that for Mixed category there was no available audio data in the test database.

### 2.2 The English IWSLT Dataset

The IWSLT dataset consists of English TED talks transcripts, and has recently became a benchmark for evaluating English punctuation recovery models [4, 15, 24, 25]. We use the same training, validation and test sets as the studies above, containing 2.1 M, 296 K and 13 K words respectively. This dataset deals with only three types of punctuations: comma, period and question mark.

## 3 Experimental Setups

### 3.1 MaxEnt Model

The maximum entropy (MaxEnt) model was suggested by Ratnaparkhi for POS Tagging [19]. In his framework, each sentence is described as a token (word) sequence. Each classified token is described with a set of unique features. The system learns the output labels based on these. In supervised learning, the output labels are hence assigned to the token series. To determine the set of features, the MaxEnt model defines a joint distribution through the available tags and the current context, which can be controlled with a radius parameter. Pre-defined features such as word forms, capitalization, etc. can also be added.

We use the MaxEnt model only with word form-related input features, and all tokens are represented in lower case. To obtain these input features, we use *Huntag*, an open-source, language independent Maximum Entropy Markov Model-based Sequential tagger for both Hungarian and English data [20].

The radius parameter of the MaxEnt tagger determines the size of the context considered. By default, left (past) and right (future) context is taken into account. We will refer to this setup as *off-line mode*. As taking future context into account increases latency, we consider the limit of it, which we will refer to by *on-line mode*. In the experiments we use round brackets to specify left and right context, respectively. Hence (5,1) means that we are considering 5 past and 1 future token actually.

### 3.2 Recurrent Neural Networks

We split the training, validation and test corpus into short, fixed-length subsequences, called chunks (see the optimized length in Table 2), without overlapping, i.e. such that every token appears once. A vocabulary is built from the k-most common words from the training set, by adding a garbage collector "*Unknown*" entry to map rare words. Incomplete sub-sequences were padded with zeros. An embedding weight matrix was added based on pre-trained embedding weights and the tokens of the vocabulary.

We investigate the performance of an unidirectional and a bidirectional RNN model in our experiments. Our target slot for punctuation prediction is preceding the actual word. The used architectures are presented in Fig. 1.

Our RNN models (WE-LSTM and WE-BiLSTM, named after using "Word Embedding") are built up in the following way: based on the embedding matrix, the preprocessed sequences are projected into the embedding space ($x_t$ represents the word vector $x$ at time step $t$). These features are fed into the following layer composed of LSTM or BiLSTM hidden cells, to capture the context of $x_t$. The output is obtained by applying a softmax activation function to predict the $y_t$ punctuation label for the slot preceding the current word $x_t$. We chose this simple and lightweight structure to allow for real-time operation with low latency.

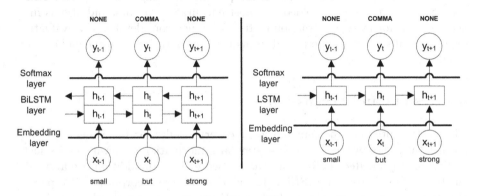

**Fig. 1.** Structure of WE-BiLSTM (left) and WE-LSTM (right) RNN model

The Hungarian punctuation models were trained on the 100 K most frequent words in the training corpus, by mapping the remaining outlier words to a shared "*Unknown*" symbol. RNN-based recovery models use 600-dimensional pre-trained Hungarian word embeddings [13]. This relative high dimensionality of the embeddings comes from the highly agglutinating nature of Hungarian. In our English RNN-models, a 100-dimensional pre-trained "GloVe" word embedding [17] is used for projection. During training, we use categorical cross-entropy cost function and also let the imported embeddings updated.

We performed a systematic grid search optimization for hyperparameters of the RNNs on the validation set: length of chunks, vocabulary size, number of hidden states, mini-batch size, optimizers. We also use early stopping to prevent overfitting, controlled with patience. Table 2 summarizes the final values of each hyperparameter used in the Hungarian and the English WE-BiLSTM and WE-LSTM models, also including those ones which were inherited from [25], to ensure a partial comparability.

**Table 2.** Hyperparameters of WE-BiLSTM and WE-LSTM models

| Language | Model | Chunk length (#words) | Vocab. Size (#words) | Word embedding dimension | #Hidden states | Batch size | Optimizer | Patience |
|---|---|---|---|---|---|---|---|---|
| HUN | WE-BiLSTM | 200 | 100 000 | 600 | 512 | 128 | RMSProp | 3 |
| HUN | WE-LSTM | | | | 256 | | | 2 |
| EN | WE-BiLSTM | 200 | 27 244 (by [25]) | 100 (by [25]) | 256 | | | 2 |
| EN | WE-LSTM | 250 | | | | | | |

As for the MaxEnt setup, we differentiate low latency and lightweight on-line mode, and robust off-line mode using the future context. All RNN models for punctuation recovery were implemented with the Keras library [6], trained on GPU. The source code of the RNN models is publicly available[1].

We briefly mention that beside word forms, we were considering other textual features too: lemmas, POS-tags (also suggested by [26]) and morphological analysis. The latter were extracted using the *magyarlánc* toolkit, designed for morphological analysis and dependency parsing in Hungarian [28]. Nevertheless, as using word forms yielded the most encouraging results, and also as further analysis for feature extraction increases latency considerably, the evaluated experimental systems rely on word forms features only, input to the embedding layers.

## 4    Results and Discussion

This section presents the punctuation recovery results for the Hungarian and English tasks. For evaluation, we use standard information retrieval metrics such as Precision (Pr), Recall (Rc), and the F1-Score (F1). In addition, we also calculate the Slot Error Rate (SER) [12], as it is able to incorporate all types of punctuation errors – insertions (Ins), substitutions (Subs) and deletions (Dels) – into a single measure:

$$SER = \frac{C(Ins) + C(Subs) + C(Del)}{C(totalslots)}, \qquad (1)$$

for slots considered following each word in the transcription (in (1) $C(.)$ is the count operator).

### 4.1    Hungarian Overall Results

First, we compare the performance of the baseline MaxEnt sequence tagger (see Subsect. 3.1) to the RNN-based punctuation recovery system (see Subsect. 3.2) on the Hungarian broadcast dataset. Both approaches are presented in two configurations. In the *on-line mode* punctuations are predicted for the slot preceding the current word in the input sequence resulting in a low latency system, suitable for real-time application. In the *off-line mode*, aimed at achieving the best

---

[1]  https://github.com/tundik/HuPP.

result with the given features and architecture, the future word context is also exploited. Please note that hyperparameters of all approaches and configurations were optimized on the validation set as explained earlier (see Sect. 3).

The test evaluations are presented in Table 3 for the reference and in Table 4 for the automatic (ASR) transcripts, respectively. In the notation of MaxEnt models $(i, j)$, $i$ stands for the backward (past), whereas $j$ stands for the forward (future) radius. As it can be seen, the prediction results for comma stand out from the others for all methods and configurations. This can be explained by the fact that Hungarian has generally clear rules for comma usage. In contrast to that, period prediction may also benefit from acoustic information, which assumption is supported by the results in [14], showing robust period recovery with less effective comma restoration.

**Table 3.** Punctuation restoration results for Hungarian reference transcripts

| Reference transcript | Model | Comma | | | Period | | | Question | | | Exclamation | | | SER |
|---|---|---|---|---|---|---|---|---|---|---|---|---|---|---|
| | | Pr | Rc | F1 | Pr | Rc | F1 | Pr | Rc | F1 | Pr | Rc | F1 | |
| Off-line mode | MaxEnt-(19, 19) | 72.5 | 59.6 | 65.5 | 52.1 | 40.0 | 45.2 | **55.7** | 21.8 | 31.3 | 31.1 | 31.5 | 31.3 | 63.5 |
| | WE-BiLSTM | **72.9** | **71.2** | **72.0** | **59.1** | **56.1** | **57.6** | 52.4 | **38.7** | **44.5** | **51.3** | 36.1 | **42.4** | **50.1** |
| On-line mode | MaxEnt-(25, 1) | 71.8 | 58.1 | 64.2 | 47.5 | 35.7 | 40.8 | 50.4 | 16.2 | 24.5 | 29.3 | **33.3** | 31.2 | 66.9 |
| | WE-LSTM | **72.7** | **69.5** | **71.1** | **56.2** | **48.3** | **52.0** | **60.4** | **31.1** | **41.1** | **61.1** | 29.4 | **39.7** | **53.6** |

As Table 3 shows, switching to the RNN-based punctuation restoration for Hungarian reference transcripts reduces SER by around 20% relative compared to the baseline MaxEnt approach. The WE-BiLSTM and WE-LSTM are especially beneficial in restoring periods, question marks and exclamation marks as they are able to exploit large contexts much more efficiently than the MaxEnt tagger. Limiting the future context in on-line configuration causes much less deterioration in results than we had expected. The features from the future word sequence seem to be useful if task requires maximizing recall, otherwise the WE-LSTM is an equally suitable model for punctuation recovery.

**Table 4.** Punctuation restoration results for Hungarian ASR transcripts

| ASR transcript | Model | Comma | | | Period | | | Question | | | Exclamation | | | SER |
|---|---|---|---|---|---|---|---|---|---|---|---|---|---|---|
| | | Pr | Rc | F1 | Pr | Rc | F1 | Pr | Rc | F1 | Pr | Rc | F1 | |
| Off-line mode | MaxEnt-(19, 19) | **64.5** | 55.8 | 59.9 | 41.1 | 31.2 | 35.6 | **41.2** | 8.8 | 14.4 | 48.8 | 17.1 | 25.4 | 79.2 |
| | WE-BiLSTM | 63.9 | **67.7** | **65.7** | **50.5** | **49.0** | **49.8** | 37.7 | **24.1** | **29.4** | **60.9** | **24.0** | **34.4** | **70.1** |
| On-line mode | MaxEnt-(25, 1) | **64.3** | 54.9 | 59.2 | 38.9 | 29.4 | 33.5 | 36.0 | 7.1 | 11.9 | 47.1 | 20.6 | 28.6 | 81.3 |
| | WE-LSTM | 63.8 | **65.1** | **64.4** | **47.8** | **42.0** | **44.7** | **48.5** | **20.5** | **28.9** | **61.8** | **21.7** | **32.1** | **73.1** |

As outlined in the introduction, limiting the future context and propagation of ASR errors into the punctuation recovery pipeline are considered to be the most important factors hindering effective recovery of punctuations in live TV streams. Results confirm that a large future context is less crucial for robust recovery of punctuations, contradictory to our expectations. In contrast, ASR errors seem to be more directly related to punctuation errors: switching from reference transcripts to ASR hypotheses resulted in 15–20% increase in SER (see

Table 4). Although the performance gap is decreased between the two approaches in case of input featuring ASR hypothesis, RNN still outperforms MaxEnt baseline by a large margin.

## 4.2 Hungarian Results by Genre

The Hungarian test database can be divided into 6 subsets based on the genres of the transcripts (see Table 1). We also analyzed punctuation recovery for these subsets, hypothesizing that more informal and more spontaneous genres are harder to punctuate, in parallel to the more ASR errors seen in these scenarios. Some of the punctuation marks for specific genres were not evaluated (see "N/A" in Table 1), if the Precision or Recall was not possible to be determined based on their confusion matrix.

As the RNN-based approach outperformed the MaxEnt tagger for every genre, we decided to include only results of WE-BiLSTM and WE-LSTM systems in Tables 5 and 6 for better readability.

**Table 5.** Hungarian reference transcript results by genres

| Reference transcript | Genre | Comma | | | Period | | | Question | | | Exclamation | | | SER |
|---|---|---|---|---|---|---|---|---|---|---|---|---|---|---|
| | | Pr | Rc | F1 | Pr | Rc | F1 | Pr | Rc | F1 | Pr | Rc | F1 | |
| RNN Off-line mode | Weather | 61.2 | 54.3 | 57.5 | 46.7 | 46.7 | 46.7 | N/A | N/A | N/A | 90.0 | 45.0 | 60.0 | 69.3 |
| | Brc.-News | 89.9 | 84.4 | 87.1 | 84.3 | 90.7 | 87.3 | 91.7 | 50.0 | 64.7 | 83.9 | 56.5 | 67.5 | 20.0 |
| | Sport news | 68.3 | 60.6 | 64.2 | 49.4 | 51.4 | 50.4 | N/A | N/A | N/A | 75.0 | 30.0 | 42.9 | 67.0 |
| | Brc.-Conv. | 80.4 | 74.5 | 77.3 | 63.9 | 64.9 | 64.4 | 63.0 | 46.4 | 53.5 | 88.9 | 18.5 | 30.6 | 38.7 |
| | Sport mag. | 61.2 | 61.1 | 61.1 | 43.9 | 49.3 | 46.5 | 55.2 | 37.5 | 44.7 | 38.5 | 9.4 | 15.2 | 73.1 |
| | Magazine | 67.6 | 67.6 | 67.6 | 45.1 | 46.3 | 45.7 | 50.5 | 29.7 | 37.5 | 50.0 | 5.6 | 10.1 | 58.6 |
| RNN On-line mode | Weather | 60.2 | 57.5 | 58.8 | 45.7 | 37.9 | 41.4 | N/A | N/A | N/A | 87.5 | 35.0 | 50.0 | 70.6 |
| | Brc.-News | 88.4 | 83.1 | 85.7 | 86.6 | 81.3 | 83.9 | 75.0 | 40.9 | 52.9 | 100.0 | 67.4 | 80.5 | 24.1 |
| | Sport news | 68.7 | 57.2 | 62.4 | 42.4 | 37.5 | 39.8 | N/A | N/A | N/A | 90.0 | 60.0 | 72.0 | 74.2 |
| | Brc.-Conv. | 80.1 | 74.0 | 76.9 | 66.7 | 54.8 | 60.1 | 63.0 | 45.6 | 52.9 | 77.6 | 29.2 | 42.5 | 40.8 |
| | Sport mag. | 60.8 | 59.7 | 60.3 | 42.3 | 34.8 | 38.2 | 53.3 | 38.3 | 44.5 | 20.0 | 7.5 | 11.0 | 77.3 |
| | Magazine | 67.6 | 65.1 | 66.3 | 43.5 | 32.8 | 37.4 | 57.3 | 27.2 | 36.9 | 36.4 | 11.3 | 17.2 | 61.5 |

If we compare the results to the statistics in Table 1, it can be seen that the punctuation recovery system performed best on those genres (broadcast news, broadcast conversations, magazine), for which we had the most training samples. However, the relatively large difference among these three, well-modeled genres suggests that there must be another factor in the background, as well, which is the predictability of the given task. Analogous to language modeling, the more formal, the task is, the better is the predictability of punctuations (see broadcast news results). Obviously, conversational (broadcast conversations) and informal (magazine) speech styles (characterized with less constrained wording and increased number of disfluencies and ungrammatical phrases) make prediction more difficult and introduce punctuation errors compared to more formal styles.

The relatively high SER of the weather forecast and the sport programs genres point out the importance of using a sufficient amount of in-domain training data. Besides collecting more training data, adaptation techniques could be utilized to improve results for these under-resourced genres.

**Table 6.** Hungarian ASR transcript results by genres

| ASR transcript | Genre | Comma | | | Period | | | Question | | | Exclamation | | | SER | WER |
|---|---|---|---|---|---|---|---|---|---|---|---|---|---|---|---|
| | | Pr | Rc | F1 | Pr | Rc | F1 | Pr | Rc | F1 | Pr | Rc | F1 | | |
| RNN Off-line mode | Weather | 64.7 | 54.3 | 59.1 | 45.9 | 42.0 | 43.9 | N/A | N/A | N/A | 100.0 | c50.0 | 66.7 | 70.0 | 6.8 |
| | Brc.-News | 79.5 | 80.0 | 79.7 | 74.7 | 82.6 | 78.4 | 50.0 | 14.3 | 22.2 | 80.0 | 46.2 | 58.5 | 37.0 | 10.1 |
| | Sport news | 47.7 | 54.8 | 51.0 | 32.1 | 40.5 | 35.8 | N/A | N/A | N/A | 100.0 | 50.0 | 66.7 | 107.5 | 21.4 |
| | Brc.-Conv. | 70.6 | 67.5 | 69.0 | 56.8 | 51.3 | 53.9 | 43.8 | 28.9 | 34.8 | 84.6 | 13.6 | 23.4 | 60.2 | 24.7 |
| | Sport mag. | 55.9 | 59.8 | 57.8 | 39.2 | 43.5 | 41.3 | 48.0 | 16.2 | 24.2 | N/A | N/A | N/A | 87.5 | 30.3 |
| | Magazine | 58.6 | 60.2 | 59.5 | 35.6 | 28.4 | 31.6 | 31.2 | 14.9 | 20.2 | N/A | N/A | N/A | 83.1 | 38.7 |
| RNN On-line mode | Weather | 65.3 | 58.7 | 61.8 | 37.8 | 34.6 | 36.1 | N/A | N/A | N/A | 100.0 | 12.5 | 22.2 | 72.7 | 6.8 |
| | Brc.-News | 76.5 | 79.3 | 77.9 | 76.8 | 70.7 | 73.6 | N/A | N/A | N/A | 100.0 | 50.0 | 66.7 | 42.9 | 10.1 |
| | Sport news | 48.9 | 54.3 | 51.5 | 28.6 | 30.1 | 29.3 | N/A | N/A | N/A | 75.0 | 60.0 | 66.7 | 108.9 | 21.4 |
| | Brc.-Conv. | 70.3 | 66.8 | 68.5 | 57.4 | 41.2 | 48.0 | 37.0 | 24.8 | 29.7 | 86.7 | 16.0 | 27.1 | 62.2 | 24.7 |
| | Sport mag. | 53.6 | 56.4 | 55.0 | 37.8 | 30.4 | 33.4 | 42.1 | 21.6 | 28.6 | 14.3 | 4.5 | 6.9 | 91.0 | 30.3 |
| | Magazine | 57.6 | 59.4 | 58.5 | 36.5 | 20.8 | 26.5 | 42.1 | 11.9 | 18.7 | N/A | N/A | N/A | 83.9 | 38.7 |

By comparing punctuation recovery error of the reference and ASR transcripts, we can draw some interesting conclusions. For the well-modeled genres (Brc.-News, Brc.-Conv., magazine) the increase in SER correlates with the word error rate (WER) of the ASR transcript. However, for the remaining genres (weather, sport news, sport magazine), this relationship between SER and WER is much less predictable. It is particularity difficult to explain the relatively poor results for the sport news genre. Whereas the WER of the ASR transcript is moderate (24.7%), the SER of punctuation is almost doubled for it (67% to 107%). We assume that this phenomenon is related to the high number of named entities in the sport news program, considering that the highest OOV Rate (10%) can be spotted for this genre among all the 6 tested genres.

### 4.3  English Results

In this subsection, we compare our solutions for punctuation recovery with some recently published models. For this purpose, we use the IWSLT English dataset, which consists of TED Talks transcripts and is a considered benchmark for English punctuation recovery. For complete comparability, we used the default training, validation and test datasets. However, the hyperparameters were optimized for this task (see Table 2). Please note that the IWSLT dataset does not contain samples for exclamation marks.

We present the English punctuation recovery results in Tables 7 and 8. As it can be seen, in on-line mode, the proposed RNN (WE-LSTM) significantly outperformed the so-called T-LSTM configuration presented in [25], which had the best on-line results on this dataset so far to the best of our knowledge. Without using pre-trained word embedding (noWE-LSTM) our results are getting very close to the T-LSTM configuration.

Although in this paper we primarily focused on creating a lightweight, low latency punctuation recovery system, we also compared our WE-BiLSTM system to the best available off-line solutions. As it is shown in Tables 7 and 8, both T-BRNN-pre from [25] configuration and Corr-BiRNN form [15] outperformed

**Table 7.** Punctuation restoration results for English reference transcripts

| Reference transcript | Model | Comma | | | Period | | | Question | | | SER |
|---|---|---|---|---|---|---|---|---|---|---|---|
| | | Pr | Rc | F1 | Pr | Rc | F1 | Pr | Rc | F1 | |
| Off-line mode | MaxEnt-(6, 6) | 45.6 | 26.7 | 33.7 | 59.4 | 57.0 | 58.2 | 52.4 | 23.9 | 32.8 | 77.2 |
| | WE-BiLSTM | 55.5 | 45.1 | 49.8 | 65.9 | **75.1** | 70.2 | 57.1 | 52.2 | 54.5 | 59.8 |
| | T-BRNN-pre [25] | **65.5** | 47.1 | 54.8 | 73.3 | 72.5 | 72.9 | **70.7** | **63.0** | **66.7** | **49.7** |
| | Corr-BiRNN [15] | 60.9 | **52.4** | **56.4** | **75.3** | 70.8 | **73.0** | 70.7 | 56.9 | 63.0 | 50.8 |
| On-line mode | MaxEnt-(10, 1) | 44.9 | 23.7 | 31.0 | 53.4 | 50.1 | 51.7 | 50.0 | 21.7 | 30.8 | 83.2 |
| | noWE-LSTM | 47.3 | **42.7** | 44.9 | 60.9 | 50.4 | 55.2 | **68.2** | 32.6 | 44.1 | 76.4 |
| | WE-LSTM | 56.3 | 40.3 | **47.0** | **61.2** | **60.5** | **60.8** | 55.5 | **43.5** | 48.8 | **68.1** |
| | T-LSTM [24] | 49.6 | 41.1 | 45.1 | 60.2 | 53.4 | 56.6 | 57.1 | 43.5 | **49.4** | 74.0 |

**Table 8.** Punctuation restoration results for English ASR transcripts

| ASR transcript | Model | Comma | | | Period | | | Question | | | SER |
|---|---|---|---|---|---|---|---|---|---|---|---|
| | | Pr | Rc | F1 | Pr | Rc | F1 | Pr | Rc | F1 | |
| Off-line mode | MaxEnt-(6, 6) | 40.6 | 23.9 | 30.1 | 56.2 | 53.5 | 54.8 | 31.6 | 17.1 | 22.2 | 84.0 |
| | WE-BiLSTM | 46.8 | 39.6 | 42.9 | 60.7 | 70.3 | 65.1 | 44.4 | 45.7 | 45.0 | 72.5 |
| | T-BRNN-pre [25] | **59.6** | 42.9 | 49.9 | **70.7** | 72.0 | **71.4** | 60.7 | 48.6 | 54.0 | **57.0** |
| | Corr-BiRNN [15] | 53.5 | **52.5** | **53.0** | 63.7 | 68.7 | 66.2 | **66.7** | **50.0** | **57.1** | 65.4 |
| On-line mode | MaxEnt-(10, 1) | 42.6 | 23.9 | 30.7 | 53.2 | 48.9 | 51.0 | 33.3 | 17.1 | 23.0 | 87.0 |
| | noWE-LSTM | 40.2 | **39.3** | 39.7 | 56.2 | 46.6 | 51.0 | **76.5** | 38.2 | **51.0** | 86.5 |
| | WE-LSTM | **48.8** | 37.1 | **42.2** | **57.6** | **57.3** | **57.4** | 41.2 | 41.2 | 41.2 | **78.3** |
| | T-LSTM [24] | 41.8 | 37.8 | 39.7 | 56.4 | 49.3 | 52.6 | 55.6 | **42.9** | 48.4 | 83.7 |

our WE-BiLSTM mainly due to their better performance for commas and question marks. However, these punctuation recovery systems are using much more complex structure and it is questionable whether they would be able to operate in real time scenarios. We consider the high recall of periods by our WE-BiLSTM models as a nice achievement both in reference and ASR transcripts.

## 5  Conclusions

In this paper, we introduced a low latency, RNN-based punctuation recovery system, which we evaluated on Hungarian and English datasets and compared its performance to a MaxEnt sequence tagger. Both approaches were tested in off-line mode, where textual features could be used from both forward and backward directions; and also in on-line mode, where only backward features were used to allow for real-time operation. The RNN-based approach outperformed the MaxEnt baseline by a large margin in every test configuration. However, what is more surprising, on-line mode causes only a small drop in the accuracy of punctuation recovery.

By comparing results on different genres of the Hungarian broadcast transcripts, we found (analogous to language modeling) that the accuracy of text based punctuation restoration mainly depends on the amount of available training data and the predictability of the given task. Note, that we are not aware of any prior work in the field of text based punctuation recovery of Hungarian speech transcripts.

In order to compare our models to state-of-the-art punctuation recovery systems, we also evaluated them on the IWSLT English dataset in both on-line and off-line modes. In on-line mode, our WE-LSTM system achieved the overall best result. In off-line mode, however, some more complex networks turned out to perform better than our lightweight solution.

For future work, we are mainly interested in merging of our word-level system and the prosody-based approach outlined in [14] for Hungarian. Extending the English model with further textual or acoustic features is also a promising direction, as we keep our focus on low latency for both languages.

All in all, we consider as important contributions of our work that (1) we use a lightweight and fast RNN model by closely maintained performance; (2) we target real-time operation with little latency; (3) we use the approach for the highly agglutinating Hungarian which has a much less constrained word order than English, as grammatical functions depend much less on the word order than on suffixes (case endings), which makes sequence modeling more difficult due to higher variation seen in the data.

**Acknowledgements.** The authors would like to thank the support of the Hungarian National Research, Development and Innovation Office (NKFIH) under contract ID *OTKA-PD-112598*; the Pro Progressio Foundation; NVIDIA for kindly providing a Titan GPU for the RNN experiments.

# References

1. Batista, F., Moniz, H., Trancoso, I., Mamede, N.: Bilingual experiments on automatic recovery of capitalization and punctuation of automatic speech transcripts. IEEE Trans. Audio Speech Lang. Process. **20**(2), 474–485 (2012)
2. Batista, F.: Recovering capitalization and punctuation marks on speech transcriptions. Ph.D. thesis. Instituto Superior Técnico (2011)
3. Beeferman, D., Berger, A., Lafferty, J.: Cyberpunc: A lightweight punctuation annotation system for speech. In: Proceedings of ICASSP, pp. 689–692. IEEE (1998)
4. Che, X., Wang, C., Yang, H., Meinel, C.: Punctuation prediction for unsegmented transcript based on word vector. In: Proceedings of LREC, pp. 654–658 (2016)
5. Cho, E., Niehues, J., Kilgour, K., Waibel, A.: Punctuation insertion for real-time spoken language translation. In: Proceedings of the Eleventh International Workshop on Spoken Language Translation (2015)
6. Chollet, F.: Keras: Theano-based deep learning library. Code: https://github.com/fchollet. Documentation: http://keras.io (2015)
7. Chung, J., Gulcehre, C., Cho, K., Bengio, Y.: Empirical evaluation of gated recurrent neural networks on sequence modeling. arxiv preprint arXiv:1412.3555 (2014)
8. Gravano, A., Jansche, M., Bacchiani, M.: Restoring punctuation and capitalization in transcribed speech. In: Proceedings of ICASSP, pp. 4741–4744. IEEE (2009)

9. Hochreiter, S., Schmidhuber, J.: Long short-term memory. Neural Comput. **9**(8), 1735–1780 (1997)

10. Huang, J., Zweig, G.: Maximum entropy model for punctuation annotation from speech. In: Proceedings of Interspeech, pp. 917–920 (2002)

11. Lu, W., Ng, H.T.: Better punctuation prediction with dynamic conditional random fields. In: Proceedings of EMNLP, pp. 177–186. ACL (2010)

12. Makhoul, J., Kubala, F., Schwartz, R., Weischedel, R.: Performance measures for information extraction. In: Proceedings of DARPA broadcast news workshop, pp. 249–252 (1999)

13. Makrai, M.: Filtering wiktionary triangles by linear mapping between distributed models. In: Proceedings of LREC, pp. 2770–2776 (2016)

14. Moró, A., Szaszák, G.: A phonological phrase sequence modelling approach for resource efficient and robust real-time punctuation recovery. In: Proceedings of Interspeech (2017)

15. Pahuja, V., Laha, A., Mirkin, S., Raykar, V., Kotlerman, L., Lev, G.: Joint learning of correlated sequence labelling tasks using bidirectional recurrent neural networks. arxiv preprint arXiv:1703.04650 (2017)

16. Paulik, M., Rao, S., Lane, I., Vogel, S., Schultz, T.: Sentence segmentation and punctuation recovery for spoken language translation. In: Proceedings of ICASSP, pp. 5105–5108. IEEE (2008)

17. Pennington, J., Socher, R., Manning, C.D.: Glove: global vectors for word representation. In: Proceedings of EMNLP, pp. 1532–1543 (2014)

18. Povey, D., et al.: The Kaldi speech recognition toolkit. In: Proceedings of ASRU, pp. 1–4. IEEE (2011)

19. Ratnaparkhi, A., et al.: A maximum entropy model for part-of-speech tagging. In: Proceedings of EMNLP, pp. 133–142 (1996)

20. Recski, G., Varga, D.: A Hungarian NP chunker. Odd Yearb. **8**, 87–93 (2009)

21. Renals, S., Simpson, M., Bell, P., Barrett, J.: Just-in-time prepared captioning for live transmissions. In: Proceedings of IBC 2016 (2016)

22. Schuster, M., Paliwal, K.K.: Bidirectional recurrent neural networks. IEEE Trans. Sig. Process. **45**(11), 2673–2681 (1997)

23. Shriberg, E., Stolcke, A., Baron, D.: Can prosody aid the automatic processing of multi-party meetings? evidence from predicting punctuation, disfluencies, and overlapping speech. In: ISCA Tutorial and Research Workshop (ITRW) on Prosody in Speech Recognition and Understanding (2001)

24. Tilk, O., Alumäe, T.: LSTM for punctuation restoration in speech transcripts. In: Proceedings of Interspeech, pp. 683–687 (2015)

25. Tilk, O., Alumäe, T.: Bidirectional recurrent neural network with attention mechanism for punctuation restoration. In: Proceedings of Interspeech, pp. 3047–3051 (2016)

26. Ueffing, N., Bisani, M., Vozila, P.: Improved models for automatic punctuation prediction for spoken and written text. In: Proceedings of Interspeech, pp. 3097–3101 (2013)

27. Varga, Á., Tarján, B., Tobler, Z., Szaszák, G., Fegyó, T., Bordás, C., Mihajlik, P.: Automatic close captioning for live Hungarian television broadcast speech: a fast and resource-efficient approach. In: Ronzhin, A., Potapova, R., Fakotakis, N. (eds.) SPECOM 2015. LNCS, vol. 9319, pp. 105–112. Springer, Cham (2015). doi:10.1007/978-3-319-23132-7_13

28. Zsibrita, J., Vincze, V., Farkas, R.: magyarlanc: A toolkit for morphological and dependency parsing of Hungarian. In: Proceedings of RANLP, pp. 763–771 (2013)

# Speech: Paralinguistics and Synthesis

# Unsupervised Speech Unit Discovery Using K-means and Neural Networks

Céline Manenti$^{(\boxtimes)}$, Thomas Pellegrini, and Julien Pinquier

IRIT, Université de Toulouse, UPS, Toulouse, France
{celine.manenti,thomas.pellegrini,julien.pinquier}@irit.fr

**Abstract.** Unsupervised discovery of sub-lexical units in speech is a problem that currently interests speech researchers. In this paper, we report experiments in which we use phone segmentation followed by clustering the segments together using k-means and a Convolutional Neural Network. We thus obtain an annotation of the corpus in pseudo-phones, which then allows us to find pseudo-words. We compare the results for two different segmentations: manual and automatic. To check the portability of our approach, we compare the results for three different languages (English, French and Xitsonga). The originality of our work lies in the use of neural networks in an unsupervised way that differ from the common method for unsupervised speech unit discovery based on auto-encoders. With the Xitsonga corpus, for instance, with manual and automatic segmentations, we were able to obtain 46% and 42% purity scores, respectively, at phone-level with 30 pseudo-phones. Based on the inferred pseudo-phones, we discovered about 200 pseudo-words.

**Keywords:** Neural representation of speech and language · Unsupervised learning · Speech unit discovery · Neural network · Sub-lexical units · Phone clustering

## 1 Introduction

Annotated speech data abound for the most widely spoken languages, but the vast majority of languages or dialects is few endowed with manual annotations. To overcome this problem, unsupervised discovery of linguistic pseudo-units in continuous speech is gaining momentum in the recent years, encouraged for example by initiatives such as the *Zero Resource Speech Challenge* [15].

We are interested in discovering pseudo-units in speech without supervision at phone level ("pseudo-phones") and at word level ("pseudo-word"). Pseudo-words are defined by one or more speech segments representing the same phonetic sequence. These are not necessarily words: they may not start/end at the beginning/ending of a true word and may contain several words, such as "I think that". The same applies for pseudo-phones that may be shorter or longer than true phones, and one pseudo-phone may represent several phones.

To find speech units, one can use dotplots [4], a graphical method for comparing sequences, and Segmental Dynamic Time Warping (S-DTW) with the

© Springer International Publishing AG 2017
N. Camelin et al. (Eds.): SLSP 2017, LNAI 10583, pp. 169–180, 2017.
DOI: 10.1007/978-3-319-68456-7_14

use of the cosine similarity that gives distances between phonetic acoustic models [1,11]. In [17], DTW and Hidden Markov Models are also used on posteriorgrams (posterior distribution over categorical units (e.g. phonemes) as a function of time) to find pseudo-words.

During our own experiments, we were able to see the usefulness of the posteriorgrams, which are data obtained by supervised learning. We therefore sought to obtain these posteriorgrams phones in an unsupervised way.

To obtain phone posteriorgrams, clustering can be used. In [17], k-means are used on parameters generated by an auto-encoder (AE), also called Bottleneck Features (BnF), after binarization. k-means are similarly used in [14], with AEs and graph clustering. Increasingly used in speech research, neural networks come in several unsupervised flavors. AEs learn to retrieve the input data after several transformations performed by neuron layers. The interesting parameters lie in the hidden layers. AEs can have several uses: denoising with the so-called denoising AEs [16], or creating new feature representations, such as Bottleneck features using a hidden layer with a number of neurons that is markedly lower than that of the other layers. The information is reformulated in a condensed form and the AE is expected to capture the most salient features of the training data. Studies have shown that, in some cases, AE posteriorgrams results are better than those of GMM [2,7]. In the context of unsupervised speech unit discovery, AE variants have emerged, such as correspondence AEs (cAEs) [13]. cAEs no longer seek to reconstruct the input data but other data, previously mapped in a certain way. They therefore require a first step of grouping segments of speech into similar pairs (pseudo-words, etc.) found by a DTW. There is another type of AEs, which avoids the DTW step by forcing to reconstruct neighboring data frames, using the speech stability properties: the so-called segmental AEs [2].

In our work, we first performed tests with different AEs. Their results mainly helped to separate the voiced sounds from the unvoiced sounds but gave poor results for our task of pseudo-phone discovery (less than 30% purity on the BUCKEYE corpus, see Sect. 3). We decided to design an alternative approach coupling k-means and supervised neural networks.

This paper is organized as follows. Section 2 presents the system architecture, then the speech material used to validate our approach on three languages is described. Finally, results both at phone- and word-levels are reported and discussed in Sect. 4.

## 2    System Description

Figure 1 shows the schema of the system. First, pseudo-phones are discovered by a k-means algorithm using log F-banks as input. Second, a CNN classifier is trained to predict these pseudo-units taken as pseudo-groundtruth. The probabilities outputted by the CNN are then used as features to run k-means once again. These last two steps (supervised CNN and k-means) are iterated as long as the CNN training cost decreases.

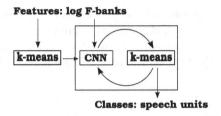

**Fig. 1.** General architecture: the system is trained in an iterative manner

## 2.1  Features

As input, we use 40 log Filterbank coefficients extracted on 16-ms duration windows, with a 3/4 hop size. 40 Mel-filters is a conventional number of filters used for speech analysis tasks.

Currently, our model needs to know the boundaries of the phones in order to standardize the input feature at segment level. We use and compare two different segmentations: the manual segmentation provided with the corpus and a segmentation derived automatically based on our previous work on cross-language automatic phone-level segmentation [8].

## 2.2  Class Assignment for CNN Learning

For initialization, the k-means algorithm uses frames of log F-bank coefficients as input and each input feature window is concatenated with its 6 neighborhood windows. It assigns a single class per window and we propagate this result on the segments delimited by the phone boundaries by a majority vote. Figure 2 illustrates the majority voting strategy used to choose the single pseudo-phone number 7 on a given segment. In the following iterations, the k-means algorithm takes as input the phone posteriorgrams generated by the CNN.

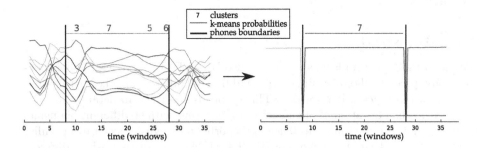

**Fig. 2.** Majority voting strategy

As we require the CNN to output the same class for all the windows comprising a segment, the model learns to output rather stable probabilities on the segments. We therefore accelerate the k-means step by using as input a single window, which is the average of the windows comprising a given segment and we obtain directly a class by segment. This simplification was shown in preliminary experiments to have no impact on the results. As a result, we have a single class assigned by segment.

### 2.3   CNN Architecture

Supervised neural networks require to know the classes of the training data. In our case, the true manually annotated phones are not available, thus, we use pseudo-phones clusters previously inferred by the k-means algorithm based on a previous segmentation of the input data.

We use a CNN with two convolution layers followed by a fully connected layer and a final output layer. The nonlinearity function used is the hyperbolic tangent. The first convolution layer is comprised of thirty $4 \times 3$ filters followed by a layer of $2 \times 2$ maxpooling and the second one of sixty $3 \times 3$ filters followed by a layer of $1 \times 2$ maxpooling. The dense layer has 60 neurons and we use dropout (0.5) before the last layer. A 0.007 learning rate was used with Nesterov momentum parameter updates.

Our experiments showed that the iterative process using pseudo-phones inferred by the k-means algorithm gives better results than those attributed by the first k-means iteration. Moreover, the CNN can also give us for each window the class probabilities. After experiments, these posteriorgrams outperformed the F-bank coefficients. We have chosen to retain these probabilities, on which we apply the consecutive k-means iterations. Our model is therefore an iterative model.

## 3   Speech Material

For our experiments, we used three corpus of different languages, sizes and conditions.

### 3.1   BUCKEYE

We used the American English corpus called BUCKEYE [12], composed of spontaneous speech (radio recordings) collected from 40 different speakers with about 30 min of time speech per speaker. This corpus is described in detail in [6].

The median duration of phonemes is about 70 ms, with 60 different phonemes annotated. It is more than the 40 usually reported for English, because of peculiar pronunciations that the authors of BUCKEYE chose to distinguish in different classes, particularly for nasal sounds.

We used 13 h of recordings of 26 different speakers, corresponding to the part of training according to the subsets defined in the *Zero Resource Speech* 2015 challenge [15].

## 3.2    BREF80

BREF80 is a corpus of read speech in French. As we are interested in less-resourced languages, we only took one hour of speech, recorded by eight different speakers. The French phone set we considered is the standard one comprised of 35 different phones, with a median duration of 70 ms.

## 3.3    NCHLT

The Xitsonga corpus [5], called NCHLT, is composed of short read sentences recorded on smart-phones, outdoors. We used nearly 500 phrases, with a total of 10,000 examples of phonemes annotated manually, from the same challenge database than the one used in the *Zero Resource Speech challenge*. The median duration of the phones is about 90 ms and there are 49 different phones.

# 4    Experiments and Results

In this section, we first report results with manual phone segmentations in order to evaluate our approach on the pseudo-phone discovery task only. Results at phone and word levels are given. In the last Subsect. 4.3, we evaluate the system under real conditions, namely with our automatic phone-level segmentations. We evaluate our system on different languages and speaking styles.

## 4.1    Results at Phone Level

To evaluate our results, we compute the standard purity metric of the pseudo-phones [9].

Let $N$ be the number of manual segments at phone level, $K$ the number of pseudo-phones, $C$ the number of phones and $n_j^i$ the number of segments labeled with phone $j$ and automatically assigned to pseudo-phone $i$. Then, the clustering purity obtained is defined as:

$$\frac{1}{N} \sum_{i=1}^{K} \arg \max_{j \in [1,C]} \left( n_j^i \right)$$

First, we sought to optimize the results of the first k-means' iteration, the one used to initialize the process by assigning class numbers to segments for the first time. The parameters that influence its results are the input features (log F-bank coefficients), the context size in number of frames and the number of means used.

We tested different context sizes and found that the influence of this parameter was at most one percent on the results. The best value is around six windows.

The choice of the number of clusters is ideally in the vicinity of the number of phones sought, that is to say generally about thirty. This is an average value of course, and there are languages that comprise many more phones, such as the Khoisan language with 141 different phones. We will look at the influence of the mean number on the search for pseudo-words in Sect. 4.2.

It is interesting to compare the results with the ones obtained in a supervised learning setting. Table 1 shows the results in terms of purity. As expected, results obtained with the supervised CNN are much better than with the clustering approach. Figure 3 shows the improvement provided by the use of a neural network. It is for a small number of pseudo-phones that the CNN improves the results the most (almost + 10% for 15 clusters).

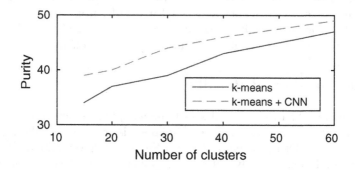

**Fig. 3.** Improvement in percent purity thanks to the neural network

One of the possible applications of this work is to help manual corpus annotation. In the case where a human would label each of the clusters attributed by the model with the real phonetic labels, thus regrouping the duplicates, we can consider using more averages than the number of phones present in the language considered. We have therefore looked at the evolution of purity as a function of the number of clusters in Fig. 4.

We see that, for few clusters, the results improve rapidly. But, starting from a hundred clusters, purity begins to evolve more slowly: in order to gain about 4% in purity, the number of averages needs to be multiplied by a factor of 10.

Table 1 gives the following pieces of information to evaluate the quality of the results:

- The percentage of purity obtained in supervised classification by the same CNN model as the one we use in the unsupervised setting. We did not try to build a complicated model to maximize the scores but rather a model adapted to our unsupervised problem. In comparison, the state of the art is around 80% of phone accuracy on the corpus TIMIT [3].
- The percentage of purity obtained by our unsupervised model. Scores are calculated for 30 pseudo-phones. We see that there are almost 20% of difference between the small corpus of read Xitsonga and the large spontaneous English corpus.

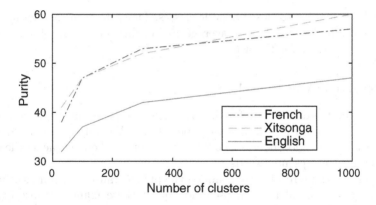

**Fig. 4.** Influence of the number of clusters on purity

– To further evaluate these results, we assign to each group the phonetic class most present among the grouped segments. Several groups can thus correspond to the same phone. The last line in the table indicates the number of different phones allocated for 30 groups found by our model. We see that we have only about fifteen different phones allocated, which is less than the number of phones present in these languages. So there are more than half of the phones that are not represented. This value changes slightly when we increase the number of clusters.

**Table 1.** Purity by segment (%) obtained for each corpus: English (En, BUCKEYE), French (Fr, BREF80) and Xitsonga (Xi, NCHLT)

| Language | En | Fr | Xi |
| --- | --- | --- | --- |
| Purity (%) supervised learning | 60 | 62 | 66 |
| Purity (%) for 30 clusters | 29 | 43 | 46 |
| Number of $\neq$ phonetic classes | 16 | 18 | 11 |

With the French and Xitsonga corpora, we obtained the best results, whether supervised or not. This can be easily explained: they are comprised of read speech, are the smallest corpora and with the least numbers of different speakers. These three criteria strongly influence the results. It is interesting to note that we get almost equivalent results with English if we only take 30 min of training data from a single speaker.

By studying in details the composition of the clusters, we found that having 30% (respectively 40%) purity scores does not mean that we have 70% (respectively 60%) of errors due to phones that differ from the phonetic label attributed during the clustering. The clusters are generally made up of two or three batches of examples belonging to close phonetic classes. Thus, the three phones most frequent in each cluster represent on average 70% of their group samples for French or Xitsonga and 57% for English.

## 4.2   Results at Word Level

To find pseudo-words, we look for the sequences comprised of the same pseudo-phones. A pseudo-word must at least appear twice. We only consider sequences of more than 5 pseudo-phones. Using shorter pseudo-phone sequences leads to too many incorrect pseudo-words.

To evaluate these results, we compare the phone transcripts constituting the different realizations of a given pseudo-word. If these manual transcripts are identical, then the pseudo-word is considered as correct. Otherwise, we count the number of phone differences. For a pseudo-word with only two realizations, we accept up to two differences in their phone sequences. For a pseudo-word with more than two examples, we rely on a median pseudo-word as done in [10], and again tolerate two differences maximum.

The results may depend on the number of groups selected. If we consider a larger number of distinct clusters, we get less pseudo-phone sequences that are the same, and thus we discover less pseudo-words. But by doing so, the groupings are purer, as shown in the Fig. 5.

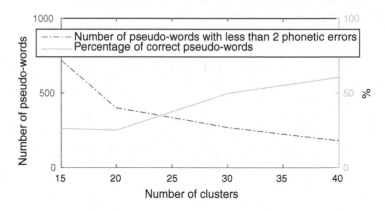

**Fig. 5.** Influence of the number of pseudo-phones on quantity and purity of pseudo-words found.

In Table 2, we look at three characteristics of the identified pseudo-words to evaluate our results:

- Number of pseudo-words found,
- Number of pseudo-words whose manual phonetic transcription between group examples has at most two differences,

**Table 2.** Pseudo-words statistics with manual and automatic segmentations

| Language | En | Fr | Xi |
|---|---|---|---|
| # Hours | 13 | 1 | 1/2 |
| # Phone examples | 586k | 36k | 10k |
| Manual segmentation | | | |
| # pseudo-words | 3304 | 671 | 231 |
| # Pseudo-words ≤ 2 differences | 1171 | 415 | 172 |
| # Identical pseudo-words | 334 | 188 | 76 |
| Automatic segmentation | | | |
| # Pseudo-words | 3966 | 540 | 200 |
| # Pseudo-words ≤ 2 differences | 843 | 269 | 120 |
| # Identical pseudo-words | 40 | 32 | 25 |

– Number of pseudo-words in which all lists representing them have exactly the same phonetic transcription.

The French corpus allows to obtain 671 pseudo-words, out of which 188 are correct and 227 with one or two differences in their phonetic transcriptions. We thus find ourselves with 415 pseudo-words with at most two differences with their manual phonetic transcriptions of the examples defining them. The results obtained on the other two corpora are worse. Proportionally, we find about ten times less pseudo-words than for with the French corpus.

In comparison, in a work performed on four hours of the corpus ESTER, 1560 pseudo-words were found, out of which 672 of them were sufficiently accurate according to their criteria, with an optimized pseudo-word search algorithm based on the DTW and self-similarities [10].

### 4.3 Towards a Fully Automatic Approach: With Automatic Segmentations

Until now we have used manual phones segmentation. To deal with real conditions when working on a few resourced language, we will now use our model without input handwritten data. We therefore use automatic phones segmentation to train our model. This automatic segmentation can be learned in other languages, with more resources.

In a previous work, we used a CNN to perform automatic segmentation task [8]. It is a supervised model, but we have demonstrated its portability to languages other than those learned.

The CNN takes as input F-bank coefficients and outputs probabilities of the presence/absence of a boundary at frame-level. The diagram of this model is represented in the Fig. 6. The output is a probability curve evolving in time whose summits are the locations of probable boundaries. To avoid duplicates due to noisy peaks, the curve is smoothed and all summits above a certain threshold are identified as phones boundaries.

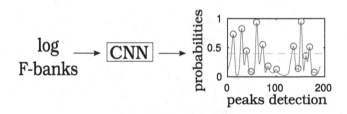

**Fig. 6.** Illustration of our automatic segmentation approach, based on a CNN

Previously, we showed that this network was portable to other languages than the one used for training and this is very useful in the present work. Indeed, for a language with few resources, we can use a corpus from other well-resourced languages. In addition, it is an additional step towards a fully unsupervised setting. Table 3 shows that the network gets a good F-measure on languages other than those used for training.

**Table 3.** F-measure (%) obtained with the automatic cross-language segmentation with the results on the columns according to the test corpus and on the lines according to the two training corpora, for 20 ms.

| Languages | Test    | En | Fr | Xi |
|-----------|---------|----|----|----|
|           | En + Fr | 73 | 74 | **53** |
| Train     | Fr + Xi | **64** | 80 | 64 |
|           | En + Xi | 73 | **63** | 58 |

Concerning the pseudo-word discovery, we get the results displayed in Table 2. The number of pseudo-words found is similar to that found with the manual segmentation but the purity score is lower with this automatic segmentation, as expected. For French and Xitsonga, half of the pseudo-words found have less than two errors, for English it is less than a quarter.

## 5   Conclusions

In this paper, we reported our experiments on speech unit discovery based first on a simple approach using the k-means algorithm on acoustic features, second on an improved version, in which a CNN is trained on the pseudo-phones clusters inferred by k-means. This solution differs from the standard approach based on AEs reported in the literature.

Our model is not yet fully unsupervised: it needs a pre-segmentation at phone level and obviously the best results were obtained with a manual segmentation. Fortunately, the loss due to the use of automatic segmentation is small and we have shown in a previous work that this segmentation can be done using a segmentation model trained on languages for which we have large manually annotated corpora. This allows us to apply our approach to less-resourced languages without any manual annotation, with the audio signal as input only. In the present work, the automatic segmentation system was trained for English, language with a lot of resources.

We tested our approach on three languages: American English, French and the less-represented language called Xitsonga. Concerning the results, there are differences according to the target language, and especially according to their characteristics. In all our experiments, the results on the BUCKEYE corpus, which is comprised of conversational speech, are worse than for the other two corpora, which are made up of read speech. The increase in the number of speakers also can be a factor of performance decrease.

With the Xitsonga corpus, for instance, with manual and automatic segmentations, we were able to obtain 46% and 42% purity scores, respectively, at phone-level with 30 pseudo-phones. Based on the inferred pseudo-phones, we discovered about 200 pseudo-words.

Our next work will focus on use DPGMM instead of k-means and on use unsupervised segmentation to have a fully unsupervised model. Furthermore, we presented first results on pseudo-word discovery based on mining similar pseudo-phone sequences. The next step will be to apply pseudo-word discovery algorithms to audio recordings, such as dotplots.

# References

1. Towards spoken term discovery at scale with zero resources. In: INTERSPEECH, pp. 1676–1679. International Speech Communication Association (2010)
2. Badino, L., Canevari, C., Fadiga, L., Metta, G.: An auto-encoder based approach to unsupervised learning of subword units. In: ICASSP, pp. 7634–7638 (2014)
3. Badino, L.: Phonetic context embeddings for DNN-HMM phone recognition. In: Interspeech 2016, 17th Annual Conference of the International Speech Communication Association, San Francisco, CA, USA, September 8–12, pp. 405–409 (2016)
4. Church, K.W., Helfman, J.I.: Dotplot: a program for exploring self-similarity in millions of lines of text and code. J. Comput. Graph. Stat. 2(2), 153–174 (1993)
5. van Heerden, C., Davel, M., Barnard, E.: The semi-automated creation of stratified speech corpora (2013)
6. Kiesling, S., Dilley, L., Raymond, W.D.: The variation in conversation (vic) project: creation of the buckeye corpus of conversational speech. In: Language Variation and Change, pp. 55–97 (2006)
7. Lyzinski, V., Sell, G., Jansen, A.: An evaluation of graph clustering methods for unsupervised term discovery. In: INTERSPEECH, pp. 3209–3213. ISCA (2015)
8. Manenti, C., Pellegrini, T., Pinquier, J.: CNN-based phone segmentation experiments in a less-represented language (regular paper). In: INTERSPEECH, p. 3549. ISCA (2016)

9. Manning, C.D., Raghavan, P., Schütze, H.: Introduction to Information Retrieval. Cambridge University Press, Cambridge (2008)

10. Muscariello, A., Bimbot, F., Gravier, G.: Unsupervised Motif acquisition in speech via seeded discovery and template matching combination. IEEE Trans. Audio Speech Lang. Process. **20**(7), 2031–2044 (2012). https://doi.org/10.1109/TASL.2012.2194283

11. Park, A.S., Glass, J.R.: Unsupervised pattern discovery in speech. IEEE Trans. Audio Speech Lang. Process. **16**(1), 186–197 (2008)

12. Pitt, M., Dilley, L., Johnson, K., Kiesling, S., Raymond, W., Hume, E., Fosler-Lussier, E.: Buckeye corpus of conversational speech (2nd release) (2007). www.buckeyecorpus.osu.edu

13. Renshaw, D., Kamper, H., Jansen, A., Goldwater, S.: A comparison of neural network methods for unsupervised representation learning on the zero resource speech challenge. In: INTERSPEECH, pp. 3199–3203 (2015)

14. Tian, F., Gao, B., Cui, Q., Chen, E., Liu, T.Y.: Learning deep representation for graph clustering, pp. 1293–1299 (2014)

15. Versteegh, M., Thiollire, R., Schatz, T., Cao, X.N., Anguera, X., Jansen, A., Dupoux, E.: The zero resource speech challenge 2015. In: INTERSPEECH, pp. 3169–3173 (2015)

16. Vincent, P., Larochelle, H., Bengio, Y., Manzagol, P.A.: Extracting and composing robust features with denoising autoencoders. In: ICML, pp. 1096–1103. ACM (2008)

17. Wang, H., Lee, T., Leung, C.C.: Unsupervised spoken term detection with acoustic segment model. In: Speech Database and Assessments (Oriental COCOSDA), pp. 106–111. IEEE (2011)

# Noise and Speech Estimation as Auxiliary Tasks for Robust Speech Recognition

Gueorgui Pironkov[1]([✉]), Stéphane Dupont[1], Sean U.N. Wood[2],
and Thierry Dutoit[1]

[1] Circuit Theory and Signal Processing Lab, University of Mons,
Boulevard Dolez 31, 7000 Mons, Belgium
{gueorgui.pironkov,stephane.dupont,thierry.dutoit}@umons.ac.be
[2] NECOTIS, Department of Electrical and Computer Engineering,
University of Sherbrooke, 2500 Boulevard de l'Université,
QC, Sherbrooke J1K 2R1, Canada
sean.wood@usherbrooke.ca

**Abstract.** Dealing with noise deteriorating the speech is still a major problem for automatic speech recognition. An interesting approach to tackle this problem consists of using multi-task learning. In this case, an efficient auxiliary task is clean-speech generation. This auxiliary task is trained in addition to the main speech recognition task and its goal is to help improve the results of the main task. In this paper, we investigate this idea further by generating features extracted directly from the audio file containing only the noise, instead of the clean-speech. After demonstrating that an improvement can be obtained through this multi-task learning auxiliary task, we also show that using both noise and clean-speech estimation auxiliary tasks leads to a 4% relative word error rate improvement in comparison to the classic single-task learning on the CHiME4 dataset.

**Keywords:** Speech recognition · Multi-task learning · Robust ASR · Noise estimation · CHiME4

## 1 Introduction

In recent years, Deep Neural Networks (DNN) have proven their efficiency in solving a wide variety of classification and regression tasks [14]. In particular, DNNs have been used as acoustic models for Automatic Speech Recognition (ASR), significantly outperforming the previous state-of-the-art methods based on Gaussian Mixture Models (GMM) [9]. Improvements brought by neural networks have progressively reduced the Word Error Rate (WER) to a level where some studies argue that ASR can now achieve near human-level performance [31]. Despite these recent improvements, dealing with noisy and reverberant conditions is still a major challenge for ASR [29]. Several techniques have been developed to address this problem, including feature enhancement for example, where features are *cleaned* at the front-end of the ASR system.

© Springer International Publishing AG 2017
N. Camelin et al. (Eds.): SLSP 2017, LNAI 10583, pp. 181–192, 2017.
DOI: 10.1007/978-3-319-68456-7_15

In this work, we use Multi-Task Learning (MTL) to improve ASR performance in the noisy and reverberant acoustic context. MTL consists of training a single system, specifically a DNN, to solve multiple tasks that are different but related, as opposed to the traditional Single-Task Learning (STL) architecture where the system is trained on only one task [2]. MTL has previously been applied in a variety of situations where ASR is the main task and different auxiliary tasks are added. In most cases, however, few MTL auxiliary tasks have been found to be helpful for the main ASR task when speech is corrupted by noise and reverberation. Generating the clean-speech feature as an auxiliary task is one of the most efficient such approaches [5,15,17,23]. We explore this idea further here by generating the noise features alone as an auxiliary task, as well as generating the noise and clean-speech features separately as two additional auxiliary tasks. The core idea is to increase the acoustic model's awareness of the noisy environment, and how it corrupts speech. To evaluate these auxiliary tasks, we use the simulated part of the CHiME4 dataset [29]. While the CHiME4 dataset contains both real and simulated data, only the simulated part may be used here since we need to extract clean-speech and noise features to train the MTL system.

This paper is organized as follows. First, we present the state-of-the-art in MTL for ASR in Sect. 2. We then describe the MTL mechanism in depth in Sect. 3. Details of the experimental setup used to evaluate the noise estimation auxiliary task are presented in Sect. 4, with the results and analysis presented in Sect. 5. Finally, the conclusion and ideas for future work are discussed in Sect. 6.

## 2    Related Work

Many speech and language processing problems including speech synthesis [10,30], speaker verification [4], and spoken language understanding [16] have benefited form MTL training. In the case of ASR, whether applying an STL or MTL architecture, the main task consists of training the acoustic model to estimate the phone-state posterior probabilities. These probabilities are then fed as input to a Hidden-Markov Model (HMM) that deals with the temporality of speech. The use of MTL for ASR has already been tested with a variety of auxiliary tasks. Early studies used MTL with gender classification as an auxiliary task [17,26], the goal being to increase the acoustic model's awareness of the impact of the speaker gender on the speech. As explained previously, the goal of the main task is to predict phone-state probabilities; some studies investigate a broader level of classes as the auxiliary task, as they try to directly predict the phone probability instead of the probability of the HMM state [1,25]. A related auxiliary task consists of classifying even broader phonetic classes (e.g. fricative, plosive, nasal,. . . ) but has shown poor performance [26]. Another approach consists of classifying graphemes as auxiliary task, where graphemes are the symbolic representation of speech (e.g. any alphabet), as opposed to the phonemes that directly describe the sound [3,26]. In order to increase the generalization ability of the network, recent studies have also focused on increasing its speaker-awareness. This is done by recognizing the speaker or by estimating the associated i-vector [6] of each speaker as auxiliary task [19, 20, 27, 28], instead of concatenating the i-vector to the input features. Adapting the acoustic model to a particular speaker can also benefit from MTL [11]. Additional information about these methods can be found in [18].

Most of the previously cited methods do not particularly focus on ASR in noisy and reverberant conditions, nonetheless robust ASR is a field of interest as well. Some studies have focused solely on improving ASR in reverberant acoustic environment by generating de-reverberated speech as auxiliary task, using reverberated speech as input during training [8, 22]. Another approach that tackles the noise problem in ASR with MTL consists of recognizing the type of noise corrupting the speech, where a single noise type among several possible types is added for each sentence of the clean speech [12, 24]. This approach does not seem to have a real positive impact on the main ASR task, however. The MTL task that shows the highest improvement consists of generating the clean-speech features as auxiliary task [15, 17, 23]. Of course, in order to generate the targets needed to train this auxiliary task, access to the clean speech is required to extract the features, and this can only be done with simulated noisy and reverberant data. It is also possible to use an MTL system as a feature extractor for robust ASR, where a bottle-neck layer is used, the goal being to use the activations of the bottle-neck layer as input of a traditional STL/ASR system [13].

Though previous studies have proposed recognizing the type of noise, or generating the clean-speech features, to the best of our knowledge, there have been no attempts to estimate the noise features alone as an auxiliary task, or to estimate both the noise and speech features separately in an MTL setup.

## 3   Multi-Task Learning

Initially introduced in 1997, the core idea of multi-task learning consists of training a single system (a neural network here) to solve multiple tasks that are different but still related [2]. In the MTL nomenclature, the *main task* is the principal task, i.e. the task that would be initially used for a STL architecture, whereas at least one *auxiliary task* is added to help improve the network's convergence to the benefit of the main task. An MTL architecture with one main task and $N$ auxiliary tasks is shown in Fig. 1 as an example.

All MTL systems share two essential characteristics: (a) The same input features are used for training both the main and the auxiliary tasks. (b) The parameters (weights and biases) of all neurons, and more generally the internal structure of the network, are shared among the main and auxiliary tasks, with the exception of the output layer. Furthermore, these parameters are updated by backpropagating a mixture of the error associated with each task, with a term:

$$\epsilon_{MTL} = \epsilon_{Main} + \sum_{n=1}^{N} \lambda_n * \epsilon_{Auxiliary_n}, \tag{1}$$

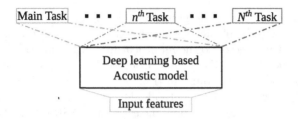

**Fig. 1.** A Multi-Task Learning system with one main task and $N$ auxiliary tasks.

where $\epsilon_{MTL}$ is the sum of all the task errors to be minimized, $\epsilon_{Main}$ and $\epsilon_{Auxiliary_n}$ are the errors obtained from the *main* and *auxiliary* tasks respectively, $\lambda_n$ is a nonnegative weight associated with each of the auxiliary tasks, and $N$ is the total number of auxiliary tasks added to the main task. The value $\lambda_n$ controls the influence of the auxiliary task with respect to the main task. If the $n^{th}$ auxiliary task has a $\lambda_n$ close to 1, the main task and the auxiliary task will contribute equally to the error estimation. On the other hand, if $\lambda_n$ is close to 0, a single-task learning system could be obtained due to the very small (or nonexistent) influence of the auxiliary task. The auxiliary task is frequently removed during testing, keeping only the main task. Selecting a relevant auxiliary task with respect to the main task is the crucial point leading to convergence of the main task. Instead of computing and training each task independently, sharing the parameters of the system among multiple tasks may lead to better results than an independent processing of each task [2].

## 4    Experimental Setup

In this section, we will present the tools and methods used to evaluate the new auxiliary task that we propose for robust ASR.

### 4.1    Database

In order to evaluate noise estimation as an auxiliary task for robust ASR, we use the CHiME4 database [29]. This database was released in 2016 for a speech recognition and separation challenge in reverberant and noisy environments. This database is composed of 1-channel, 2-channel, and 6-channel microphone array recordings. Four different noisy environments (café, street junction, public transport, and pedestrian area) were used to record real acoustic mixtures through a tablet device with 6-channel microphones. The WSJ0 database [7] is used to create simulated data. WSJ0 contains clean-speech recordings to which noise is added. The noise is recorded from the four noisy environments described above. For the noise estimation auxiliary task, we use features extracted from these recordings containing only noise as targets for training. As we cannot obtain these targets for real data, we only use the simulated data in this study.

All datasets (training, development, and test sets) consist of 16 bit wav files sampled at 16 kHz. The training set consists of 83 speakers uttering 7138 simulated sentences, which is the equivalent of ~15 h of training data. The development set consists of 1640 utterances (~2.8 h) uttered by 4 speakers. Finally, 4 additional speakers compose the test set with 1320 utterances corresponding to approximately 4.5 h of recordings.

In this work, we investigate noise and clean-speech estimation as auxiliary tasks, therefore we use only the noise recorded from a single channel during training (channel no 5). The test and development set noises are randomly selected from all channels, making the task harder but also challenging the generalization ability of the setup.

## 4.2 Features

The features used as input for training the MTL system as well as targets for the noise and/or clean-speech estimation tasks are obtained through the following traditional ASR pipeline:

1. Using the raw audio wav files, 13-dimensional Mel-Frequency Cepstral Coefficients (MFCC) features are extracted and normalized through Cepstral Mean-Variance Normalization (CMVN).
2. For each frame, the adjacent ±3 frames are spliced.
3. These 91-dimensional feature vectors are reduced through a Linear Discriminative Analysis (LDA) transformation to a 40-dimensional feature space.
4. The final step consists of projecting the features through a feature-space speaker adaptation transformation known as feature-space Maximum Likelihood Linear Regression (fMLLR).

Finally, the 40-dimensional features that are computed through this pipeline are spliced one more time with the surrounding ±5 frames for the input features fed to the acoustic model, thus giving additional temporal context to the network during training. For the auxiliary tasks' targets, the same pipeline is followed to generate the clean-speech and noise features but there is no ±5 splicing at the final stage. Alignments from the clean-speech are reused for the transformations applied on noisy features.

## 4.3 Training the Acoustic Model

Training and testing this MTL auxiliary tasks was done using the *nnet3* version of the Kaldi toolbox [21].

We use a classic feed-forward deep neural network acoustic model to evaluate the performance of this new auxiliary task. The DNN is composed of 4 hidden layers, each of them consisting of 1024 neurons activated through Rectified Linear Units (ReLU). The main task used for STL and MTL computes 1972 phone-state posterior probabilities after a softmax output layer. The training of the DNN is done through 14 epochs using the cross-entropy loss function for the main task, and quadratic loss function for the auxiliary tasks (as they are regression issues), with an initial learning rate starting at 0.0015 that is progressively reduced to 0.00015. Stochastic gradient descent (SDG) is used to update the parameters of the network through the backpropagation of the error derivatives. The size of the mini-batch used to process the input features is set 512. These parameters were selected through empirical observations.

The same experiments were also conducted using other deep learning algorithms including Recurrent Neural Networks (RNN) with Long Short-Term Memory (LSTM) cells and Time-Delay Neural Networks (TDNN). However, the feed-forward DNN showed similar or better results than these more complex architectures on the simulated data of CHiME4. Also, the computational time for the RNN-LSTM network was much higher than for the feed-forward DNN. While the complexity and temporarily of the main and auxiliary tasks did not require a more complex acoustic model here, we note that for some auxiliary tasks, having a more complex network can be crucial for the convergence of the auxiliary task, as is the case for speaker classification for instance [19].

During decoding, the most likely transcriptions are obtained through the phone-state probabilities estimated by the feed-forward network, and used by the HMM system and associated with a language model. The language model is the 3-gram KN language model trained on the WSJ 5K standard corpus.

## 4.4 Baseline

The baseline of our system is obtained by training the setup presented in the previous section in single-task learning manner. We compute the word error rate for both the development and test sets over all four noisy environments for the simulated data of CHiME4. The results are shown in Table 1. A very significant mismatch coming from the recording environments between the development and test set can be noticed, explaining the higher WER for the test set. For the rest of this paper we display only the *Average* results as the trends and evolutions of the WER are similar over all four noisy environments.

**Table 1.** Word error rate in % on the development and test sets of CHiME4 dataset used as baseline. *Average* is the mean WER of all 4 environmental noises and *Overall* is the mean WER over the development and test sets.

|          | Average | Bus   | Café  | Pedestrian | Street |
|----------|---------|-------|-------|------------|--------|
| Dev set  | 18.54   | 16.55 | 22.05 | 15.03      | 20.52  |
| Test set | 26.82   | 21.44 | 30.99 | 26.90      | 27.96  |
| Overall  | 22.68   | 19.00 | 26.52 | 20.97      | 24.24  |

## 5   Results

In this section, we investigate the improvement brought by the new MTL auxiliary task, namely regenerating the noise contained in the corrupted sentence, in comparison to STL. We also combine this auxiliary task with the more traditional clean-speech generation auxiliary task.

### 5.1   Noise Features Estimation

In order to evaluate the impact of estimating the noise features as an auxiliary task in our MTL setup, we vary the value of $\lambda_{noise}$, thus varying the influence of this auxiliary task with respect to main ASR task. The obtained results for values of $\lambda_{noise}$ varying between 0 (STL) and 0.5 are presented in Table 2. There is a small but persistent improvement of the WER for $\lambda_{noise} = 0.05$, over both the development and test sets. For smaller values ($\lambda_{noise} = 0.01$), the improvement is nearly insignificant as the value of $\lambda_{noise}$ brings the training too close to STL ($\lambda_{noise} = 0$), while for values of $\lambda_{noise}$ too high ($\lambda_{noise} \geq 0.15$), the WER is worse than for STL as the influence of the auxiliary task overshadows the main ASR task.

**Table 2.** Average word error rate (in %) of the Multi-Task Learning architecture when the auxiliary task is noise feature estimation, where $\lambda_{noise}$ is the weight attributed to the noise estimation auxiliary task during training. The baseline, which is the Single-Task Learning architecture, is obtained for $\lambda_{noise} = 0$. The *Overall* values are computed over both datasets.

| $\lambda_{noise}$ | *0 (STL)* | 0.01 | 0.05 | 0.1 | 0.15 | 0.2 | 0.3 | 0.5 |
|---|---|---|---|---|---|---|---|---|
| Dev set | 18.54 | 18.43 | **18.19** | 18.31 | 18.65 | 18.82 | 19.59 | 20.83 |
| Test set | 26.82 | 26.63 | **26.50** | 26.55 | 26.85 | 27.08 | 28.01 | 29.89 |
| Overall | 22.68 | 22.53 | **22.35** | 22.43 | 22.75 | 22.95 | 23.80 | 25.36 |

In order to further highlight these observations, we present the relative WER improvement brought by MTL in comparison to STL in Fig. 2. An improvement is obtained for values of $\lambda_{noise}$ between 0.01 and 0.1. The highest improvement is obtained for $\lambda_{noise} = 0.05$, with a relative improvement in comparison to STL going up to 1.9% on the development set for instance. Larger values of $\lambda_{noise}$ degrade performance on the main speech recognition task.

As discussed in Sect. 4.3, training is done over 14 epochs. In order to prove the ASR improvement is not only the result of the introduction of a small noise into the system, but rather that both tasks are converging, we present the error over these 14 epochs in Fig. 3, highlighting in this way the error reduction obtained on both tasks loss functions over time.

Despite the persistence of the relative improvement for small values of $\lambda_{noise}$, it can be noted that this improvement is quite small. This can be explained by

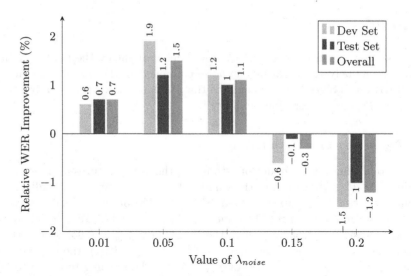

**Fig. 2.** Evaluation of the relative improvement of the word error rate brought by multi-task learning in comparison to single-task learning, with $\lambda_{noise}$ the weight attributed to the noise estimation auxiliary task. The *Overall* values are computed over both the development and test datasets.

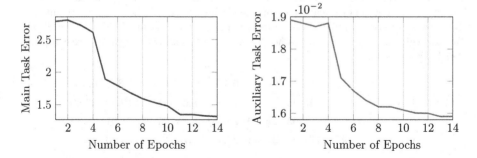

**Fig. 3.** Evolution of the tasks errors over training epochs. The *Main Task* is the speech recognition error computed through the cross-entropy loss function, whereas the *Auxiliary Task* corresponds to the noise estimation error obtained through the quadratic loss function.

several considerations. First, this auxiliary task is less directly related to the main task than for instance clean speech generation, meaning that the convergence of the auxiliary task may not significantly help the main task. Another consideration is that the auxiliary task is in fact quite a hard task here as the Signal-to-Noise Ratio (SNR) is always in favor of the clean-speech and not the noise, making it hard to estimate the noise alone. Finally, the suitability of the features extracted following the pipeline presented in Sect. 4.2, as well as using fMLLR transformation in this context, is most likely not optimal for noise.

Despite these considerations, using noise estimation as auxiliary task seems to be helpful for the main ASR task when $\lambda_{noise}$ is properly selected. Additionally, using a MTL setup is easy to implement and does not require extensive computational time in comparison to STL (as the same network is trained for both tasks). Finally, the targets for this particular auxiliary task, noise estimation, are easy to get as we have access to the noise when generating the simulated data.

## 5.2 Combining Noise and Clean-Speech Features Estimation

Instead of separately generating clean-speech or noise as auxiliary tasks, we investigate here the combination of both tasks in the MTL framework. In order to do that, we first repeat the same experiment as in Sect. 5.1 but where we generate only the clean-speech features as the auxiliary task. After varying the

**Table 3.** Average word error rate in % on the development and test sets of CHiME4 dataset, when different auxiliary tasks are applied. *Overall* is the mean WER over the development and test sets data.

| Auxiliary task(s) | Dev set | Test set | Overall |
|---|---|---|---|
| None (STL) | 18.54 | 26.82 | 22.68 |
| Noise estimation ($\lambda_{noise} = 0.05$) | 18.19 | 26.50 | 22.35 |
| Clean-speech estimation ($\lambda_{speech} = 0.15$) | 17.99 | 26.06 | 22.03 |
| Noise + clean-speech estimation | **17.79** | **25.78** | **21.79** |

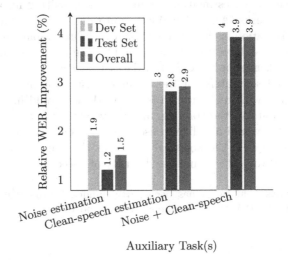

**Fig. 4.** Evaluation of the relative improvement of the word error rate brought by multi-task learning in comparison to single-task learning, with different auxiliary tasks. The *Overall* values are computed over both the development and test datasets.

value of $\lambda_{speech}$ we found the best WER is obtained for $\lambda_{speech} = 0.15$. The obtained results are depicted in Table 3 and, as in the previous section, we compute the relative improvement brought by the different auxiliary tasks (plus their combination) in comparison to STL in Fig. 4.

The results show that, as expected, a better WER is obtained when using clean-speech estimation as auxiliary task in comparison to noise estimation, with an overall relative improvement of 2.9% (while it was 1.5% in the previous experiment). Interestingly however, using both the clean-speech and noise estimation auxiliary tasks lead to even better performance, with 3.9% overall relative improvement and more than 1% absolute improvement on the test set. This result highlights the fact that the network is learning different and valuable information from both auxiliary tasks in order to improve the main task. Once again, implementing these auxiliary tasks is simple and does not require significant additional computational time in comparison to classic single-task learning architectures.

## 6    Conclusion

In this paper, we have studied multi-task learning acoustic modeling for robust speech recognition. While most previous studies focus on clean-speech generation as auxiliary task, we propose and investigate here another different but related auxiliary task: noise estimation. This auxiliary task consists of generating the features extracted from the audio file containing only the noise that is later added to the clean-speech to create the simulated noisy data. After showing that an improvement can be obtained with this auxiliary task, we combined it with the clean-speech estimation auxiliary task, resulting in one main task and two auxiliary tasks. A relative WER improvement of 4% can be obtained thanks to the association of these two auxiliary tasks in comparison to the classic single-task learning architecture. Training and testing here was done only on the simulated data taken from the CHiME4 dataset, as the clean-speech and noise audio are required separately for the auxiliary tasks training, thus making it impossible to train with real data. In future work, we would like to find a way to integrate real data to the training, and re-evaluate the impact of these two auxiliary tasks. We would also like to use other types of features which may be more suitable to capture the noise variations, as the features we are currently using are designed to best capture the diversity of speech.

**Acknowledgments.** This work has been partly funded by the Walloon Region of Belgium through the SPW-DGO6 Wallinov Program no 1610152.

## References

1. Bell, P., Renals, S.: Regularization of context-dependent deep neural networks with context-independent multi-task training. In: IEEE International Conference on Acoustics, Speech and Signal Processing (ICASSP), pp. 4290–4294. IEEE (2015)

2. Caruana, R.: Multitask learning. Mach. learn. **28**(1), 41–75 (1997)
3. Chen, D., Mak, B., Leung, C.C., Sivadas, S.: Joint acoustic modeling of triphones and trigraphemes by multi-task learning deep neural networks for low-resource speech recognition. In: IEEE International Conference on Acoustics, Speech and Signal Processing (ICASSP), pp. 5592–5596. IEEE (2014)
4. Chen, N., Qian, Y., Yu, K.: Multi-task learning for text-dependent speaker verification. In: Sixteenth Annual Conference of the International Speech Communication Association (2015)
5. Chen, Z., Watanabe, S., Erdogan, H., Hershey, J.R.: Speech enhancement and recognition using multi-task learning of long short-term memory recurrent neural networks. In: INTERSPEECH, pp. 3274–3278. ISCA (2015)
6. Dehak, N., Kenny, P., Dehak, R., Dumouchel, P., Ouellet, P.: Front-end factor analysis for speaker verification. IEEE Trans. Audio Speech Lang. Process. **19**(4), 788–798 (2011)
7. Garofolo, J., Graff, D., Paul, D., Pallett, D.: CSR-I (WSJ0) Complete LDC93S6A. Web Download. Linguistic Data Consortium, Philadelphia (1993)
8. Giri, R., Seltzer, M.L., Droppo, J., Yu, D.: Improving speech recognition in reverberation using a room-aware deep neural network and multi-task learning. In: IEEE International Conference on Acoustics, Speech and Signal Processing (ICASSP), pp. 5014–5018. IEEE (2015)
9. Hinton, G., Deng, L., Yu, D., Dahl, G.E., Mohamed, A.R., Jaitly, N., Senior, A., Vanhoucke, V., Nguyen, P., Sainath, T.N., et al.: Deep neural networks for acoustic modeling in speech recognition: the shared views of four research groups. Sig. Process. Mag. **29**(6), 82–97 (2012)
10. Hu, Q., Wu, Z., Richmond, K., Yamagishi, J., Stylianou, Y., Maia, R.: Fusion of multiple parameterisations for DNN-based sinusoidal speech synthesis with multi-task learning. In: Proceedings of Interspeech (2015)
11. Huang, Z., Li, J., Siniscalchi, S.M., Chen, I.F., Wu, J., Lee, C.H.: Rapid adaptation for deep neural networks through multi-task learning. In: Sixteenth Annual Conference of the International Speech Communication Association (2015)
12. Kim, S., Raj, B., Lane, I.: Environmental noise embeddings for robust speech recognition (2016). arxiv preprint arXiv:1601.02553
13. Kundu, S., Mantena, G., Qian, Y., Tan, T., Delcroix, M., Sim, K.C.: Joint acoustic factor learning for robust deep neural network based automatic speech recognition. In: IEEE International Conference on Acoustics, Speech and Signal Processing (ICASSP), pp. 5025–5029. IEEE (2016)
14. LeCun, Y., Bengio, Y., Hinton, G.: Deep learning. Nature **521**(7553), 436–444 (2015)
15. Li, B., Sainath, T.N., Weiss, R.J., Wilson, K.W., Bacchiani, M.: Neural network adaptive beamforming for robust multichannel speech recognition. In: Proceedings of Interspeech (2016)
16. Li, X., Wang, Y.Y., Tur, G.: Multi-task learning for spoken language understanding with shared slots. In: Twelfth Annual Conference of the International Speech Communication Association (2011)
17. Lu, Y., Lu, F., Sehgal, S., Gupta, S., Du, J., Tham, C.H., Green, P., Wan, V.: Multitask learning in connectionist speech recognition. In: Proceedings of the Australian International Conference on Speech Science and Technology (2004)
18. Pironkov, G., Dupont, S., Dutoit, T.: Multi-task learning for speech recognition: an overview. In: Proceedings of the 24th European Symposium on Artificial Neural Networks (ESANN) (2016)

19. Pironkov, G., Dupont, S., Dutoit, T.: Speaker-aware long short-term memory multi-task learning for speech recognition. In: 24th European Signal Processing Conference (EUSIPCO), pp. 1911–1915. IEEE (2016)

20. Pironkov, G., Dupont, S., Dutoit, T.: Speaker-aware multi-task learning for automatic speech recognition. In: 23rd International Conference on Pattern Recognition (ICPR) (2016)

21. Povey, D., Ghoshal, A., Boulianne, G., Burget, L., Glembek, O., Goel, N., Hannemann, M., Motlicek, P., Qian, Y., Schwarz, P., et al.: The kaldi speech recognition toolkit. In: IEEE 2011 Workshop on Automatic Speech Recognition and Understanding. IEEE Signal Processing Society (2011)

22. Qian, Y., Tan, T., Yu, D.: An investigation into using parallel data for far-field speech recognition. In: IEEE International Conference on Acoustics, Speech and Signal Processing (ICASSP), pp. 5725–5729. IEEE (2016)

23. Qian, Y., Yin, M., You, Y., Yu, K.: Multi-task joint-learning of deep neural networks for robust speech recognition. In: IEEE Workshop on Automatic Speech Recognition and Understanding (ASRU), pp. 310–316. IEEE (2015)

24. Sakti, S., Kawanishi, S., Neubig, G., Yoshino, K., Nakamura, S.: Deep bottleneck features and sound-dependent i-vectors for simultaneous recognition of speech and environmental sounds. In: Spoken Language Technology Workshop (SLT), pp. 35–42. IEEE (2016)

25. Seltzer, M.L., Droppo, J.: Multi-task learning in deep neural networks for improved phoneme recognition. In: IEEE International Conference on Acoustics, Speech and Signal Processing (ICASSP), pp. 6965–6969. IEEE (2013)

26. Stadermann, J., Koska, W., Rigoll, G.: Multi-task learning strategies for a recurrent neural net in a hybrid tied-posteriors acoustic model. In: INTERSPEECH, pp. 2993–2996 (2005)

27. Tan, T., Qian, Y., Yu, D., Kundu, S., Lu, L., Sim, K.C., Xiao, X., Zhang, Y.: Speaker-aware training of LSTM-RNNS for acoustic modelling. In: IEEE International Conference on Acoustics, Speech and Signal Processing (ICASSP), pp. 5280–5284. IEEE (2016)

28. Tang, Z., Li, L., Wang, D.: Multi-task recurrent model for speech and speaker recognition (2016). arxiv preprint arXiv:1603.09643

29. Vincent, E., Watanabe, S., Nugraha, A.A., Barker, J., Marxer, R.: An analysis of environment, microphone and data simulation mismatches in robust speech recognition. Computer Speech & Language (2016)

30. Wu, Z., Valentini-Botinhao, C., Watts, O., King, S.: Deep neural networks employing multi-task learning and stacked bottleneck features for speech synthesis. In: IEEE International Conference on Acoustics, Speech and Signal Processing (ICASSP), pp. 4460–4464. IEEE (2015)

31. Xiong, W., Droppo, J., Huang, X., Seide, F., Seltzer, M., Stolcke, A., Yu, D., Zweig, G.: Achieving human parity in conversational speech recognition (2016). arxiv preprint arXiv:1610.05256

# Unified Approach to Development of ASR Systems for East Slavic Languages

Radek Safarik[✉] and Jan Nouza

SpeechLab, Technical University of Liberec, Studentska 2,
461 17 Liberec, Czech Republic
{radek.safarik,jan.nouza}@tul.cz
https://www.ite.tul.cz/speechlab/

**Abstract.** This paper deals with the development of language specific modules (lexicons, phonetic inventories, LMs and AMs) for Russian, Ukrainian and Belarusian (used by 260M, 45M and 3M native speakers, respectively). Instead of working on each language separately, we adopt a common approach that allows us to share data and tools, yet taking into account language unique features. We utilize only freely available text and audio data that can be found on web pages of major newspaper and broadcast publishers. This must be done with large care, as the 3 languages are often mixed in spoken and written media. So, one component of the automated training process is a language identification module. At the output of the complete process there are 3 pronunciation lexicons (each about 300K words), 3 partly shared phoneme sets, and corresponding acoustic (DNN) and language (N-gram) models. We employ them in our media monitoring system and provide results achieved on a test set made of several complete TV news in all the 3 languages. The WER values vary in range from 24 to 36%.

**Keywords:** Speech recognition · Multi-lingual · Cross-lingual · East slavic languages · Language identification

## 1 Introduction

Since 2000s we have been developing ASR systems that would suit the needs of Slavic languages, i.e. those with a high degree of inflection, rich morphology and more or less free word order. We started with Czech (our native tongue) and designed a fully automated framework which includes also our own real-time decoder. Currently, it has been used in several applications, such as broadcast monitoring [11], voice dictation, or spoken archive processing [13]. Later, we have adapted it to other Slavic languages: Slovak [12], Polish [14], Croatian, Serbian, Slovene, Macedonian, and Bulgarian [10].

There is a big interest from media monitoring companies to apply the ASR technology for their business, namely for automatic transcription, sub-titling, indexation, or on-line alerting. We collaborate with a company that is applying our ASR engine and the above mentioned language specific components for

© Springer International Publishing AG 2017
N. Camelin et al. (Eds.): SLSP 2017, LNAI 10583, pp. 193–203, 2017.
DOI: 10.1007/978-3-319-68456-7_16

broadcast monitoring in the countries where Slavic languages are used. To cover all, we had to include also the East Slavic family of languages, i.e. Russian, Ukrainian and Belarusian. They have several common features, the most visible one being the Cyrillic alphabet (azbuka), although each has several unique letters (in orthography) and phonemes (in pronunciation). Instead of working on each language separately, we have built a development platform that allows to combine the sharable data and tools with those components that are language dependent. The research and the practical implementation is supported by the Technology Agency of the Czech Republic within a large project called MultiLinMedia.

We started our work on the East Slavic languages with an initial study focused on Russian, as it is the largest one and also because it offers some real challenges, such as the crucial role of stress in pronunciation or a large number of palatalized phonemes. In [19], we presented an approach that tried to solve these issues in an efficient way and yielded rather fair results. Therefore, we decided to take Russian as the pilot language, from which the other two are bootstrapped.

While getting enough text and audio data is not that hard for Russian, for Ukrainian, and namely for Belarusian, it is more difficult and less straightforward. One reason is that (from historical reasons) Russian is frequently used in Ukraine and Belarus, and all the 3 languages are often mixed in newspapers and in broadcasting. Hence, a good language identifier must be a part of the development platform. Like in some other under-resourced languages, we had to consider also combination of training data from multiple languages.

## 2    Related Works

Let us recall that the first ASR system for Russian was proposed already in 1960s. It used dynamic programming and worked with some 300 words [3]. In next decades, many small/middle vocabulary ASR have been reported for different applications, such as phone call routing, voice commanding, etc. The first real LVCSR system for Russian was designed by IBM researchers in 1996 [4]. It had 36k words in its lexicon and achieved 5% WER, but only with short read sentences. Later, many other ASR systems were created using different approaches to deal with Russian specific features, such as high degree of inflection or rich phonology. In [21], grapheme-based acoustic modelling was used and achieved 32.8% WER on the GlobalPhone data-base [22]. To cope with rich inflection, in [6] syntactic/morphemic analysis was used to 'compress' the lexicon and language model. It was tested on 78k and 208k lexicons and yielded 44% WER. A similar approach was applied also in [5] with 26.9% WER on the SPIIRAS database [2] with statistical 3 gram model interpolated with syntactic-statistical 2 gram model and 204k words in the lexicon. In [23], a fully automated scheme (RLAT) for the rapid development of LMs was presented. It was tested also on the Russian part of GlobalPhone and yielded 36.2% of WER.

For the Ukrainian language, published works have a shorter record. First LVCSR systems were reported in 2000s. Among them, there was e.g., a computerized stenographer for the Ukrainian Parliament [17] achieving 28.5% WER, or systems built on the Ukrainian Broadcast Speech corpus [16] with about 40% WER for spontaneous speech [7] and 10% for dictation [18] (with a 100k lexicon). Another work utilizes the GlobalPhone and RLAT for Ukrainian (in the same way as in [23]) and reports 21.6% WER [20]. There is also a paper that deals with the detection of and code switching between Russian and Ukrainian in spoken documents [8]. It reports 24 and 31% WER for Ukrainian and Russian, respectively.

As to Belarusian, we have found only one paper [9] which describes an HTK based voice-command demo system. The authors reported 43% WER with 1838 commands and 8% WER with a subset of 460 specially selected ones.

Our work differs from the mentioned ones by (a) focusing on a practical real-world application (on-line transcription of broadcast programs), (b) utilizing only freely available data, (c) employing an own decoder optimized and already deployed for 10 other languages, (d) making almost all the development process fully automated, which also means that no native speakers needed to be involved.

## 3  East Slavic Languages

The East Slavic language family consists of 3 main official languages (Russian, Ukrainian and Belarusian) and several smaller regional ones, sometimes considered as dialects, such as Rusyn or Polesian. While Russian is the language with the highest number of speakers among all the Slavic languages, Ukrainian and Belarusian have less speakers and they are not always native tongues for a large part of population in their respective countries. Due to historical reasons, there was and still is a big Russian influence in the neighbouring post-Soviet states. About one third of Ukrainian people consider Russian as their native language and for Belarus it is even about two thirds. This a true challenge for automatic gathering and processing of Ukrainian and Belarusian text and speech data.

All the 3 languages use the Cyrillic alphabet but each has several unique letters. Fortunately, this enables to distinguish between their written texts. Phonetics is the most complex one among the whole Slavic group. Most consonants have two versions, hard and soft pronounced (palatalized) ones. Syllable stress plays an important role both in pronunciation and in perception. It does not have a fixed position in a word and it is not marked in standard texts. A grammatical case or even a meaning is changed by moving it (e.g. zamók – lock and zámok – castle). The stress also causes vowel reduction which means that the unstressed phonemes significantly change their quality (e.g. unstressed /o/ is reduced to /a/). However, this applies mainly to Russian. In the other two languages, the vowel reduction is not that strong and it is already reflected in the orthography (e.g. odin [ad$^j$in] in Russian and adzin [ad$^j$in] in Belarusian for word 'one'). Therefore, from practical reasons, we distinguish between stressed and unstressed vowels only in the Russian phoneme inventory.

As in the other Slavic languages, the degree of inflection is rather high, which results in larger ASR vocabularies if we want to get an acceptable coverage rate (>97%) needed for the general broadcast task. Large and multi-domain text corpora are necessary not only for reliable word frequency lists but also for representative LMs because the Slavic languages have a relatively free word order in a sentence (Table 1).

**Table 1.** East Slavic languages

| Language | Abbreviation | Speakers |
|----------|--------------|----------|
| Russian | RU | 260 million |
| Ukrainian | UK | 45 million |
| Belarusian | BE | 3.2 million |

## 4    Language Specific ASR Modules

For each language, three specific modules need to be made: a pronunciation lexicon, a language model (LM) and an acoustic model (AM). The first two require large text corpora, for the third we have to collect enough audio data containing speech of many speakers in the target language.

**Table 2.** Statistics on corpora and ASR vocabularies

| Language | RU | UK | BE |
|----------|-----|-----|-----|
| # Web | 12 | 28 | 11 |
| Downloaded text | 2.83 GB | 4.39 GB | 2.56 GB |
| After processing | 998 MB | 2.37 GB | 814 MB |
| After lang. filtering | 998 MB | 758 MB | 280 MB |
| # Words | 149M | 111M | 42M |
| Lexicon size | 326k | 324k | 293k |
| Pronunciations | 408k | 372k | 353k |
| Phonemes | 53 | 39 | 36 |

### 4.1    Text Corpus, Lexicon and Language Model

**Text Corpus.** The best free sources of texts are web pages of newspapers and broadcasters. They contain multi-domain texts in large quantities and with a relatively small number of typos and other errors. We have made a tool that crawls through a given web source, process its HTML code and extracts text blocks that meet several basic criteria (e.g. constrains on the number and length of strings, number of digits in a sequence, length of supposed-to-be sentences, etc.).

This is to avoid downloading of captions, tables, advertisements and other non-verbal data. After that, the corpus is further processed to remove or unify punctuation, URLs, specific characters, etc. Repetitive sentences that may have origin in multi-page web layout are also removed. Table 2 shows the amount of downloaded data and its size after processing and language filtering.

As mentioned earlier, there is a high number of Russian speaking people in Ukraine and Belarus. Russian texts may occur everywhere in these two countries, on official web pages and documents, in interviews, in citations, etc. Luckily, the three languages differ in few characters, especially in those representing sounds /i/ and /y/ that are very common letters. We proposed a simple filter that counts these unique letters in a sentence for each pair of languages and classifies it to that with the highest count. This method achieved 96.9% accuracy on 500 test sentences labelled by native speakers. Table 3 shows which letters of the language in the row are not present in the language in the column.

In our scheme, the last step is a conversion of the texts into Latin script. For this purpose we have created a 1-to-1 mapping of Cyrillic letters to the Latin ones; to those with the same or similar pronunciation (see [19]). It significantly simplifies reading, typing, editing non-Latin texts to those who are not familiar with azbuka. Moreover, it allows us to use the same string manipulation routines (needed, e.g., for digit transcription or G2P conversion) for all Slavic languages.

**Table 3.** Character differences in three alphabets

|      | RU             | UK           | BE            |
|------|----------------|--------------|---------------|
| RU   | /              | э, ё, ы, ъ   | и, ъ, щ       |
| UK   | г, е, i, ï, '  | /            | г, е, и, ï, щ |
| BE   | г, i, ў, '     | э, ё, ы, ў   | /             |

**Pronunciation Lexicon.** The ASR lexicon is based on the frequency of strings found in the corpus for the given language. Not all are true words, and we need to apply filtering, e.g. to remove strings with digits inside, or to omit 1- and 2-letter strings that are not valid words or abbreviations. The lexicon is made of those words that reach some minimum count (usually 5). This leads to lexicons with about 300K words and about 3% OOV rate.

All Slavic languages have fairly straightforward relation between orthography and pronunciation and this applies even more for those using Cyrillic alphabet where foreign words and names are phonetically transliterated.

For Russian, we use a set of 53 phonemes, including the palatalized consonants and pairs of stressed/unstressed vowels [19]. For UK and BE languages, we omit the unstressed ones and several specific phonemes, and get subsets with 39 and 36 phonemes, respectively. Pronunciation is generated according to the rules, from which many are common to all Slavic languages (e.g. consonant assimilation). The specific ones are related mainly to the palatalization phenomenon.

The method how we deal the stress in Russian is described in detail in [19]. Special care is also given to abbreviations were we allow alternative pronunciation. The size of the lexicons and some other statistics on them are summarized in Table 2.

**Language Model.** From the corpora we compute N-gram statistics that allow us to identify the most frequently occurring collocated words. These are added to the lexicon and after that a bigram LM is made using Knesser-Ney smoothing algorithm. Due to the added multi-words it has a larger span than a standard 2-gram model yet with lower computation demands.

## 4.2   Speech Data and Acoustic Model

Within the MultiLinMedia project we use only freely available speech data that can be gathered from audio and video sources on the Internet. There exist some dedicated speech databases, like e.g., GlobalPhone [22] with Russian and recently added Ukrainian parts, but they are not free, their size is limited, usually they cover just read speech, and they are not available for all the target languages. The AM training is done using HTK speech recognition toolkit and Torch library to train the neural networks.

**Speech Data Harvesting Scheme.** We have developed our own approach that allows us to gather (and annotate) enough speech from such sources, like broadcast or parliament archives. We rely on the fact that these archives contain both spoken records and some related text in form of, e.g. printed articles, subtitles for videos or stenograms from parliament sessions. Our goal is to detect automatically those documents or their parts (we call them chunks) where the spoken content agrees with the attached text. The estimate of the spoken content is provided by the best ASR system available at that stage and its output is matched with the provided text. The chunks where the former perfectly matches the latter are considered as correctly transcribed and their orthographic and phonetic transcriptions are used to build up a regular training database. The scheme proved to be robust. If there is no correspondence between the text and audio/video documents, the system will produce no training data in this case and it will cost us nothing except of the used CPU time. On the other side, if for a given document or its chunk, the ASR output is exactly same as the provided text, we can be sure that the transcription is trustworthy and that also the phonetic annotation provided by the system is precise enough for the AM training. The amount of the data gathered in this way depends on the nature of the sources. E.g., the parliament archives have a high degree of agreement between spoken and written content, while broadcasters' web pages can have a rather loose relation between them. However, because these sources are usually very large, there is always a chance to gather hours or even tens of hours of training data. The system and its efficiency improves in time. It is lower at the initial phase when the ASR system need to use an AM borrowed from other language(s) but it increases after several iterations when the newly acquired training data is added to the previous one.

The scheme is applied in several iterative steps:

1. Take a document with a related text from web
2. Transcribe it with the best available AM
3. Find chunks with closely matched text and transcript
4. Compare transcription with related text:
   (a) If match score = 100%, add it to training set
   (b) (Optional) If match score > threshold (e.g. 90%), check and correct by human and add to training set
5. Train new AM from the current training set
6. If any data left, go back to 1., 2., or 4., accordingly.

A more detailed description can be found in [10,14].

In each language, we start the data gathering process by searching for webs of major TV and radio stations and also the national parliaments. We have to find if they contain audio or video data, and a text that could be somehow related. (Usually this can be guessed without any particular knowledge of the language). The documents are automatically downloaded and the process described above is launched. It can run almost without any human supervision but it is useful to include also the optional human check (step 4b). We use a tool that replays the audio chunks whose score is above the specified threshold. The annotator also sees the original text and ASR output with highlighted differences and just clicks on that choice that sounds more appropriate to him/her. This human intervention is used mainly during initial iterations to boost the progress and also to discover potential errors in pronunciation.

**Application to East Slavic Languages.** In Table 4, there are some relevant statistics from the data harvesting process. We show the number of used web sources (it was all TV and radio stations, and in case of Russian also the national parliament Duma), the total size of downloaded audio data, the size of the extracted chunks with text match, and the real amount of speech data gathered for AM training.

**Table 4.** Statistics on acoustic data

| Language | RU | UK | BE |
|---|---|---|---|
| # Web sources | 7 | 6 | 3 |
| Downloaded audio [hours] | 4627 | 4015 | 955 |
| Total size of chunks [hours] | 192 | 161 | 48 |
| Gathered and annotated data [hours] | 58.3 | 48.0 | 16.4 |

**Language Identification.** As mentioned earlier, there is a lot of Russian speakers in Belarus. In TV news, Belarusian is often mixed with Russian and Ukrainian. If we wanted to get a pure training database for BE, we had to incorporate a spoken language identification (LID) module. Classic LID methods based on phonotactics do not work well for closely related languages. In such a situation, it is better to use a method that takes into account a longer context, it means words and their combinations, i.e. the LM. We used the approach described in [15]. We took our ASR system and let it work with special components: (a) an AM trained on the RU and UK speech data (obviously, BE data was not available at that stage), a lexicon composed of 100K most frequent words from RU, UK and BE (with a language label attached to each word), and (c) an LM trained on the joint RU+UK+BE corpus. The ASR system processed the audio data from the BE sources, segmented them into chunks and for each chunk it made a decision about the language based on the majority labels and a threshold. The latter was tuned on a small development set to get a balance between precision and recall values. When applied to the all Belarusian sources, it approved about 55% audio chunks as those belonging to BE. This is the main reason why we gathered only 16.4 h of BE training data as shown in Table 4. Anyway, in the following section we demonstrate that an AM based on this pure data works better than a mixed AM.

## 5    Evaluation on Real Broadcast Data

For evaluation we have prepared a standard test set used in the whole MultiLinMedia project. For each language, we took 3 full TVR main news shows, each about 30 min long. The shows were complete and contained jingles, headlines, clean studio speech, interviews in streets, large background noise, dubbed speech and also utterances in other languages (denoted as out-of-language, OOL). Precise orthographic annotations were made by native speakers. It should be also noted that the data was recorded at the beginning of 2017, while all the data used in the AM and LM training comes from the period 2010–2016. Like in our previous projects, we make them publicly available. They can be downloaded from *gitlab.ite.tul.cz/SpeechLab/EastSlavicTestData*. In all the experiments, we employed the same LVCSR system as in [10]. It uses 39-dimensional log-filter banks and DNN-HMM architecture. All the DNNs have 5 hidden layers (1024-768-768-512-512) with ReLU activation functions.

The results for the 3 languages are summarized in Table 5. We present two WER values. The first represents the total number of errors made by the system trained for the target language. It includes also those errors caused by talks in other languages (OOL passages). If we skip them in evaluation, we get the second WER value, which reflects the performance measured only on the target language speech. We can see that this plays a significant role mainly in UK and BE news.

The results for Russian (22%) are on the similar level like we got for most other Slavic languages. Ukrainian is worse but still acceptable for the broadcast monitoring task. Belarusian shows the worst performance, which is most likely caused by the small AM training set (16.4 h). We run an experiment in which the BE test set was recognized using either the UK acoustic model or that trained on the UK and BE data. The results are shown in Table 6. They are worse compared to the case when the genuine Belarusian AM was used. We have also analysed the main sources of ASR errors. In case of clean speech, most are due to misrecognized word-forms belonging to the same lemma and differing just in acoustically (and phonetically) similar suffixes. In these languages, the suffixes often have unstressed and hence reduced pronunciation. Obviously, more serious errors occur in noisy and spontaneous speech passages. One important issue that we are currently working on is the implementation of the LID module that runs on-line and allow for switching between the languages.

**Table 5.** Broadcast news test set and performance. The last column shows WER after excluding OOL passages, i.e. those in other than target language

| Language | Duration | #words | OOV | WER | WER excl. OOL |
|---|---|---|---|---|---|
| Russian | 93 min | 12277 | 2.18% | 23.02% | 22.08% |
| Ukrainian | 75 min | 9440 | 2.75% | 34.25% | 30.15% |
| Belarusian | 82 min | 1716 | 3.12% | 44.52% | 35.95% |

## 6  Discussion

One may ask, how good is an AM created in the presented way compared to that made of a dedicated speech database. As we have the Russian part of the Global-Phone (GP) dataset, we run several experiments. We trained an AM on the GP train set (22 h) and compared it to our AM based on 58 h gathered automatically. Two test sets were used: the test part of GP (10 speakers) and the broadcast set mentioned in Sect. 5. The results are summarized in Table 7 and show that the AM from the free sources outperforms the GP based one in both the tasks. We assume that it is mainly because the former is larger in size, it includes more speakers and, in particular, a larger variety of speaking styles and acoustic conditions.

**Table 6.** Results achieved on Belarusian test set with cross-lingual (UK) and multi-lingual (BE+UK) AMs

| AM | WER axcl. OOL |
|---|---|
| UK | 47.02% |
| BE+UK | 36.36% |

**Table 7.** Comparison of 2 train & test sets in Russian (The values are WERs in %)

| Test set\Train set | GlobalPhone | Automatically gathered |
|---|---|---|
| GlobalPhone | 18.21% | 14.30% |
| Broadcast News | 50.74% | 23.02% |

# 7    Conclusions

In this paper, we show that both the linguistic as well as the acoustic part of a state-of-the-art LVCSR system can be trained on data that is freely available on the Internet. The data can be acquired and processed in an almost automatic way, with minimum human intervention and without any particular knowledge of the target language.

We spent almost 2 years working on the Russian system, as it was our first experience with an East Slavic language, azbuka, the stress issue, etc. Anyway, it helped us to get a deeper insight. The other two languages were processed in a faster and more efficient way, using Russian as a start point, well suited for bootstrapping. We have also benefited from the multi-lingual platform and its tools built during the previous work on other Slavic languages. Recently, we have a portfolio that contains phonetic inventories, grapheme-to-phoneme converters, digit-to-text transducers [1], lexicons, acoustic models and language models for all 13 Slavic languages.

**Acknowledgments.** The research was supported by the Technology Agency of the Czech Republic (project $TA04010199$) and by the Student Grant Scheme (SGS) at the Technical University of Liberec.

# References

1. Chaloupka, J.: Digits to words converter for slavic languages in systems of automatic speech recognition. In: Karpov, A., Potapova, R., Mporas, I. (eds.) Speech and Computer. LNCS, vol. 10458, pp. 312–321. Springer, Cham (2017). doi:10.1007/978-3-319-66429-3_30
2. Jokisch, O., et al.: Multilingual speech data collection for the assessment of pronunciation and prosody in a language learning system. In: Proceeding of SPECOM (2009)
3. Vintsyuk, T.: Speech discrimination by dynamic programming. Kibernetica **1**, 15–22 (1968)
4. Kanevsky, D., Monkowski, M., Sedivy, J.: Large vocabulary speaker-independent continuous speech recognition in Russian language. In: Proceeding of SPECOM, vol. 96, pp. 28–31 (1996)
5. Karpov, A., et al.: Large vocabulary Russian speech recognition using syntactico-statistical language modeling. Speech Commun. **56**, 213–228 (2014)
6. Karpov, A., Kipyatkova, I., Ronzhin, A.: Very large vocabulary ASR for spoken Russian with syntactic and morphemic analysis. In: Proceeding of Interspeech (2011)

7. Lyudovyk, T., Robeiko, V., Pylypenko, V.: Automatic recognition of spontaneous Ukrainian speech based on the Ukrainian broadcast speech corpus. In: Dialog 2011 (2011)
8. Lyudovyk, T., Pylypenko, V.: Code-switching speech recognition for closely related languages. In: SLTU (2014)
9. Nikalaenka K., Hetsevich, Y.: Training algorithm for speaker-independent voice recognition systems using HTK. In: Pattern recognition and information processing (2016)
10. Nouza J., Safarik R., Cerva, P.: Asr for south slavic languages developed in almost automated way. In: LNCS (LNAI) (2017)
11. Nouza, J., et al.: Continual on-line monitoring of czech spoken broadcast programs. In: Proceeding of Interspeech, pp. 1650–1653 (2006)
12. Nouza, J., et al.: Czech-to-slovak adapted broadcast news transcription system. In: Proceeding of Interspeech (2008)
13. Nouza, J., et al.: Speech-to-text technology to transcribe and disclose 100,000+ hours of bilingual documents from historical czech and czechoslovak radio archive. In: Proceeding of Interspeech (2014)
14. Nouza, J., Cerva, P., Safarik, R.: Cross-lingual adaptation of broadcast transcription system to polish language using public data sources. In: Proceeding of Interspeech (2016)
15. Nouza, J., Cerva, P., Silovsky, J.: Dealing with bilingualism in automatic transcription of historical archive of czech radio. In: Petrosino, A., Maddalena, L., Pala, P. (eds.) ICIAP 2013. LNCS, vol. 8158, pp. 238–246. Springer, Heidelberg (2013). doi:10.1007/978-3-642-41190-8_26
16. Pylypenko, V., et al.: Ukrainian broadcast speech corpus development. In: Proceeding of SPECOM (2011)
17. Pylypenko, V., Robeyko, V.: Experimental system of computerized stenographer for Ukrainian speech. In: Proceeding of SPECOM (2009)
18. Robeiko, V., Sazhok, M.: Real-time spontaneous Ukrainian speech recognition system based on word acoustic composite models. In: Proceeding of UkrObraz (2012)
19. Safarik, R., Nouza, J.: Methods for rapid development of automatic speech recognition system for Russian. In: Proceeding of IEEE Workshop ECMS (2015)
20. Schlippe, T., et al.: Rapid bootstrapping of a Ukrainian large vocabulary continuous speech recognition system. In: Acoustics, Speech and Signal Processing (ICASSP) (2013)
21. Stüker, S., Schultz, T.: A grapheme based speech recognition system for Russian. In: Proceeding of International Conference SPECOM 2004 (2004)
22. Schultz, T.: Globalphone: a multilingual speech and text database developed at karlsruhe university. In: Proceeding of Interspeech (2002)
23. Vu, N.T., et al.: Rapid bootstrapping of five eastern European languages using the rapid language adaptation toolkit. In: Proceeding of Interspeech (2010)

# A Regularization Post Layer: An Additional Way How to Make Deep Neural Networks Robust

Jan Vaněk[(✉)], Jan Zelinka, Daniel Soutner, and Josef Psutka

Department of Cybernetics, New Technologies for the Information Society,
University of West Bohemia in Pilsen, Univerzitní 22,
301 00 Pilsen, Czech Republic
vanekyj@kky.zcu.cz

**Abstract.** Neural Networks (NNs) are prone to overfitting. Especially, the Deep Neural Networks in the cases where the training data are not abundant. There are several techniques which allow us to prevent the overfitting, e.g., L1/L2 regularization, unsupervised pre-training, early training stopping, dropout, bootstrapping or cross-validation models aggregation. In this paper, we proposed a regularization post-layer that may be combined with prior techniques, and it brings additional robustness to the NN. We trained the regularization post-layer in the cross-validation (CV) aggregation scenario: we used the CV held-out folds to train an additional neural network post-layer that boosts the network robustness. We have tested various post-layer topologies and compared results with other regularization techniques. As a benchmark task, we have selected the TIMIT phone recognition which is a well-known and still favorite task where the training data are limited, and the used regularization techniques play a key role. However, the regularization post-layer is a general method, and it may be employed in any classification task.

**Keywords:** Speech recognition · Phone recognition · Acoustic modeling · Neural networks · Regularization · Neural networks ensemble

## 1 Introduction

A usual problem that occurs during neural network training is called overfitting: the accuracy on the training set is driven to be a very nice, but when new data is presented to the network the accuracy drops. The network has nicely memorized the training data, but it has not learned to generalize the new ones.

A method for improving network generalization is to use a network that is just large enough to provide an adequate fit. The larger/deeper network we use, the more complex the functions the network can create. If we use a small enough network, it will not have enough power to overfit the data. Unfortunately, it is hard to know beforehand how large a network should be for a specific task. In the automatic speech recognition (ASR) domain, too shallow networks do not have an ability to learn the complex dependencies in the speech data. Therefore, it

© Springer International Publishing AG 2017
N. Camelin et al. (Eds.): SLSP 2017, LNAI 10583, pp. 204–214, 2017.
DOI: 10.1007/978-3-319-68456-7_17

is preferred to use a deep enough network in combination with some techniques that prevent the overfitting.

## 2 Related Work

A brief overview of the most popular techniques that prevent the overfitting follows.

### 2.1 Early Stopping

A possible way to regularize a neural network is *early stopping*, meaning that the training procedure is monitored by a validation set. When the validation error increases for a specified number of iterations, the training is stopped.

### 2.2 Stochastic Gradient Descent (SGD)

Despite the fact that the main benefit SGD is a training process acceleration, SGD could be also seen as an overfitting avoiding technique because it does not let a training process use the same examples over and over again.

### 2.3 L1, L2 Regularization

The other way to better neural networks generalization ability is to add an extra term to the error function that will penalize complexity of the network [1]. Penalizing the number of nonzero weights is called *L1* regularization: a coefficient times a sign of weight values are added to the error gradient function. *L2* regularization uses a sum of the squares: a fraction of weights values is added to the error gradient function. We can set the L1 and L2 coefficients for an entire network or on a per-layer basis.

### 2.4 Dropout

The term *dropout* refers to dropping out units (hidden and visible) in a neural network [2]. The dropping a unit out means temporarily removing it from the network, along with all its incoming and outgoing connections. The choice of which units to drop is random with a fixed probability $p$ independent of other units, where $p$ can be chosen using a validation set or can simply be set in an interval between 0 and 0.5. For the input units, however, $p$ is usually closer to 0 than to 0.5. At training time, a new random set of weights is sampled and trained for each training sample. At test time, a very simple approximate averaging method works well in practice. The weights of this network are scaled-down versions of the trained weights. The outgoing weights are multiplied by $1 - p$ at test time. This ensures that for any hidden unit the expected output is the same as the actual output at test time [2].

The dropout technique described above prevents feature co-adaptation by encouraging independent contributions from different features. However, repeatedly sampling a random subset of active weights makes training much slower. Wang et al. in [3] experimented with a Gaussian approximation of integrated dropout sampling. This approximation, justified by the central limit theorem and empirical evidence, gives an order of magnitude speedup and more stability.

## 2.5 Shakeout

From the statistic point of view, the dropout works by implicitly imposing an $L2$ regularizer on the weights. Kang et al. presented in [4] a new training scheme: *shakeout*. Instead of randomly discarding units as the dropout does at the training stage, the shakeout method randomly chooses to enhance or inverse the contributions of each unit to the next layer. This scheme leads to a combination of L1 regularization and L2 regularization imposed on the weights of the model. The technique was called *shakeout* because of the randomly shaking process and the L1 regularization effect pushing network weights to zero. The models trained by shakeout generalize better than the standardly trained networks. Moreover, in adverse training conditions, such as with limited data, the shakeout training scheme outperforms the dropout training.

## 2.6 Unsupervised Pre-training

Since gradient-based optimization often appears to get stuck in poor solutions starting from random initialization, Hinton et al. in [5] proposed a greedy layer-wise unsupervised pre-training learning procedure. This procedure relies on the training algorithm of restricted Boltzmann machines (RBM) and initializes the parameters of a deep belief network (DBN), a generative model with many layers of hidden causal variables. The greedy layer-wise unsupervised training strategy helps the optimization by initializing weights in a region near a good local minimum, but also implicitly acts as a sort of regularization that brings better generalization and encourages internal distributed representations that are high-level abstractions of the input [6]. The main idea is a using unsupervised learning at each stage of a deep network as a part of a training procedure for DBN. Upper layers of a DBN are supposed to represent more abstract concepts that explain the input observation, whereas lower layers extract low-level features from the input. In other words: this model first learns simple concepts, on which it builds more abstract concepts. In combination with the early stopping technique, the DBN pre-trained networks provide better generalization ability.

## 2.7 NNs Ensemble: Models Combination and Aggregation

Ensemble method or combining multiple classifiers is another way to improve the generalization ability [7–9]. By averaging or voting the prediction results from multiple networks, we can significantly reduce the model classification variance. The motivation of combining several neural networks is to improve a new data classification performance over individual networks. Perrone et. al showed, theoretically, in [10] that the performance of the ensemble can not be worse than any single model used separately if the predictions of the individual classifier are unbiased and uncorrelated.

**Bootstrap Aggregation - Bagging.** Bootstrap aggregation, also called bagging, is a method for generating multiple versions of a predictor and using these to get an aggregated predictor [11]. The aggregation averages over the versions when predicting a numerical outcome and does a plurality vote when predicting a class. The multiple versions are formed by making bootstrap replicates of the learning set and using these as new learning sets. The vital element is the instability of the prediction method. If perturbing the training set can cause significant changes in the predictor constructed, then bagging can improve accuracy.

**Cross-Validation Aggregation - Crogging.** In classification, cross-validation (CV) is widely employed to estimate the expected accuracy of a predictive algorithm by averaging predictive errors across mutually exclusive sub-samples of the data. Similarly, bootstrapping aims to increase the validity of estimating the expected accuracy by repeatedly sub-sampling the data with replacement, creating overlapping samples of the data. Beyond error estimation, bootstrap aggregation or bagging is used to make a NNs ensemble. Barrow et al. considered in [12] similar extensions of cross-validation to create diverse models. By bagging, it was proposed to combine the benefits of cross-validation and prediction aggregation, called crogging. In [12], the crogging approach significantly improved prediction accuracy relative to the bagging.

# 3   Regularization Post-Layer (RPL)

The RPL technique is based on cross-validation aggregation; however, it is more advanced. Cross-validation ensures that all observations are used for both training and validation, though not simultaneously, and each observation is guaranteed to be used for model estimation and validation the same number of times. Furthermore, the validation set available in CV can be used to control for overfitting in neural network training using early stopping. A $k$-fold cross-validation allows the use of all $k$ validation sets in performing early stopping, and this potentially further reduces the risk of overfitting. Moreover, prediction values for all $k$ folds can be obtained in validation mode. These validation-predictions for all folds – all training data – make a new valuable training set for an additional NN layer that we called regularization post-layer.

The RPL input dimension is equal to the number of classes, and the output dimension is the same. We assume that there is softmax function in the last layer of a standard NN. To be compatible, RPL must end by the softmax layer too. In principle, an application of the proposed regularization is not restricted on a post-layer or on only one single layer. But an output of the last layer has a straightforward influence on a criterion value. Thus a robustness of the layer is crucial. Any regularization of some deeper layer might make a training process less stable. This danger is always present but it is clearly minimal in the case of the last layer. Furthermore, dropout as a technique that works well for hidden layers can be applied technically also on the last layers but, for obvious reasons, is inappropriate. We prefer using of log-softmax values of the main NN as an input to RPL because RPL ends by additional softmax layer that assumes log-domain inputs. Because of log-domain processing, a multiplication in RPL corresponds to power at the output/probability domain, and an addition in RPL corresponds to a multiplication at the output. We lacked in a function that corresponds to an addition at the output. Therefore, we proposed to add a log-addition part to RPL.

Let $x_i$ be a log-sofmax output vector of the main NN, where $i = 1 \ldots N$. N is number of training samples. In general case, we proposed the RPL topology as follows:

$$y_i = \text{softmax}\Big(\text{addlog}(x + Wx + b, c)\Big), \tag{1}$$

where addlog is a log-addition function

$$\text{addlog}(\alpha, \beta) = \log\Big(\exp(\alpha) + \exp(\beta)\Big), \tag{2}$$

that is illustrated on Fig. 1 and can be computed in a robust way as follows:

$$\text{addlog}(\alpha, \beta) = \max(\alpha, \beta) + \log\Big(1 + \exp\Big(\min(\alpha, \beta) - \max(\alpha, \beta)\Big)\Big), \tag{3}$$

$W$ is a square matrix, $b$ is a vector of offsets, and $c$ is a logarithm of a vector of offsets in non-log-domain. Using $x + Wx$ (i.e. $(I + W)x$) instead of a simpler Wx keeps values of gradients significantly non-zero in the beginning of a training process when items of W and b are still small numbers. In this case, $(I + W)$ is a regularization of W. Therefore, the matrix W and the vector $b$ may be initialized close to zero. For initialization of the vector $c$, values close to the neutral element of the operation addlog are required. The neutral element is not 0 in this case but negative infinity. Thus the initial values were logarithms of small positive numbers e.g. $\log(1e-6)$. We trained RPL in the same way as the main NN: By SGD with identical training parameters.

As Fig. 1 shows, the addlog operation used in described way (see Eq. (1)) is similar to rectified linear unit (ReLU) function (precisely, it is the so-called softplus function for $c = 0$) but the difference is that the addlog function infimum (i.e. $c$) is not fixed as zero. The set infimum possibility is not relevant in hidden layer because a subsequent layer can do that too by means of its biases. In the last layer, especially when softmax function is applied, it is crucial because a posterior close to zero could be done only by means increasing other posteriors (before softmax) and that could make a NN and its training process less effective. The withc have the greatest influence on rare classes where the variance of predictions in log-domain is high.

When then the number of classes is high (i.e. thousands of classes as it is the typical situation in acoustic modeling for speech recognition), W is too large and some regularization or other treatment is necessary. Besides L1/L2

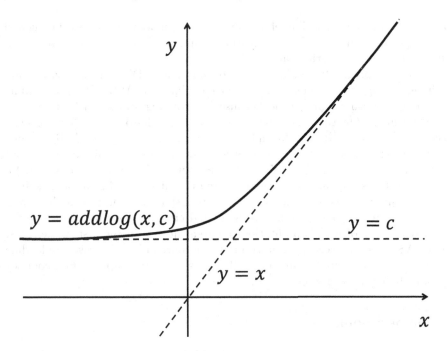

**Fig. 1.** Addlog function behavior

regularization, one may try a diagonal matrix $W$ only or decomposition the square matrix $W$ to a multiplication of two rectangular matrices $W = AB$, where an additional tunable hidden dimension parameter of matrices $A$ and $B$ is available.

In the testing phase, $k$ outputs of main NNs are available. We follow the crogging technique: we aggregate all $k$ outputs first and then proceed to the RPL that produces the single final output.

## 4  Regularization Techniques in Automatic Speech Recognition (ASR)

Training of NN as an acoustical model for automatic speech recognition (ASR) has some unique issues. The main fundamental issue is the data dependency. Usually, the training data samples are defined to be independent, but it does not hold in ASR. In a stream of processed audio signal, subsequent samples are mutually dependent and very similar. Also, data produced by the same speaker or under the same condition (recording hardware, codec, environment) are dependent. It is well known in the community and it could be partially avoided by splitting between training/validation parts at speaker-level. Cross-validation and bootstrapping need to be done at speaker-level also. The dependency in data lowers the "effective" amount of training data and we need to process a higher number of training samples during training relatively to another task where the data samples are independent. Therefore, ASR NNs are more prone to overfitting and the proper combination of regularization techniques belongs to key parts that affect the total ASR performance.

Early stopping, L2 regularization, dropout, and DBN unsupervised pre-training belongs in the most often used techniques in ASR. The pre-training also reduces total training time because the training starts much closer to a local optimum. On the other hand, pre-training is hard to use for the most recent NN types as (B) LSTM and TDNN. Using of dropout in ASR is sometimes tricky; $p$-values closer to one is used, or dropout is utilized in the first several training epochs only. While the dropout method was not designed for noise robustness, in [13] it was demonstrated that it was useful for noisy speech as it produced a network that was highly robust to variabilities in the input. NNs ensemble methods including bagging and crogging have not been popular in ASR because of high computation requirement also in the test phase. Most of the ASR systems need to keep computation burden low. However, offloading most of the NN computation to a GPU offers a space to apply more complex NNs, including NNs ensemble.

## 5   Experiments

As a benchmark task, we have selected the TIMIT phone recognition which is a well-known and still popular task where the training data are limited, and the used regularization technique plays a key role. As a baseline system, we used Kaldi and its TIMIT training example which is publicly available and offer easy experiments repeatability.

The TIMIT corpus contains recordings of phonetically-balanced prompted English speech. It was recorded using a Sennheiser close-talking microphone at 16 kHz rate with 16 bit sample resolution. TIMIT contains a total of 6300 sentences (5.4 h), consisting of 10 sentences spoken by each of 630 speakers from 8 major dialect regions of the United States. All sentences were manually segmented at the phone level.

The prompts for the 6300 utterances consist of 2 dialect sentences (SA), 450 phonetically compact sentences (SX) and 1890 phonetically-diverse sentences (SI).

The training set contains 3696 utterances from 462 speakers. The core test set consists of 192 utterances, 8 from each of 24 speakers (2 males and 1 female from each dialect region). The training and test sets do not overlap.

### 5.1   Speech Data, Processing, and Test Description

As mentioned above, we used TIMIT data available from LDC as a corpus LDC93S1. Then, we ran the Kaldi TIMIT example script, which trained various NN-based phone recognition systems with a common HMM-GMM tied-triphone model and alignments. The common baseline system consisted of the following methods: It started from MFCC features which were augmented by $\Delta$ and

$\Delta\Delta$ coefficients and then processed by LDA. Final feature vectors dimension was 40. We obtained final alignments by HMM-GMM tied-triphone model with 1909 tied-states. We trained the model with MLLT and SAT methods, and we used fMLLR for the SAT training and a test phase adaptation. We dumped all training, development and test fMLLR processed data, and alignments to disk. Therefore, it was easy to do compatible experiments from the same common starting point. We employed a bigram language/phone model for final phone recognition. A bigram model is a very weak model for phone recognition; however, it forced focus to the acoustic part of the system, and it boosted benchmark sensitivity. The training, as well as the recognition, was done for 48 phones. We mapped the final results on TIMIT core test set to 39 phones (as it is usual by processing TIMIT corpora), and phone error rate (PER) was evaluated by the provided NIST script to be compatible with previously published works. In contrast to the Kaldi recipe, we used a different phone decoder to be able to test novel types of NNs. The decoder is simple, but it does not tie mapped triphones and it process entire "full-triphone" lattice. It gives slightly better results than Kaldi standard WFST decoder. To be comparable, we used our full-triphone decoder for all results in this paper.

## 5.2 NNs Topology and Initialization

We used a common NN topology for all tests because the focus of this article is NN regularization and generalization ability. It followed the standard Kaldi recipe: we stacked input fMLLR feature 11 frames long window to 440 NN input dimension. All the input vectors were transformed by an affine transform to normalize input distribution. The net had 6 layers with 1024 sigmoid neurons and the final softmax layer with 1909 neurons. We used the DBN pre-training. In this point, we exported the NN parameters from Kaldi to our Theano-based training tool and made 10 epochs of SGD. Ten epochs are not enough to train a good model, but it was our common starting point for all our experiments. This way, we reduced the total computation time. We trained all NNs with L2 regularization with value 1e–4. Note that we also tested more recent types of NNs: LSTM NN and TDNN. However, we were not able to obtain better results compared to above mentioned deep NN (DNN).

## 5.3 Results

In Table 1, we show the summary of our results together with a couple state-of-the-art recently published results. *DNN* means the above described NN randomly initialized and trained by SGD with the L2 regularization and early stopping. *DBN_DNN* is identical to *DNN* but with DBN unsupervised pre-training. *DBN_DNN_sMBR* is *DBN_DNN* followed by a sequence-discriminative training [14]. *DBN_DNN_Dropout* means *DBN_DNN* trained with dropout (probability of dropping out $p = 0.1$). *DBN_DNN_Bagging_5* is an aggregation of 5 bootstraped NNs, all trained from the same DBN initialization. *DBN_DNN_Crogging_5* is an aggregation of 5 NNs done by a 5-fold cross-validation, all NNs were trained from

**Table 1.** Phone error rate [%] on TIMIT for various acoustic model NNs

| NNs | Dev | Test |
|---|---|---|
| DNN | 21.6 | 23.0 |
| DBN_DNN_Dropout | 18.2 | 19.7 |
| DBN_DNN_Bagging_5 | 16.5 | 17.3 |
| DBN_DNN_sMBR | 16.4 | 17.1 |
| DBN_DNN | 16.3 | 17.0 |
| DBN_DNN_Crogging_5 | 16.5 | 16.9 |
| DBN_DNN_RPL$_{Full}$ | 16.4 | 16.8 |
| DBN_DNN_RPL$_{Diag}$ | 16.2 | 16.5 |
| RNNDROP [15] | 15.9 | 16.9 |
| CNN_Hierarchical_Maxout_Dropout [16] | 13.3 | 16.5 |

the same DBN initialization. *DBN_DNN_RPL$_{Full}$* and *DBN_DNN_RPL$_{Diag}$* are based on *DBN_DNN_Crogging_5* but the RPL with full- or diagonal-matrix $W$ was added and trained. The results in Table 1 show that the DBN unsupervised pre-training was a key to get a low PER. Progress over *DBN_DNN* baseline was rather small. The sequence-discriminative training did not help on this task. A big surprise was the dropout result. Here the dropout results were unsatisfactory, and it was a problem to train the dropout NN in Kaldi successfully. However, other published more positive result with dropout on TIMIT [15,17]. Bagging produced slightly higher PER that the *DBN_DNN* baseline. We have also done an experiment with a change of a count of aggregated NNs in the bagging method. We show the accurate bagging results in Fig. 2. From the bagging experiment analysis, it is clear that 5 NNs was enough for aggregating and more NNs did not brings any further improvement. The greatest difference was between single NN and 2-NNs bagging. Crogging itself produced similar results as the baseline. We obtained an improvement with RPL only. With 1909 classes (tied-states of HMM), the diagonal RPL worked better than the full matrix variant, and we obtained PER that matched the best published TIMIT core test results [15,16].

## 5.4    Using of RPL in Image Processing

Besides using of RPL in phone recognition, we successfully applied it in a Kaggle image classification competition. We utilized RPL in the National Data Science Bowl competition 2015 (www.kaggle.com/c/datasciencebowl). Thanks to RPL, we achieved a very low drop of accuracy between development and test sets (the public and private leaderboard, respectively) and our team was moved up by 7 places on the leaderboard and we have finished at 19th place from 1,049 teams.

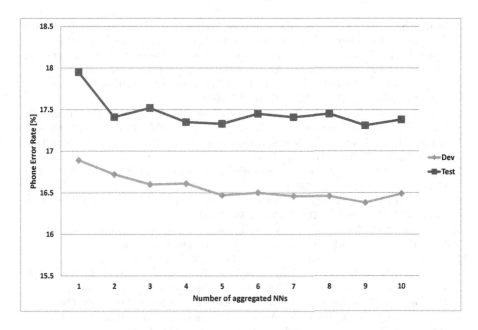

**Fig. 2.** Bagging PER for various number of NNs aggregated

# 6  Conclusion

In this paper, we propose the regularization post layer as an additional way to improve a deep neural networks generalization ability. It may be combined with other techniques as L1/L2 regularization, early stopping, dropout and others. It is based on cross-validation training and aggregation of NNs ensemble; therefore, the main drawback is higher computational requirements in the test phase. However, this drawback may be overcome by using of GPUs to accelerate NNs evaluation.

On TIMIT benchmark task, only a using of regularization post layer gives better results than DNN with DBN pre-training and we obtained PER that matched the best published TIMIT core test results.

**Acknowledgments.** This research was supported by the Grant Agency of the Czech Republic, project No. GAČR GBP103/12/G084.

# References

1. Girosi, F., Jones, M., Poggio, T.: Regularization theory and neural networks architectures. Neural Comput. **7**(2), 219–269 (1995)
2. Srivastava, N., Hinton, G.E., Krizhevsky, A., Sutskever, I., Salakhutdinov, R.: Dropout: a simple way to prevent neural networks from overfitting. J. Mach. Learn. Res. (JMLR) **15**, 1929–1958 (2014)

3. Wang, S.I., Manning, C.D.: Fast dropout training. In: Proceedings of the 30th International Conference on Machine Learning, vol. 28, pp. 118–126 (2013)
4. Kang, G., Li, J., Tao, D.: Shakeout: a new regularized deep neural network training scheme. In: Proceedings of the AAAI Conference, pp. 1751–1757 (2016)
5. Hinton, G.E., Osindero, S., Teh, Y.W.: A fast learning algorithm for deep belief nets. Neural Comput. **18**(7), 1527–1554 (2006)
6. Larochelle, H., Bengio, Y., Louradour, J., Lamblin, P.: Exploring strategies for training deep neural networks. J. Mach. Learn. Res. (JMLR) **1**, 1–40 (2009)
7. Hansen, L.K., Salamon, P.: Neural network ensembles. IEEE Trans. Pattern Anal. Mach. Intell. **12**(10), 993–1001 (1990)
8. Zhang, G.: Neural networks for classification: a survey. IEEE Trans. Syst. Man Cybern. **30**(4), 451–462 (2000)
9. Zhou, Z.H., Wu, J., Tang, W.: Ensembling neural networks: many could be better than all. Artif. Intell. **137**(1–2), 239–263 (2002)
10. Perrone, M.P., Cooper, L.N.: When networks disagree: ensemble methods for hybrid neural networks. Technical report, DTIC Document (1992)
11. Breiman, L.: Bagging predictors. Mach. Learn. **24**(2), 123–140 (1996)
12. Barrow, D.K., Crone, S.F.: Crogging (cross-validation aggregation) for forecasting - a novel algorithm of neural network ensembles on time series subsamples. In: Proceedings of the International Joint Conference on Neural Networks (2013)
13. Seltzer, M.L., Yu, D., Wang, Y.: An investigation of deep neural networks for noise robust speech recognition. In: Proceedings of the ICASSP (2013)
14. Vesely, K., Ghoshal, A., Burget, L., Povey, D.: Sequence-discriminative training of deep neural networks. In: Proceedings of the INTERSPEECH, pp. 2345–2349 (2013)
15. Moon, T., Choi, H., Lee, H., Song, I.: RNNDROP: a novel dropout for RNNs in ASR. In: Proceedings of the ASRU (2015)
16. Tóth, L.: Convolutional deep maxout networks for phone recognition. In: Proceedings of the INTERSPEECH, pp. 1078–1082 (2014)
17. Tóth, L.: Combining time- and frequency-domain convolution in convolutional neural network-based phone recognition. In: Proceedings of the ICASSP (2014)

# Speech Recognition: Modeling and Resources

# Detecting Stuttering Events in Transcripts of Children's Speech

Sadeen Alharbi[1(✉)], Madina Hasan[1], Anthony J. H. Simons[1],
Shelagh Brumfitt[2], and Phil Green[1]

[1] Department of Computer Science, The University of Sheffield, Sheffield, UK
{ssmalharbi1,m.hasan,a.j.simons,s.m.brumfitt,p.green}@sheffield.ac.uk
[2] Department of Human Communication Sciences, The University of Sheffield,
Sheffield, UK

**Abstract.** Stuttering is a common problem in childhood that may persist into adulthood if not treated in early stages. Techniques from spoken language understanding may be applied to provide automated diagnosis of stuttering from children speech. The main challenges however lie in the lack of training data and the high dimensionality of this data. This study investigates the applicability of machine learning approaches for detecting stuttering events in transcripts. Two machine learning approaches were applied, namely HELM and CRF. The performance of these two approaches are compared, and the effect of data augmentation is examined in both approaches. Experimental results show that CRF outperforms HELM by 2.2% in the baseline experiments. Data augmentation helps improve systems performance, especially for rarely available events. In addition to the annotated augmented data, this study also adds annotated human transcriptions from real stuttered children's speech to help expand the research in this field.

**Keywords:** Stuttering event detection · Speech disorder · Human-computer interaction · CRF · HELM

## 1 Introduction

Stuttering, sometimes referred to as 'stammering', is a speech disorder problem that starts in childhood and may result in severe emotional, communicational, educational and social maladjustment. Inadequate diagnoses and intervention at an early age may increase the risk that the condition may become chronic and has negative consequences on children with stuttering and their families [2,5]. Thus, clinical intervention should take place as early as the preschool years because later intervention does not help. Also, it is not possible to determine a child's chance of naturally recovering from stuttering. Moreover, children are less tractable as they get older due to the reduction of neural plasticity [11]. During the assessment phase, clinicians need to carefully measure the stuttering events to determine the severity of stuttering. This measurement is usually conducted by counting the number of stuttering events in the child's speech. This process is extremely dependent on the clinician's experience [1]. In another approach,

© Springer International Publishing AG 2017
N. Camelin et al. (Eds.): SLSP 2017, LNAI 10583, pp. 217–228, 2017.
DOI: 10.1007/978-3-319-68456-7_18

the clinician transcribes a recorded session and classifies each spoken term into one of several normal, disfluent or stuttering categories [4]. This process takes a long time because of the need to write every spoken word which takes time and effort, requires knowledge of the relevant categories. An automated speech transcription of the recorded speech using Automatic Speech Recognition (ASR) could help clinicians speed up the assessment process and store the data for further investigations. However, understanding children's speech is well known to be a challenge even for humans, due to several factors, such as speech spontaneity, slow rate of speech and variability in the vocal effort [13]. Therefore, a large amount of data is required to train an ASR with an acceptable word error rate (WER) and to process the ASR output to automatically identify the stuttering events in the transcription.

Research in this area investigate three main approaches to detect stuttering events. The first area of study attempts to detect stuttering events from recorded speech signals. Howell and Sackin [9], for example, proposed the first attempt at stuttering recognition. Their study applied Artificial Neural Network (ANNs) and focused on identifying repetitions and prolongations. The basic idea is that the input vector of ANNs are the autocorrelation function and envelope. Their best accuracy was 80%. Geetha et al. [3] presented an objective method of differentiating stuttering disfluencies. They used ANN techniques on two groups of disfluent children. Several features were chosen to discriminate between normal and stuttering speech. They reported that ANN classifiers could predict the classifications of normal and stuttering with 92% accuracy. Another approach detects stuttering events from transcriptions. Mahesha and Vinod [15] is used a lexical Rule-Based (RB) algorithm to detect and estimate the severity of 4 types of stuttering events: Interjection (I), word repetition (W), syllable repetition (S) and prolongation (P), in orthographic transcripts from University College London's Archive of Stuttered Speech (UCLASS) [8]. In particular, they use prior domain knowledge to construct expert-based sets of rules to count the number of occurrences of each of the 4 stuttering events. The third approach is a combination of the previous two approaches. An automatic speech recognition approach has been proposed by Heeman et al. [6,7] in an attempt to merge a clinician's annotations with an ASR transcript to produce an annotated transcript of audio files (between 1 and 2 min duration) of read speech. Three types of stuttering were considered in [6]; revisions, interjections, and phrase, word and sound repetitions. However, the proposed system relied on the availability of the clinician's annotations of the read recordings.

This work investigates the detection of stuttering events in orthographic transcripts from UCLASS corpus. Traditional RB algorithm, for event detection tasks, is powerful in transferring the experiences of domain experts to make automated decisions. For offline applications where time and effort are not concerns and it can work with high accuracy for limited target data. However, this approach depends on the expert's knowledge [14], which means it only works if all situations of stuttering events are considered. This condition cannot be satisfied in practice due to the continuous variability in data volume and complexity.

Moreover, this knowledge based approach is deterministic as it uses rules like "If word $W$ is preceded by word $Z$, within $C$ number of words, trigger the event $Y$", and if such scenarios are missed false decisions will be made without giving probability that evaluates those decisions.

Alternative probabilistic approaches are therefore required to learn the rules from the structure embedded in the data (i.e. the stuttering pattern encapsulated in the stuttering sentences). Machine learning classifiers such as Hidden Event Language Model (HELM) and Conditional Random Fields (CRF) can actually help build data driven rules, and furthermore, as we find more data, these classifiers can be easily and frequently retrained. As a precursor to developing ASR for children with stuttering, this work investigates the applicability of machine learning approaches; particularly HELM and CRF, for automatically detecting stuttering events in transcripts of children's speech. Moreover, it is well known that the main limitation in children's speech related research is the lack of large publicly available corpora. To slightly alleviate the lack of training data in this field, additional recordings (from the children recordings in Release One of UCLASS has been transcribed and annotated with the stuttering events to support the research in this field. This study also examines the effect of augmenting the training data with artificially generated data. The rest of the paper is organised as follows. The guidelines and methodology used for producing the stuttering data transcriptions and annotations are described in Sect. 2. Section 3 presents the process of data normalisation and extraction of classification features. The two classification approaches are then described in Sect. 4. The data augmentation design and process is presented in Sect. 5. Section 6 explains the common measures used in stuttering events detection. Section 7 presents the experiments used in this study. Finally, the conclusion and future work are discussed in Sect. 8.

## 2    Data Transcription and Annotation

### 2.1    Data Transcription

This study uses the 31 publicly available orthographic transcriptions of children's speech monologue in Release One of UCLASS [8]. The transcription method in this release adopting certain conventional orthographies to indicate stuttering disfluencies. For example, "This is is a a a amazing". In addition to those transcriptions, this study adds the orthographic transcriptions of another 32 files from the same release following the same transcription guidelines. The data consists of 45 males and 18 females between 7 and 17 years of age. The 63 transcription files were then annotated to include the stuttering type for each word using the annotation approach described in Sect. 2.2.

### 2.2    Data Annotation Approach

The annotation approach followed in this study is the one proposed by Yairi and Ambrose [21] and used by Juste and De Andrade [16]. In this approach, eight types of stuttering are considered: (1) sound repetitions, which include phoneme repetition (e.g.,'c c c complex'), (2) part-word repetitions, which consider a repetition of less than a word and more than a sound (e.g., 'com com complex'), (3) word repetitions that count the whole word repeated (e.g., 'mommy mommy'), (4) dysrhythmic prolongations, which involve an inappropriate duration of a phoneme sound (e.g., 'mmmmommy'), (5) phrase repetitions that repeat at least two complete words (e.g., 'this is this is'), (6) interjections, which involve the inclusion of meaningless words (e.g., 'ah', 'umm'), (7) revisions that attempt to fix grammar or pronunciation mistakes (e.g., 'I ate I prepared dinner'). (8) The block type includes inappropriate breaks in different parts of the sentence in between or within words. In this study, all types of stuttering were considered except the revision and block types. All stuttering types examined in the study are listed with their corresponding abbreviations in Table 1. Illustrative examples of the 6 different stuttering types are given in Fig. 1. The annotation methodology was reviewed by a speech language pathologist (SLP), who is one of the co-authors[4] of this paper. The distribution of each type of stuttering event, as well as the number of words in the training and testing data, are summarised in Table 2.

**Table 1.** Stuttering types

| Label | Stuttering type |
|-------|-----------------|
| I | Interjection |
| S | Sound repetitions |
| PW | Part-word repetitions |
| W | Word repetitions |
| PH | Phrase repetitions |
| P | Prolongation |
| NS | Non Stutter |

*Mommy mommy I want I want t t t to* go to *mmmy* school and *umm pla play*
    HW          PH        S           P           I    PW

**Fig. 1.** Stuttering examples

## 3     Data Normalisation and Features Extraction

Text normalisation is a very important step for the detection of stuttering events. It is also considered to be a prerequisite step for lots of downstream speech and language processing tasks. Text normalisation categorises text entities like dates,

**Table 2.** Data statistics

| Set | Words | %I | %W | %PW | %S | %PH | %P | %NS |
|---|---|---|---|---|---|---|---|---|
| Train | 11204 | 4.6 | 2.7 | 2.2 | 11.8 | 1.1 | 1.6 | 76 |
| Test | 2501 | 3.8 | 2.7 | 2.0 | 12.3 | 1.8 | 0.6 | 76.8 |
| All Data | 13705 | 4.5 | 2.6 | 2.1 | 11.9 | 1.2 | 1.4 | 76.3 |

numbers, times and currency amounts, and transforms those entities into words. For our experiments, we normalised the transcriptions and extracted word level based features to be used in the classification approaches used in this work. These features included n-grams for n = 2, 3 and 4, and up to two following words, referred to as post words.

## 4  Classification Approaches

### 4.1  Hidden Event Language Model

The Hidden Event Language Model (HELM) technique was adopted in this work, since it is an appropriate model to use when events of interest are not visible in every training context [17]. Stuttering events may be treated as hidden events, within a context that normally expects regular words. Standard language models are normally used to predict the next word and give word history. However, the language model here is applied to measure the probability of the appearance of each stuttering event at the end of each observed word, given its context. The inter-words events sequence are predicted by the model, E = e0, e1, e2,... en, based on given of a sequence of words, W =w0, w1, w2,... wn, using a quasi-HMM technique. The states of the model are represented as Word/event pairs, while the hidden state is represented as the stuttering event type. A standard language model provides the observations of previous words, and the probabilities.

### 4.2  Conditional Random Fields

Linear-Chain Conditional Random Fields (CRFs) are discriminative models that have been intensively used for sequence labelling and segmentation purposes [19]. The model aims to estimate and directly optimise the posterior probability of the label sequence, given a sequence of features (hence the frequently used term direct model). The CRF++ [12] toolkit was used in this work.

## 5  Data Augmentation

Data augmentation is a technique used for machine learning tasks in which there are too few training resources and usually not enough for training a model with reasonable performance. In speech processing, for example, the data augmentation is performed by adding perturbation from different sources such as artificial background noise, vocal tract length perturbation [10] and changing speaking rate of the spoken utterance. For this study, we used a language model that was trained on the stuttering data (the training set), to generate additional sentences to supplement the original training data. The SRILM toolkit [18] was used to generate random sentences from a word-list, weighted by the probability of word-distribution in a language model. The word list was designed to include stuttering versions of the words in the publicly available word list (lm-csr-64k-vp-3gram) [20], in addition to the original word list.

The generated sentences (416,456 words) are of nonsense and not grammatically correct, most of the time, just like children's speech. Despite this fact, those generated sentences tend to exhibit feasible stuttering patterns, including less-common ones.

In order to automatically annotate the generated sentences, before it can be used for training the classifiers, an RB algorithm was built through several attempts with human annotators interventions. The annotation rules described in Sect. 2.2 were followed in this offline annotation process. A subset of 3000 words, was taken from the generated data and manually annotated as a reference. This reference was used to improve the performance of the RB algorithm. To further improve the labels on the generated data, some samples were revised and edited by human annotators. However, it is important to clarify that the RB annotation of the generated data is not fully revised by human annotators. Table 6 presents the labels distribution in the generated data.

## 6   Metrics

In this work, the conventional metrics: precision $Prec$, recall $Rec$, F1 score and accuracy $Acc$ are used to evaluate the performance of the classifiers. The definitions of these metrics are given below.

$$\text{Prec} = \frac{TP}{TP + FP}, \qquad \text{Rec} = \frac{TP}{TP + FN}$$

$$\text{F1} = 2\frac{Precision * Recall}{Precision + Recall}, \quad \text{Acc} = \frac{TP + TN}{TP + TN + FP + FN}.$$

$TP$, $FP$ and $FN$ refer to true positive, false positive, and false negative counts, in that order.

## 7   Experiments

The following section presents our experiments on UCLASS data using the approaches discussed in Sects. 4 and 5 for detecting stuttering events.

## 7.1   Baseline Experiments

Initial experiments were conducted to determine the best order of textual features to be used for training HELM and CRF classifiers. These initial experiments were performed using 10-fold cross-validation (CV) sets, to verify the reliability of the model performance. Tables 3 and 4 show the CV results for HELM and CRF approaches, respectively. These results suggest that the best results of the HELM approach are obtained with 3-gram features, yielding an accuracy of 88%, Similarly, the best results for the CRF approach are obtained with 2-gram plus 2-post-words features with an accuracy of 90%. Generally, The CRF approach outperforms the HELM approach by **2.2%** relatively on accuracy.

Acceptable scores were obtained from both classifiers for detecting the $HW$, $I$ and $S$ classes. An important observation is the failure to detect the $PH$, $PW$, and $P$ types of stuttering events. The main reason for this failure is referred to the scarcity of these classes in the data, as shown in Table 2. Based on these results, the rest of the experiments were designed to consider the 3-gram, and 2-gram

**Table 3.** Cross-validation results using HELM approach, with **Acc = 90%**

| N-gram | Stuttering-type | Precision | Recall | f1-score |
|--------|-----------------|-----------|--------|----------|
| 2g     | I               | 0.55      | 0.15   | 0.22     |
|        | W               | 0.99      | 0.88   | 0.93     |
|        | NS              | 0.86      | 0.99   | 0.92     |
|        | P               | 0.00      | 0.00   | 0.00     |
|        | PH              | 0.00      | 0.00   | 0.00     |
|        | PW              | 0.31      | 0.04   | 0.07     |
|        | S               | 0.92      | 0.65   | 0.76     |
| 3g     | I               | 0.85      | 0.28   | 0.41     |
|        | W               | 0.99      | 0.82   | 0.90     |
|        | NS              | 0.87      | 1.00   | 0.93     |
|        | P               | 0.00      | 0.00   | 0.00     |
|        | PH              | 0.05      | 0.01   | 0.02     |
|        | PW              | 0.38      | 0.07   | 0.11     |
|        | S               | 0.96      | 0.65   | 0.78     |
| 4g     | I               | 0.87      | 0.27   | 0.40     |
|        | W               | 0.99      | 0.80   | 0.88     |
|        | NS              | 0.87      | 0.99   | 0.93     |
|        | P               | 0.00      | 0.00   | 0.00     |
|        | PH              | 0.05      | 0.01   | 0.02     |
|        | PW              | 0.39      | 0.04   | 0.07     |
|        | S               | 0.96      | 0.67   | 0.78     |

**Table 4.** Cross-validation results using CRF approach, with **Acc = 92%**

| N-gram | Stuttering-type | Precision | Recall | f1-score |
|--------|-----------------|-----------|--------|----------|
| 2g+2p  | I   | 0.78 | 0.23 | 0.34 |
|        | W   | 0.99 | 0.95 | 0.97 |
|        | NS  | 0.90 | 1.00 | 0.94 |
|        | P   | 0.00 | 0.00 | 0.00 |
|        | PH  | 0.20 | 0.02 | 0.03 |
|        | PW  | 0.25 | 0.04 | 0.07 |
|        | S   | 0.95 | 0.82 | 0.88 |
| 3g+2p  | I   | 0.84 | 0.25 | 0.38 |
|        | W   | 0.99 | 0.95 | 0.97 |
|        | NS  | 0.90 | 0.99 | 0.94 |
|        | P   | 0.00 | 0.00 | 0.00 |
|        | PH  | 0.20 | 0.04 | 0.07 |
|        | PW  | 0.26 | 0.04 | 0.07 |
|        | S   | 0.95 | 0.82 | 0.88 |
| 4g+2p  | I   | 0.91 | 0.21 | 0.34 |
|        | W   | 0.99 | 0.95 | 0.97 |
|        | NS  | 0.89 | 1.00 | 0.94 |
|        | P   | 0.00 | 0.00 | 0.00 |
|        | PH  | 0.10 | 0.03 | 0.04 |
|        | PW  | 0.33 | 0.05 | 0.08 |
|        | S   | 0.95 | 0.80 | 0.87 |

plus 2-post-words features for HELM and CRF approaches, respectively. in order
to avoid the cost of performing repeated cross-validation tests, we partitioned
the data into training (80%) and evaluation (20%) sets and we deliberately
ensured that the training and test sets had equal distributions of stuttering
events from the start. Table 6 shows the distribution of the 6 different types
of stuttering events in addition to the no stuttering event (NS). The initial
experiments described above were also repeated on the defined training and
evaluation sets, to check the generality of the defined sets. Table 5 shows the
baseline results on the evaluation set. Similar observations to the cross-validation
set of experiments are found.

## 7.2 Effect of Data Augmentation

Using the technique explained in Sect. 5, 416,456 words were generated and annotated. The distributions of the 6 stuttering events in the generated data are
presented in Table 6. The HELM and CRF models were retrained on the generated data, jointly with the original training data. The results of the retrained

**Table 5.** HELM vs CRF results on the evaluation set, with **Acc = 90%** and **Acc = 92%**, respectively

| Classifier | Stuttering-type | Precision | Recall | f1-score |
|---|---|---|---|---|
| HELM | I | 0.86 | 0.47 | 0.61 |
| | W | 0.96 | 0.85 | 0.90 |
| | NS | 0.89 | 0.99 | 0.94 |
| | P | 0.00 | 0.00 | 0.00 |
| | PH | 0.00 | 0.00 | 0.00 |
| | PW | 0.00 | 0.00 | 0.00 |
| | S | 0.98 | 0.78 | 0.87 |
| CRF | I | 0.89 | 0.35 | 0.50 |
| | W | 1.00 | 0.96 | 0.98 |
| | NS | 0.92 | 0.99 | 0.96 |
| | P | 0.00 | 0.00 | 0.00 |
| | PH | 0.00 | 0.00 | 0.00 |
| | PW | 0.00 | 0.00 | 0.00 |
| | S | 0.95 | 0.94 | 0.95 |

**Table 6.** Data statistics of generated data

| Words | %I | %W | %PW | %S | %PH | %P | %NS |
|---|---|---|---|---|---|---|---|
| 416456 | 6.5 | 8.5 | 6.8 | 27.2 | 5.3 | 1.6 | 44.1 |

HELM and CRF classifiers on detecting and classifying the 6 stuttering and non-stuttering events, on the evaluation set, are presented in Table 7. Compared to the baseline results in Table 5, the performance of both classifiers was improved, with accuracies of 92%, and 94% for HELM and CRF approaches, respectively. These results also show that the performance of the CRF classifier was improved for all labels, including for those events that were infrequent in the original training data. The improvement obtained by the retrained HELM is however less, compared to that obtained by the CRF approach. Both classifiers still fail to detect the $PH$ events. This is, however, expected due to the fact that the method used in the augmentation is based on a word list, not a list of phrases. Finally, despite the general improvements obtained by retraining using the augmented data, there is slight deterioration in the detection of $NS$, the dominant class, as shown in the CRF confusion matrix Table 8. This deterioration may due to the noisy labels of the generated data.

**Table 7.** Effect of data augmentation on the performance of HELM and CRF, when used to detect the stuttering events on the evaluation set

| Classifier | Stuttering-type | Precision | Recall | f1-score |
|---|---|---|---|---|
| HELM | I | 0.85 | 0.52 | 0.64 |
| | W | 0.97 | 0.74 | 0.84 |
| | NS | 0.91 | 0.99 | 0.95 |
| | P | 1.00 | 0.75 | 0.86 |
| | PH | 0.00 | 0.00 | 0.00 |
| | PW | 1.00 | 0.49 | 0.65 |
| | S | 0.92 | 0.84 | 0.88 |
| CRF | I | 0.96 | 0.49 | 0.65 |
| | W | 1.00 | 1.00 | 1.00 |
| | NS | 0.93 | 0.99 | 0.96 |
| | P | 1.00 | 0.57 | 0.73 |
| | PH | 0.00 | 0.00 | 0.00 |
| | PW | 0.61 | 0.32 | 0.42 |
| | S | 0.97 | 0.93 | 0.95 |

**Table 8.** CRF confusion matrix of stuttering event detection on the evaluation set: before and after augmentation

| Stuttering-type | I | W | NS | P | PH | PW | S |
|---|---|---|---|---|---|---|---|
| *CRF trained on train set* | | | | | | | |
| I | 30 | 0 | 50 | 0 | 0 | 3 | 4 |
| W | 0 | 90 | 4 | 0 | 0 | 0 | 0 |
| NS | 2 | 0 | 1910 | 1 | 0 | 0 | 7 |
| P | 0 | 0 | 13 | 0 | 0 | 0 | 2 |
| PH | 0 | 0 | 44 | 0 | 0 | 0 | 0 |
| PW | 2 | 0 | 32 | 0 | 0 | 0 | 1 |
| S | 0 | 0 | 19 | 0 | 0 | 0 | 284 |
| *CRF trained on augmented data* | | | | | | | |
| I | 46 | 0 | 48 | 0 | 0 | 0 | 0 |
| W | 0 | 94 | 0 | 0 | 0 | 0 | 0 |
| NS | 0 | 0 | 1899 | 0 | 0 | 7 | 7 |
| P | 0 | 0 | 5 | 8 | 0 | 0 | 1 |
| PH | 0 | 0 | 44 | 0 | 0 | 0 | 0 |
| PW | 2 | 0 | 21 | 0 | 0 | 11 | 0 |
| S | 0 | 0 | 21 | 0 | 0 | 0 | 285 |

# 8    Conclusions and Future Work

In this work we studied the performance of HELM and CRF approaches as alternatives to the expert-based RB approach, in detecting the stuttering events in orthographic transcripts. Experimental results show that CRF consistently outperforms the HELM approach. Baseline experiments show how low frequency stuttering events ($PW/PH/P$) fail to be detected by both HELM and CRF classifiers, because those rare events were not seen or seen infrequently in the training set. In an attempt to increase the training data to improve the performance of these classifiers, data augmentation approach was adopted to generate additional random sentences according to an n-gram distribution pattern of words with probability of some stuttering event. Despite the fact that generated sentences are only probability-weighted nonsense, they tend to exhibit feasible stuttering patterns, including less common ones. Data augmentation helped improve the performance of both classifiers, especially for infrequent events. Experimental results reflect how the augmented data helped the CRF approach to improve the recovery of most labels including the rare $P$ and $PW$ events. However, $PH$ events were still challenging to both classifiers. A phrase-based augmentation method, for sentence generation that creates realistic phrase repetition, could be a suitable solution.

Another contribution of this study has been to enlarge the corpus of human-transcribed stuttering speech data. We have approximately doubled the number of annotated sentences in the UCLASS corpus.

**Acknowledgments.** This research has been supported by the Saudi Ministry of Education, King Saud University.

# References

1. Brundage, S.B., Bothe, A.K., Lengeling, A.N., Evans, J.J.: Comparing judgments of stuttering made by students, clinicians, and highly experienced judges. J. Fluen. Dis. **31**(4), 271–283 (2006)
2. Craig, A., Calver, P.: Following up on treated stutterers studies of perceptions of fluency and job status. J. Speech Lang. Hear. Res. **34**(2), 279–284 (1991)
3. Geetha, Y., Pratibha, K., Ashok, R., Ravindra, S.K.: Classification of childhood disfluencies using neural networks. J. Fluen. Dis. **25**(2), 99–117 (2000)
4. Gregory, H.H., Campbell, J.H., Gregory, C.B., Hill, D.G.: Stuttering Therapy: Rationale and Procedures. Allyn & Bacon, Boston (2003)
5. Hayhow, R., Cray, A.M., Enderby, P.: Stammering and therapy views of people who stammer. J. Fluen. Dis. **27**(1), 1–17 (2002)
6. Heeman, P.A., Lunsford, R., McMillin, A., Yaruss, J.S.: Using clinician annotations to improve automatic speech recognition of stuttered speech. In: Interspeech 2016, pp. 2651–2655 (2016)
7. Heeman, P.A., McMillin, A., Yaruss, J.S.: Computer-assisted disfluency counts for stuttered speech. In: INTERSPEECH, pp. 3013–3016 (2011)
8. Howell, P., Davis, S., Bartrip, J.: The university college London archive of stuttered speech (uclass). J. Speech Lang. Hear. Res. **52**(2), 556–569 (2009)

9. Howell, P., Sackin, S.: Automatic recognition of repetitions and prolongations in stuttered speech. In: Proceedings of the first World Congress on fluency disorders, vol. 2, pp. 372–374 (1995)

10. Jaitly, N., Hinton, G.E.: Vocal tract length perturbation (VTLP) improves speech recognition. In: Proceeding ICML Workshop on Deep Learning for Audio, Speech and Language (2013)

11. Jones, M., Onslow, M., Packman, A., Williams, S., Ormond, T., Schwarz, I., Gebski, V.: Randomised controlled trial of the lidcombe programme of early stuttering intervention. BMJ **331**(7518), 659 (2005). http://www.bmj.com/content/331/7518/659

12. Kudoh, T.: CRF++ (2007). https://sourceforge.net/projects/crfpp/

13. Liao, H., Pundak, G., Siohan, O., Carroll, M., Coccaro, N., Jiang, Q.M., Sainath, T.N., Senior, A., Beaufays, F., Bacchiani, M.: Large vocabulary automatic speech recognition for children. In: Interspeech (2015)

14. Liu, H., Gegov, A., Cocea, M.: Complexity control in rule based models for classification in machine learning context. In: Angelov, P., Gegov, A., Jayne, C., Shen, Q. (eds.) Advances in Computational Intelligence Systems. AISC, vol. 513, pp. 125–143. Springer, Cham (2017). doi:10.1007/978-3-319-46562-3_9

15. Mahesha, P., Vinod, D.S.: Using orthographic transcripts for stuttering dysfluency recognition and severity estimation. In: Jain, L.C., Patnaik, S., Ichalkaranje, N. (eds.) Intelligent Computing, Communication and Devices. AISC, vol. 308, pp. 613–621. Springer, New Delhi (2015). doi:10.1007/978-81-322-2012-1_66

16. Juste, F.S., de Andrade, C.R.F.: Speech disfluency types of fluent and stuttering individuals: age effects. Folia Phoniatr. et Logop. **63**(2), 57–64 (2010)

17. Stolcke, A., Shriberg, E.: Statistical language modeling for speech disfluencies. In: 1996 IEEE International Conference on Acoustics, Speech, and Signal Processing, ICASSP-1996, vol. 1, pp. 405–408. IEEE (1996)

18. Stolcke, A., et al.: SRILM-an extensible language modeling toolkit. In: Interspeech, pp. 901–904 (2002)

19. Tseng, H., Chang, P., Andrew, G., Jurafsky, D., Manning, C.: A conditional random field word segmenter for sighan bakeoff 2005. In: Proceedings of the Fourth SIGHAN Workshop on Chinese Language Processing, vol. 171. Citeseer (2005)

20. Vertanen, K.: Csr lm-1 language model training recipe (2007)

21. Yairi, E., Ambrose, N.G.: Early childhood stuttering ipersistency and recovery rates. J. Speech Lang. Hear. Res. **42**(5), 1097–1112 (1999)

# Introducing AmuS: The Amused Speech Database

Kevin El Haddad[1(✉)], Ilaria Torre[2], Emer Gilmartin[3], Hüseyin Çakmak[1],
Stéphane Dupont[1], Thierry Dutoit[1], and Nick Campbell[3]

[1] University of Mons, Mons, Belgium
{kevin.elhaddad,huseyin.cakmak,stephane.dupont,
thierry.dutoit}@umons.ac.be
[2] Plymouth University, Plymouth, UK
ilaria.torre@plymouth.ac.uk
[3] Trinity College Dublin, Dublin, Ireland
{gilmare,nick}@tcd.ie

**Abstract.** In this paper we present the **AmuS** database of about three
hours worth of data related to amused speech recorded from two males
and one female subjects and contains data in two languages French and
English. We review previous work on smiled speech and speech-laughs.
We describe acoustic analysis on part of our database, and a perception
test comparing speech-laughs with smiled and neutral speech. We show
the efficiency of the data in **AmuS** for synthesis of amused speech by
training HMM-based models for neutral and smiled speech for each voice
and comparing them using an on-line CMOS test.

**Keywords:** Corpora and language resources · Amused speech · Laugh ·
Smile · Speech synthesis · Speech processing · HMM · Affective computing · Machine learning

## 1 Introduction

Recognition and synthesis of emotion or affective states are core goals of affective
computing. Much research in these areas deals with several emotional or affective
states as members of a category of human expressions – grouping diverse states
such as anger, happiness and stress together. The emotive states are either considered as discrete classes [27,33] or as continuous values in a multidimensional
space [1,31], or as dynamically changing processes over time. Such approaches
are understandable and legitimate as the goal in most cases is to build a single
system that can deal with several of the emotional expressions displayed by users.

However, such approaches often require the same features to be extracted
from all classes for modeling purposes. Emotional states vary greatly in their
expression, and in how they manifest in different subjects, posing difficulties in
providing uniform feature sets for modeling a range of emotions – for example,
happiness can be expressed with laughs, which can be a periodic expressions

© Springer International Publishing AG 2017
N. Camelin et al. (Eds.): SLSP 2017, LNAI 10583, pp. 229–240, 2017.
DOI: 10.1007/978-3-319-68456-7_19

with a certain rhythm, while disgust is usually expressed with continuous non-periodic expressions. Some studies focused on a single emotion such as stress [18] and amusement [12], with the aim to build a model of one emotional state. In this study, we explore the expression of amusement through the audio modality.

Limiting models to audio cues has several advantages. Speech and vocalization are fundamental forms of human communication. Audio is easier and less computationally costly to collect, process, and store than other modalities such as video or motion capture. Many applications rely on audio alone to collect user spoken and affective information. Conversational agents in telephony or other 'hands/eyes free' platforms use audio to "understand" and interact with the user. Many state-of-the-art robots such as NAO cannot display facial expressions, and rely on audio features to express affective states.

Amusement is very frequently present in human interaction, and therefore data are easy to collect. As amusement is a positive emotion, collection is not as complicated as the collection of more negative emotions such as fear or disgust, for ethical reasons. In addition, the ability to recognize amusement can be very useful in monitoring user satisfaction or positive mood.

Amusement is often expressed through smiling and laughter, common elements of our daily conversations which should be included in human-agent interaction systems to increase naturalness. Laugher accounts for a significant proportion of conversation – an estimated 9.5% of total spoken time in meetings [30]. Smiling is very frequent, to the extent that smiles have been omitted from comparison studies as they were so much more prevalent than other expressions [6].

Laughter and smiling have different social functions. Laughter is more likely to occur in company than in solitude [19], and punctuates rather than interrupts speech [37]; it frequently occurs when a conversation topic is ending [3], and can show affiliation with a speaker [20], while smiling is often used to express politeness [22]. In amused speech, both smiling and laughter can occur together or independently. As with laughter, smiling can be discerned in the voice when co-occurring with speech [8,40]. The phenomenon of laughing while speaking is sometimes referred to as speech-laughs [43], smiling while speaking has been called smiling voice [36,42], speech-smiles [25] or smiled speech [14]. As listeners can discriminate amused speech based on the speech signal alone [29,42], the perception of amused speech must be directly linked to these components and thus to parameters which influence them, such as duration and intensity. Mckeown and Curran showed an association between laughter intensity level and humor perception [32]. This suggests that the intensity level of laughter and, by extension, of all other amused speech components may be a particularly interesting parameter to consider in amused speech.

In this paper we present the **AmuS** database, intended for use as a resource for the analysis and synthesis of amused speech, and also for purposes such as amusement intensity estimation. **AmuS** is publicly and freely available for research purposes, and can be obtained from the first author[1]. It contains several

---

[1] **AmuS** is available at: http://tcts.fpms.ac.be/~elhaddad/AmuS/.

amused speech components from different speakers and in different languages (English and French) in sufficient quantity for corpus-based statistically robust studies. The database contains recorded sentences of actors producing smiled speech along with corresponding neutrally pronounced speech (i.e., with no specific emotion expressed) as well as laughter and speech-laughs from some of these speakers adequate for analysis and synthesis. The difference from previous work on the topic is, to the best of our knowledge, the quantity and nature of material provided.

We also describe the following experiments, which demonstrate the efficiency of the **AmuS** database for different problems:

- Acoustic study of smiled speech data from **AmuS** and comparison of results with previous findings.
- On-line perception tests, comparing speech-laughs, smiled speech and neutral sentences in terms of the arousal intensity scale.
- Evaluation results of HMM-based amused speech synthesis systems trained with the **AmuS** data.

Below we briefly review the characteristics of smiled speech and speech-laughs, outline the recording protocols we employed and describe the database of recordings. Finally, we present and discuss results of the acoustic study carried out on the recordings and results of the perception and evaluation experiments.

## 2    Motivations and Contributions

For the purposes of this study, we classify amused speech into two categories, smiled speech and speech-laughs, although finer distinctions are possible.

*Smiled Speech.* As an emotional expression, when it co-occurs with speech, smiling is formed not only by labial spreading but also by additional modifications of the vocal tract. Lasarcyk and Trouvain [29] conducted a perception test asking participants to rate the "smiliness" of vowels synthesized using an articulatory synthesizer. These were synthesized by modifying three parameters: lip spreading, fundamental frequency and the larynx length. The vowels synthesized by modifying all three parameters were perceived as most "smiley". In previous work, amused smiled speech was perceived as more amused than synthesized neutral or spread-lips speech, and the synthesized spread-lips sentences were also perceived as more amused than the synthesized neutral sentences [12]. This suggests that simply spreading the lips can give an impression of amusement. In earlier studies, Tartter found that recorded spread-lips sentences were indeed perceived as "smiled" [40]. Thus, although several factors contribute to emotional smiled speech, spreading the lips while speaking can play a strong role in its expression.

Smiled speech has also been studied at the acoustic level, with emphasis on analysis of the fundamental frequency (f0) and the first three formants (F1, F2, F3). f0 was found to be higher in smiled than neutral vowels in several studies [2,16,28,40]. Emond et al. and Drahota et al. also compared mean f0, f0 height and f0 range but did not find any systematic change between non-smiled and smiled recorded speech [8,14]. In these two studies, they compared whole sentences rather than isolated vowels or short words. However, after perception tests using these same sentences were performed, both works reported a certain correlation or relationship between f0 increase and perception of smiling.

Several acoustic studies show that the modification of the vocal tract caused by smiling during speech affects formant frequencies [8,12,16,29,38,40,42], although this effect varies somewhat from one study to another. Several authors [2,8,16,38,40,41] report higher F2 in smiled than non-smiled speech. In addition, in work dealing with smiled speech in which the speaker was asked not to express any specific positive emotion while smiling was not reported to be subject to any spread-lips stimuli, authors reported a less important or even absent increase of F1 [12,16,38,40]. Barthel and Quené [2] used naturalistic data of dyadic conversations, in the form of spontaneous smiles by speakers, to acoustically compare smiled and neutral utterances. They report increases in the first three formants but only obtained significant results for F2 and for rounded vowels. In [42], a comparison was made between smiled and non-smiled data in a dataset of naturally occurring conversation. An increase in F3 was reported from neutral to smiled speech data, as well as an average increase of formants for rounded vowels. Drahota et al. [8] also used conversational data for their study of different kinds of smile, where they compared the distance between mean formant values for smiled and neutral speech, e.g. F2-F1 for smiled speech compared versus F2-F1 neutral speech, instead of differences in formant height for smiled speech and neutral speech. They report that the more times a speaker is perceived as not smiling, the larger the difference F3-F2. In a previous study, we found variation, in the form of increase and decrease in both F1 and F2 values, for amused smiled speech with a greater effect on F2 [12].

The work cited above leads to several conclusions. f0 is reported as higher in smiled than neutral speech or to be an important parameter of smiled speech perception. Lip spreading predominantly affects F2 and has no or little effect on F1. Relationships between formants (such as their distance) should be considered as potentially important parameters for discriminating smiled from neutral speech. Finally, vocal tract deformation in smiled speech varies in different ways depending on context and speaker.

Smile detection or recognition systems often form part of a multi-class emotion classification system (smile would then be associated to positive emotions such as happiness) [24], and are also found in facial detection systems rather than systems based on audio features [23]. Recognition has been largely based on feature extraction for later classification [15]. The latest works on recognition using audio cues rely on feature learning systems using the power of deep learning to learn an internal representation of raw data with respect to a given task [17].

Smiled speech synthesis is generally reported in the literature as part of a voice adaptation or voice conversion system aiming to generate different emotions [46]. Among the very few studies on smiled speech synthesis is the articulatory synthesis approach of Lasarcyk and Trouvain [29], which only considers smiled (but not specifically amused) vowels.

The scarcity of work related to amused or even smiled speech is probably due to the lack of relevant data and the complexity of collecting it. Several databases containing emotional speech exist, such as [4,5]. However, to the best of our knowledge, these databases contain smiled speech related to happiness or joy rather than amusement, and thus might not be representative of amused smiled speech. For analysis and tasks like voice conversion it is very useful to have utterances expressed in a certain emotion (in this case amused smiled speech) and as neutral utterances; especially if the pairs came from the same speaker. Such data can be found in [4] (again for happy and not precisely amused smiled speech), but the amount for a single speaker might not be sufficient for current systems data requirements.

We have collected a database of amused smiled speech and corresponding neutral speech sentences from different speakers (male and female) in English and French. This database contains enough data for analysis and for use in machine learning-based systems. For a deeper understanding of amused smiled speech we also recorded corresponding spread-lips data.

*Speech-Laughs.* Speech-laughs are instances of co-occurrence of laughter and speech. The phenomenon has not been clearly defined in the literature. Provine reports laughter occurring mostly at the extremities of a sentences [37], while Nowkah et al. [34] report that up to 50% of conversational laughter is produced simultaneously with speech. Kohler notes the occurrence of speech - smiled speech - speech-laugh - laughter and vice versa sequencing in amusing situations in a small-scale study and mentions the need for further investigations [25]. Trouvain proposes an acoustic account of speech-laughs and describes types of speech-laughs found in the analyzed data. He mentions the intervention of breath bursts during speech and notes the presence of "tremor" in voiced segments. He also investigates whether speech-laughs and smiled speech can be placed on a continuum of amused expression, while commenting that the variety of functions possible for laughter and the individuality of laughter among subjects would lead to significant variation in the forms of speech-laugh encountered [43]. This introduces another important aspect to be considered in relation to amused speech: the continuity between smiled speech and laughter. Are these nothing more than different levels of amusement expressions? Can intensity levels of amusement be mapped to combinations of amused speech components detected or not in an utterance? No clear answer has been given concerning this aspect of amused speech, although Trouvain rejected the hypothesis of a smile-laughter continuum [43]. Dumpala et al. found that f0 was higher in laughter than in speech-laughs and higher in speech-laughs than in neutral speech [9].

There has been very little work on synthesis and recognition of speech-laughs and smiled speech. Dumpala et al. [9] present a speech-laugh/laughter discrimination system. Oh and Wang [35] tried real-time modulation of neutral speech to make it closer to speech-laughs, based on the variation of characteristics such as pitch, rhythm and tempo. However, no evaluation of the naturalness of that approach has been reported.

The **AmuS** database contains amused smiled speech, different types of speech-laughs and also laughs. We hope that the collection will thus reflect the fact that laughter can interrupt or intermingle with speech [25,34,43], or happen at the extremities of sentences [37]. Several laughter databases can be found related to isolated laughter [10], but we know of no speech-laugh database suitable for machine learning-based work or analysis. **AmuS** will facilitate research into the use of these components in amused speech and their relation with amused speech arousal/intensity, and also into the smile-laughter continuum.

## 3 Database

The **AmuS** database contains recordings of acted smiled speech and corresponding neutral utterances from three different speakers in two different languages, English and French. For some speakers, spread-lips speech was also recorded. It also contains, for some speakers, speech-laughs and laughter data. The speech-laughs were semi-acted since they were collected during the smiled speech recordings but without the speakers being explicitly asked to utter them. The laughs were spontaneously expressed since they were elicited with appropriate stimuli. The data were recorded in quiet rooms and resampled to 16 kHz for uniformity. A more detailed description is given below and the data are summarized in Table 1.

**Smiled Speech:** To provide comparative neutral and smiled speech, speakers were asked to read the same sentences neutrally (not expressing any particular emotion) and while sounding amused but without laughing. Two readers were also asked to read the same utterances while spreading their lips without trying to sound happy or amused. This was done to obtain a set of the same sentences read in different speech styles for comparison. Noisy (saturated, or containing artifacts) or wrongly pronounced data were removed from the dataset. The final dataset is shown in the "Speech Styles" columns of Table 1. Voices from two males (M) and one female (F) were collected. SpkA and SpkB are French native speakers while SpkC is a British English native speaker. The sentences used to record SpkA were phonetically balanced. The other French data were recorded using a subset of these sentences. For English, a subset of the CMU Arctic Speech Database [26] was used. All utterances were then force-aligned with their phonetic transcriptions using the HMM ToolKit (HTK) software [44]. The transcriptions were stored in label files in the HTK label format. SpkC's annotations were also manually checked. The amount of data available, to the best of our knowledge, is comparable (SpkB and SpkC) and in some cases superior (SpkA) to currently available databases.

**Speech-Laughs:** To represent the effect of laughter altering speech and thus creating speech-laughs, we collected different types of speech-laughs from SpkA and SpkC. The speech-laughs were collected from the smiled speech recordings where actors produced them intuitively when trying to sound amused. In total 161 speech-laughs were collected from SpkA and SpkC. These were divided into two groups based on how they were produced. The first group contains tremor-like sounds happening in vowels only. These were previously investigated by us and will be referred to as chuckling (or shaking) vowels [21]. The second type contains bursts of air appearing usually at the end of a syllable or between a consonant and a vowel. A total of 109 and 52 instances were collected from the first and second type respectively.

Laughs were collected from SpkB. Since these laughs are to be used in sentences, we needed them to occur during an utterance. So, SpkB was asked to sustain a vowel while watching funny videos. He eventually laughed while pronouncing a vowel. A total of 148 laughs were recorded. These types of laughs proved to be efficient to produce synthesized amused speech [13]. The perceived arousal level or affective intensity of these laughs were then annotated in an online experiment. Annotators were presented with 35 randomly picked laughs and asked to rate how amused the laughs sounded, on a scale from 0 to 4 (0 begin not amused and 4 being very amused). The annotators were free to logout at any time. A total of 22 annotators took part in this experiment and each laugh was annotated on average 5.08 times.

**Table 1. AmuS** database content description. The numbers represent the number of utterance collected. M = Male, F = Female, Lang = Language, SL = Speech-laughs, L = laughs, * = these are the same data since the laughs came from the same speaker.

| Speakers | Speech styles | | | Lang | SL | L |
|---|---|---|---|---|---|---|
| | **N** | **Sm** | **Sp** | | | |
| SpkA (M) | 1085 | 1085 | - | Fr | 48 | - |
| SpkB-Fr (M) | 249 | 199 | 199 | Fr | - | 148* |
| SpkB-Eng (M) | 180 | 213 | - | Eng | - | 148* |
| SpkC (F) | 170 | 84 | 152 | Eng | 113 | - |

## 4   Acoustic Analysis

In this section, we present data analysis on the acoustic effects of smiling on speech, comparing neutral speech vowels to amused smiled and spread-lips speech vowels. We considered 15 and 16 vowels from French and English respectively.

The f0 and first three formants (F1, F2 and F3) were calculated for each sentence in each of the three speech styles using a sliding Hamming window of 25 ms length, shifted by 10 ms using the Snack library [39]. *F2-F1* and *F3-F2* were also calculated for each vowel. Pairs were formed of the same sentences from amused smiled and neutral speech. Since the durations of two corresponding sentences in a pair were different, Dynamic Time Warping (DTW) was applied at the phoneme level to align each of the extracted parameters. Values for neutral speech were subtracted from those for amused smiled speech. The mean values of all the differences obtained were then calculated. The same method was then applied between the spread-lips and neutral speech styles. Table 2 shows the results of these mean differences. These are expressed in percentage of variation with respect to the neutral speech values (e.g. 5% represents an increase of 5% of the neutral speech value, −5% a decrease). This table shows the mean percentage obtained as well as the mean standard deviation.

**Table 2.** Mean difference of the acoustic parameters extracted between the neutral and the other two styles. These are expressed in % of variation with respect to the neutral speech, e.g. $56\% \pm 20\%$ shows and increase of 56% of and from the neutral speech values on average with 20% standard deviation.

| Speakers | Amused smiled - neutral | | | | Spread-lips - neutral | |
|---|---|---|---|---|---|---|
| | SpkA | SpkB-Fr | SpkB-Eng | SpkC | SpkB-Fr | SpkC |
| $F_0$ | $56\% \pm 20\%$ | $8\% \pm 16\%$ | $48\% \pm 21\%$ | $16\% \pm 18\%$ | $4\% \pm 15\%$ | $14\% \pm 13\%$ |
| F1 | $13\% \pm 13\%$ | $-6\% \pm 14\%$ | $-2\% \pm 15\%$ | $2\% \pm 10\%$ | $3\% \pm 10\%$ | $3\% \pm 14\%$ |
| F2 | $-1\% \pm 9\%$ | $-2\% \pm 12\%$ | $-5\% \pm 9\%$ | $0.6\% \pm 7\%$ | $2\% \pm 6\%$ | $11\% \pm 11\%$ |
| F3 | $-0.2\% \pm 10\%$ | $-5\% \pm 9\%$ | $-2\% \pm 9\%$ | $-1\% \pm 5\%$ | $-2\% \pm 6\%$ | $1\% \pm 8\%$ |
| F2 - F1 | $-4\% \pm 15\%$ | $7\% \pm 17\%$ | $-2\% \pm 15\%$ | $1\% \pm 11\%$ | $3\% \pm 11\%$ | $16\% \pm 16\%$ |
| F3 - F2 | $11\% \pm 26\%$ | $-4\% \pm 23\%$ | $13\% \pm 27\%$ | $-3\% \pm 20\%$ | $-6\% \pm 20\%$ | $-7\% \pm 23\%$ |

A 95% confidence interval paired Student's t-test was used to study the statistical significance of the results in Table 2. The mean value was calculated for each of the acoustic parameters and for all the vowels for each sentence. The set of values obtained for amused smiled and for the spread-lips styles were each compared to the neutral style for each speaker. All proved to be significantly different except for the spread-lips vs neutral F3 values of SpkB-Fr.

As can be seen from Table 2, f0 increased in all cases, congruently with previous studies. Instead, no common pattern could be noticed in the formants even for the same speaker in two different languages (SpkB-Fr and SpkB-Eng), although F3 seems to be decreasing in all cases. Regarding the spread lips, instead, the pattern observed is more consistent between speakers (all changing in the same way except for F3). Thus, amused smiles affect speech in different ways with different vocal tract modifications. Since these are read sentences, further perceptual studies should be made to compare the naturalness and amusement perceived. This might help to understand the acoustic variations better.

## 5    Perception Test

A perception test was carried out in order to compare the level of perceived amusement from neutral (N), smiled speech (Sm) and speech-laughs (Sl). For this, we used sentences from SpkA and SpkC. 16 sentences were randomly selected from SpkA, of which 6 were the same sentences for all three speech styles, while the remaining 10 were chosen randomly. From SpkC, 15 sentences were randomly picked from each speech style, since unfortunately no identical sentences could be found in the smiled and speech-laugh styles. The sentences were then paired so that each style could be compared to the other two (N vs. Sm, Sm vs. Sl and N vs. Sl). Except for the 6 same sentences from SpkA, all the others were randomly paired.

These sentences were then presented to 20 on-line raters as a Comparative Mean Opinion Score (CMOS) test. Each rater was given thirty pairs of sentences chosen randomly and asked to rate which one sounded more amused. The raters were given seven possible choices each time: three on the right in favor of the utterance on the right, three on the left in favor of the utterance on the left and one in the middle, representing neutrality (i.e., both audio utterances sound the same). Each choice was mapped to an integer ranging from −3 to +3. The scores were as follows: 0.130 in favor of Sl when compared to Sm, 0.92 for Sm when compared to N and 1.135 in favor of Sl when compared to N.

Thus, Sl sentences were perceived as more amused than both Sm and N, and Sm more amused than N. Although the scores obtained in favor of Sl when compared to Sm are not very high, this result suggests that containing the tremor and/or air outburst in speech is more likely to increase the amusement intensity level perception.

## 6  Synthesis Evaluation

The smiled and neutral speech from all voices were also used to train Hidden Markov Model (HMM)-based speech synthesis systems [45] to generate a smiled amused voice and a neutral voice for each speaker. The systems obtained from SpkA and SpkB-Fr were previously trained and evaluated in [11] and [12], respectively. The systems from SpkC and SpkB-Eng were trained and evaluated for the purpose of this study.

HMMs were trained for each of the smiled and neutral speech styles and for each speaker and language separately. Their topology was identical in each case and consisted of 5 states left-to-right HMMs with no skip. Gaussian Mixture Models (GMM) were used to model the observation probabilities for each state. These were single multivariate Gaussian distributions with diagonal covariance matrices since a unique voice was being modeled in each case. The features used for training were the Mel Generalized Cepstral Coefficients (order 35, $\alpha = 0.35$ and $\gamma = 0$) and the f0, along with their derivatives and double derivatives. The features were extracted using a 25 ms wide window shifted by 5 ms, using the Snack library. Except for SpkA, for which two HMM models were created independently for each of the smiled and neutral voices, all the other models were built using adaptation via the CMLLR algorithm [7], since SpkB-Fr, SpkB-Eng and SpkC have fewer samples. For adaptation, a large dataset was used as a source on which to adapt a smaller dataset, the target. Thus, for the French voice, the neutral data of SpkA were used to build the source model and the SpkB-Fr neutral and smiled targets. For the English voice, the RMS and SLT voices from the CMU Arctic Speech Database were used as source data respectively to target SpkB-Eng and SpkC (for both neutral and smiled). The implementation was done using the HTS toolkit [45].

Synthesized smiled and neutral sentences were used for a comparison study on an amusement level scale for each of the three voices. The same sentences were synthesized in the neutral and smiling conditions for each speaker and paired for a CMOS on-line evaluation. Participants were asked the same question as in the

CMOS of Sect. 5, and in all cases the synthesized smiled sentences generated were perceived as more amused than the neutral ones, indicating that this database is suitable to train a parametric speech synthesizer such as an HMM-based one (0.24 in favor of the synthesized amused smiled sentences).

# 7    Conclusion

In this article we presented an amused speech database containing different amused speech components which can be grouped into the broad categories of smiling and laughter. We also reviewed a state-of-the-art for the acoustic studies relative to smiled speech and previous literature concerning speech-laughs. Our database was also used for acoustic studies and well as synthesis and perceptual evaluations. Perception tests suggest that the presence of "low level" laughs occurring in amused speech, known as speech-laughs, may increase the perception of amusement of the utterance. In the future we plan on improving this database by adding naturalistic non-acted amusement expressions, which could be used for research purposes such as affective voice conversion or speech synthesis.

# References

1. Ressel, J.A.: A circumplex model of affect. J. Pers. Soc. Psychol. **39**, 1161 (1980)
2. Barthel, H., Quené, H.: Acoustic-phonetic properties of smiling revised-measurements on a natural video corpus. In: Proceedings of the 18th International Congress of Phonetic Sciences (2015)
3. Bonin, F., Campbell, N., Vogel, C.: Time for laughter. Knowl.-Based Syst. **71**, 15–24 (2014)
4. Burkhardt, F., Paeschke, A., Rolfes, M., Sendlmeier, W.F., Weiss, B.: A database of German emotional speech. In: Interspeech, vol. 5, pp. 1517–1520 (2005)
5. Busso, C., Bulut, M., Lee, C.C., Kazemzadeh, A., Mower, E., Kim, S., Chang, J., Lee, S., Narayanan, S.S.: IEMOCAP: interactive emotional dyadic motion capture database. J. Lang. Res. Eval. **42**(4), 335–359 (2008)
6. Chovil, N.: Discourse oriented facial displays in conversation. Res. Lang. Soc. Interact. **25**(1–4), 163–194 (1991)
7. Digalakis, V.V., Rtischev, D., Neumeyer, L.G.: Speaker adaptation using constrained estimation of Gaussian mixtures. IEEE Trans. Speech Audio Process. **3**(5), 357–366 (1995)
8. Drahota, A., Costall, A., Reddy, V.: The vocal communication of different kinds of smile. Speech Commun. **50**(4), 278–287 (2008)
9. Dumpala, S., Sridaran, K., Gangashetty, S., Yegnanarayana, B.: Analysis of laughter and speech-laugh signals using excitation source information. In: 2014 IEEE International Conference on Acoustics, Speech and Signal Processing (ICASSP), pp. 975–979, May 2014
10. Dupont, S., et al.: Laughter research: a review of the ILHAIRE project. In: Esposito, A., Jain, L.C. (eds.) Toward Robotic Socially Believable Behaving Systems - Volume I. ISRL, vol. 105, pp. 147–181. Springer, Cham (2016). doi:10.1007/978-3-319-31056-5_9

11. El Haddad, K., Cakmak, H., Dupont, S., Dutoit, T.: An HMM approach for synthesizing amused speech with a controllable intensity of smile. In: IEEE International Symposium on Signal Processing and Information Technology (ISSPIT), Abu Dhabi, UAE, 7–10 December 2015

12. El Haddad, K., Dupont, S., d'Alessandro, N., Dutoit, T.: An HMM-based speech-smile synthesis system: an approach for amusement synthesis. In: International Workshop on Emotion Representation, Analysis and Synthesis in Continuous Time and Space (EmoSPACE), Ljubljana, Slovenia, 4–8 May 2015

13. El Haddad, K., Dupont, S., Urbain, J., Dutoit, T.: Speech-laughs: an HMM-based approach for amused speech synthesis. In: International Conference on Acoustics, Speech and Signal Processing (ICASSP), Brisbane, Australia, 19–24 April 2015

14. Émond, C., Ménard, L., Laforest, M.: Perceived prosodic correlates of smiled speech in spontaneous data. In: Bimbot, F., Cerisara, C., Fougeron, C., Gravier, G., Lamel, L., Pellegrino, F., Perrier, P. (eds.) INTERSPEECH, pp. 1380–1383. ISCA (2013)

15. Eyben, F., Scherer, K., Schuller, B., Sundberg, J., André, E., Busso, C., Devillers, L., Epps, J., Laukka, P., Narayanan, S., Truong, K.: The geneva minimalistic acoustic parameter set (gemaps) for voice research and affective computing. IEEE Trans. Affect. Comput. **7**(2), 190–202 (2015). Open access

16. Fagel, S.: Effects of smiling on articulation: lips, larynx and acoustics. In: Esposito, A., Campbell, N., Vogel, C., Hussain, A., Nijholt, A. (eds.) Development of Multimodal Interfaces: Active Listening and Synchrony. LNCS, vol. 5967, pp. 294–303. Springer, Heidelberg (2010). doi:10.1007/978-3-642-12397-9_25

17. Fayek, H.M., Lech, M., Cavedon, L.: Evaluating deep learning architectures for speech emotion recognition. Neural Netw. **92**, 60–68 (2017)

18. Garcia-Ceja, E., Osmani, V., Mayora, O.: Automatic stress detection in working environments from smartphones' accelerometer data: a first step. IEEE J. Biomed. Health Inform. **20**(4), 1053–1060 (2016)

19. Glenn, P.: Laughter in Interaction, vol. 18. Cambridge University Press, Cambridge (2003)

20. Haakana, M.: Laughter and smiling: notes on co-occurrences. J. Pragmat. **42**(6), 1499–1512 (2010)

21. Haddad, K.E., Çakmak, H., Dupont, S., Dutoit, T.: Amused speech components analysis and classification: towards an amusement arousal level assessment system. Comput. Electr. Eng. (2017). http://www.sciencedirect.com/science/article/pii/S0045790617317135

22. Hoque, M., Morency, L.-P., Picard, R.W.: Are you friendly or just polite? – analysis of smiles in spontaneous face-to-face interactions. In: D'Mello, S., Graesser, A., Schuller, B., Martin, J.-C. (eds.) ACII 2011. LNCS, vol. 6974, pp. 135–144. Springer, Heidelberg (2011). doi:10.1007/978-3-642-24600-5_17

23. Ito, A., Wang, X., Suzuki, M., Makino, S.: Smile and laughter recognition using speech processing and face recognition from conversation video. In: 2005 International Conference on Cyberworlds (CW 2005), pp. 437–444, November 2005

24. Kim, Y., Provost, E.M.: Emotion spotting: discovering regions of evidence in audio-visual emotion expressions. In: Proceedings of the 18th ACM International Conference on Multimodal Interaction, ICMI 2016, New York, NY, USA, pp. 92–99. ACM (2016)

25. Kohler, K.J.: "Speech-smile", "speech-laugh", "laughter" and their sequencing in dialogic interaction. Phonetica **65**(1–2), 1–18 (2008)

26. Kominek, J., Black, A.W.: The CMU arctic speech databases. In: Fifth ISCA Workshop on Speech Synthesis (2004)

27. Koolagudi, S.G., Rao, K.S.: Emotion recognition from speech: a review. Int. J. Speech Technol. **15**(2), 99–117 (2012)
28. Kraut, R.E., Johnston, R.E.: Social and emotional messages of smiling: an ethological approach. J. Pers. Soc. Psychol. **37**(9), 1539 (1979)
29. Lasarcyk, E., Trouvain, J.: Spread lips+ raised larynx+ higher f0= Smiled Speech?-an articulatory synthesis approach. In: Proceedings of ISSP (2008)
30. Laskowski, K., Burger, S.: Analysis of the occurrence of laughter in meetings. In: Proceedings of the 8th Annual Conference of the International Speech Communication Association (Interspeech 2007), Antwerp, Belgium, pp. 1258–1261, 27–31 August 2007
31. Bradley, M.M., Greenwald, M.K., Petry, M.C., Lang, P.J.: Remembering pictures: pleasure and arousal in memory. J. Exp. Psychol. Learn. Mem. Cogn. **18**, 379 (1992)
32. McKeown, G., Curran, W.: The relationship between laughter intensity and perceived humour. In: Proceedings of the 4th Interdisciplinary Workshop on Laughter and Other Non-verbal Vocalisations in Speech, pp. 27–29 (2015)
33. Ming, H., Huang, D., Xie, L., Wu, J., Dong, M., Li, H.: Deep bidirectional LSTM modeling of timbre and prosody for emotional voice conversion. In: 17th Annual Conference of the International Speech Communication Association, Interspeech 2016, 8–12 September 2016, San Francisco, CA, USA, pp. 2453–2457 (2016)
34. Nwokah, E.E., Hsu, H.C., Davies, P., Fogel, A.: The integration of laughter and speech in vocal communicationa dynamic systems perspective. J. Speech Lang. Hear. Res. **42**(4), 880–894 (1999)
35. Oh, J., Wang, G.: Laughter modulation: from speech to speech-laugh. In: INTER-SPEECH, pp. 754–755 (2013)
36. Pickering, L., Corduas, M., Eisterhold, J., Seifried, B., Eggleston, A., Attardo, S.: Prosodic markers of saliency in humorous narratives. Discourse process. **46**(6), 517–540 (2009)
37. Provine, R.R.: Laughter punctuates speech: linguistic, social and gender contexts of laughter. Ethology **95**(4), 291–298 (1993)
38. Robson, J., Janet, B.: Hearing smiles-perceptual, acoustic and production aspects of labial spreading. In: XIVth Proceedings of the XIVth International Congress of Phonetic Sciences, vol. 1, pp. 219–222. International Congress of Phonetic Sciences (1999)
39. Sjölander, K.: The Snack Sound Toolkit [computer program webpage] (consulted on September, 2014). http://www.speech.kth.se/snack/
40. Tartter, V.: Happy talk: perceptual and acoustic effects of smiling on speech. Percept. Psychophys. **27**(1), 24–27 (1980)
41. Tartter, V.C., Braun, D.: Hearing smiles and frowns in normal and whisper registers. J. Acoust. Soc. Am. **96**(4), 2101–2107 (1994)
42. Torre, I.: Production and perception of smiling voice. In: Proceedings of the First Postgraduate and Academic Researchers in Linguistics at York (PARLAY 2013), pp. 100–117 (2014)
43. Trouvain, J.: Phonetic aspects of "speech laughs". In: Oralité et Gestualité: Actes du colloque ORAGE, Aix-en-Provence. L'Harmattan, Paris, pp. 634–639 (2001)
44. Young, S.J., Young, S.: The HTK hidden Markov model toolkit: design and philosophy. In: Entropic Cambridge Research Laboratory, Ltd. (1994)
45. Zen, H., Nose, T., Yamagishi, J., Sako, S., Masuko, T., Black, A., Tokuda, K.: The HMM-based speech synthesis system (HTS) version 2.0. In: Proceeding 6th ISCA Workshop on Speech Synthesis (SSW-6), August 2007
46. Zen, H., Tokuda, K., Black, A.W.: Statistical parametric speech synthesis. Speech Commun. **51**(11), 1039–1064 (2009)

# Lexical Emphasis Detection in Spoken French Using F-BANKs and Neural Networks

Abdelwahab Heba[1,2(✉)], Thomas Pellegrini[2], Tom Jorquera[1],
Régine André-Obrecht[2], and Jean-Pierre Lorré[1]

[1] Linagora, Toulouse, France
{aheba,tjorquera,jplorre}@linagora.com
[2] IRIT, Université de Toulouse, Toulouse, France
{aheba,pellegri,obrecht}@irit.fr

**Abstract.** Expressiveness and non-verbal information in speech are active research topics in speech processing. In this work, we are interested in detecting emphasis at word-level as a mean to identify what are the focus words in a given utterance. We compare several machine learning techniques (Linear Discriminant Analysis, Support Vector Machines, Neural Networks) for this task carried out on SIWIS, a French speech synthesis database. Our approach consists first in aligning the spoken words to the speech signal and second to feed classifier with filter bank coefficients in order to take a binary decision at word-level: neutral/emphasized. Evaluation results show that a three-layer neural network performed best with a 93% accuracy.

**Keywords:** Emphasized content recognition · Non verbal information in speech · SIWIS French speech synthesis database

## 1 Introduction

Speech in human communication is not only about the explicit message conveyed or the meaning of words, but also includes information, intentionally or not, which are expressed through nonverbal behaviors. Verbal and non-verbal information shape our interactions with others [4]. In [25], for instance, an appropriate use of emphasis was shown to improve the overall perception of synthesized speech. Word-level *emphasis* is considered as an important form of expressiveness in the speech synthesis field with the objective of drawing the listener attention on specific pieces of information.

A speech utterance may convey different meanings according to intonation. Such ambiguities can be clarified by emphasizing some words in different positions in a given utterance. Automatically detecting emphasized content may be useful in spoken language processing: localizing emphasized words may help speech understanding modules, in particular in semantic focus identification [15].

In speech production, various processes occur at word, sentence, or larger chunk levels. According to [6], the tonal variation, defined by pitch variation,

© Springer International Publishing AG 2017
N. Camelin et al. (Eds.): SLSP 2017, LNAI 10583, pp. 241–249, 2017.
DOI: 10.1007/978-3-319-68456-7_20

is considered as a type of pronunciation variation at the suprasegmental level. Generally, systems for automatic classification of accented words use prosody, and typically use combination of suprasegmental features such as duration, pitch, and intensity features [5,14,17,20,24]. Emphasis cues found in natural speech are more vague and heavily affected by suprasegmental features. Compared to the intensity, pitch and duration are more insensitive to the channel effects such as the distance between the speaker and the microphone. Furthermore, rather than intensity, changes in vocal loudness also affect features such as spectral balance, spectral emphasis or spectral tilt, which were explored in the detection of prominent words [3], focal accent [9], stressed and unstressed syllables [18,19, 22]. Indeed, these measures were also generally found to be more reliable than intensity.

In this paper, a statistical approach that models and detects word-level emphasis patterns is investigated. Related works are dedicated to the detection of lexical stress and pitch accent detection, in particular for Computer-Assisted Language Learning [12,13,21,27,28]. The present study differs from these works from the fact that we target at detecting acoustic emphasis at lexical level and in native speech. We plan to detect emphasis at word-level as a first step for future applications we would like to address. In particular, we would like to study if keyword detection in speech transcripts could be improved using a measure of emphasis as an additional piece of information.

Our methods consists first in aligning the speech signal to the spoken words, second in classifying each word segment as emphasized or neutral using filter-bank coefficients (F-BANKs) as input to a classifier. These acoustic features measure the energy from a number of frequency bands and take time dynamics into account. Furthermore, our preliminary experiments showed that F-BANKs outperform the use of single pitch variations (F0). We compare several types of classifiers for this task. As will be reported in this paper, neural networks performed the best.

The present article is structured as follows. Section 2 describes our methodology for word-level emphasis detection, including feature extraction and model description. In Sect. 3, we present the SIWIS French speech synthesis database, then we report a comparison of approaches and analyze the classification results.

## 2   Method

Figure 1 illustrates the global system schema for an example sentence: "ce FICHIER facilitera principalement la recherche [...]" ("this FILE will mainly ease the search for [...]"). In this sentence, the word "FICHIER" (FILE) was emphasized by the speaker. As a first step, a word alignment is carried out, which automatically aligns the expected text to the audio speech signal. Then, low-level acoustic features described hereafter are extracted and fed to a binary classifier that takes decisions on the emphasized/neutral pronunciations at word-level.

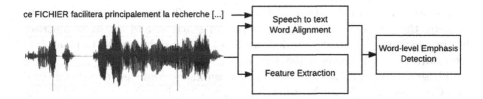

**Fig. 1.** Word-level emphasis detection

## 2.1   Word Alignment

We adopt a standard approach used in speech recognition called the time alignment procedure. This procedure is accomplished using supervised phone-based
recognition and produces phone-by-phone time markings, which are reduced to
a word-by-word format involving the following steps [23]:

- create a word-level grammar from the orthographic transcription (read
  speech);
- extract acoustic features from the speech signal;
- associate a phone transcription to each word, either extracting it from our
  pronunciation lexicon or generating them automatically with a grapheme-to-
  phoneme model; several pronunciations may be associated to a given word;
- perform the word alignment;
- extract the time markings from the aligned word segments.

We used in-house acoustical models trained with the Kaldi Speech Recognition Toolkit [16]. They were trained with the ESTER corpus [8] which consists
of 90 h of French broadcast news speech, each broadcast session contains from
20 to 40 min of spontaneous speech. Non-speech sounds, such as breath noises
and laughter are indicated in the transcriptions and we explicitly modeled them.

We followed a standard Kaldi recipe to train the models to obtain triphone
Gaussian Mixture Models/Hidden Markov Models, on 39 static, delta, and delta-
delta Mel-frequency cepstral coefficients, with LDA-MLLT and Speaker Adaptive Training (SAT). Finally, we obtain triphone with about 150 k Gaussian
mixtures and 21.2 k HMM states.

Regarding the pronunciation lexicon, we used the 105 k entry CMU-Sphinx
French dictionary. For out-of-vocabulary words, pronunciations were derived
from a grapheme-to-phoneme tool trained over the CMU-Sphinx lexicon [2].
This concerned a set of 471 words over the 33,628 different word types contained
in the SIWIS corpus used in this work.

## 2.2   Features

As input to the emphasis/neutral classifiers, we use 26 log filter-bank coefficients
(F-BANKs) extracted on 25 ms duration frames with a hop size of 10 ms. Different numbers of filter bands were tested, but the set of 26 static F-BANK features
has shown to perform well.

For each word of duration $N$ frames, a $N \times 26$ matrix is extracted. We compare the use of these length varying input matrices and the same matrices but in which some context is added: we add left and right frames to reach a $1s$ total duration, which gives a $108 \times 26$ matrix for each word.

These matrices are used in two ways, as an image (dimension: $108 \times 26$) fed to a Convolutional Neural Network (CNN), or as global statistics features (one value per filter-bank coefficient along time): the minimum, maximum, mean, median, standard deviation, skewness and kurtosis (dimension: $7 \times 26$), which are expected to characterize the behavior of each F-BANK coefficient to improve the temporal modeling.

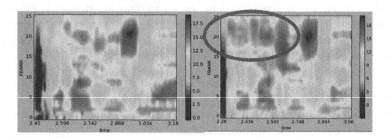

**Fig. 2.** The two figures represent the F-BANK coefficients for the word: "*courageuse-ment*" (*courageously*) with and without focus emphasis on the right and on the left, respectively. (Color figure online)

Figure 2 shows the 26 F-BANK coefficient images for both an emphasized and a neutral pronunciations of the word "courageusement" (*courageously*). We can notice that the high frequency region on the right figure (with emphasis) has higher energy values, particularly at the beginning of the word, as depicted with a blue ellipse over this area [7].

## 2.3    Models

Two types of neural networks were tested for our task: a neural network with fully-connected layers (FCNN), and a convolutional neural network (CNN).

In the case of FCNNs, the input layer is the concatenation of the global statistics on the 26 F-BANKs, i.e. a vector of size $7 \times 26 = 182$. The $k_0$ Softmax outputs estimate the emphasis of each trial. We use rectified linear (ReLU) units that have shown accurate performance in speech recognition tasks [26]. Furthermore, we experimented different number of hidden layers and different number of units. We report results with a single layer and three hidden layers each comprised of 200 units.

With CNNs, the input layer is composed of 108 frames of 26 log filter bank coefficient. Three convolution layers were respectively applied: the frequency filtering $1 \times 26$, then dynamic time filtering $108 \times 1$ and, finally, $3 \times 3$ squared filters. Followed by $2 \times 2$ down-sampling (max-pooling) layers, and produce respectively

32, 16, and 8 activation maps that serve as input parameters for three 200-unit dense hidden layers with rectified linear unit (ReLU) activation function. Finally, the output dense layer comprises 2 units with a Softmax activation function to provide a probability.

The networks were trained with the Adam optimization [11] using a cross-entropy cost function. The regularization $L^2$ was used over all hidden layers. Those models are not very deep but appear to be sufficient to get insights on emphasis detection on a small database such as SIWIS. To carry out our work, Tensorflow was used to perform the experiments on a GPU TITAN 1080 device [1].

Support Vector Machines (SVM) with a Gaussian kernel and Linear Discriminant Analysis (LDA) were also used as a baseline of our experiments.

## 3   Experiments

### 3.1   Speech Material

The SIWIS French Speech Synthesis corpus contains read speech recorded from a single native female French speaker, who reads texts selected from three different written sources: books from French novels, parliament speeches and semantically unpredictable sentences. These three written sources were divided to six subsets and serve different purposes. In our study, we only use the sentences containing emphasized words (named "part 5") and their corresponding neutral sentences (contained in parts 1 to 4). The corpus contains $1575 * 2$ sentences equivalent to 3 h 35 min duration of audio, moreover, emphasized words can be seen at different positions in the sentences (begin, middle and end). The manual annotations of emphasized phones are available in the HTS label format. Indeed, SIWIS aims at building TTS systems, investigate multiple styles, and emphasis. For more information about the corpus, the reader may refer to [10].

**Word Alignment Experiments.** Since the manual annotation provided with the SIWIS database did not allow to get time markings at word-level easily, one needed to perform word alignment as a pre-processing step for emphasis detection. We have grouped the manual time markings of emphasized phones for each word to evaluate the root mean square difference between the manual and the automatic word boundaries, according to the following formula, in which $t_m^i$ and $t_a^i$ are the manual and automatic time markers for the $i^{th}$ word, respectively, and $N$ the total number of words to be aligned:

$$\text{RMS} = \sqrt{\frac{1}{N} \sum_{i=1}^{N} (t_m^i - t_a^i)^2}$$

We worked on $1817 * 2$ words (emphasized and neutral pronunciations). The mean duration of these words is $0.372$ s ($\pm 19\%$ std), the word "*monoparentales*" (*uniparental*) has the longest duration of about 1 s, and the shortest word "*un*" (*a*) has a $0.03$ s duration.

The RMS obtained was 0.243. The smaller, the better the alignments. Nevertheless, we are aware that this value *per se* is difficult to interpret without any other reference value obtained on other speech data. Furthermore, the phone error rate is 8.7%.

A set of 1695 emphasized words were found when grouping phones (*e.g.* "temps en temps" was considered as one word by the corpus annotators), but our manual re-checking of the 1575 sentences lead to 1817 emphasized words. This increase is due to the fact that we consider each word in contiguous emphasized word sequences as several emphasized words (we consider *"temps en temps"* as three different words).

## 3.2  Classification Results

In this part, we evaluate the word-level emphasis detection method, which consists in applying the procedure shown in Fig. 1. The dataset contains $1817 * 2$ words (emphasized and neutral). The data was split into a training and a test subsets in 80%/20% proportions, respectively, and we performed a 5-fold cross-validation. We chose to keep pairs of the same words with emphasized and neutral pronunciations in the same subset, either in a training or a test fold.

In a first experiment, we focused on the global statistical features extracted over the F-BANKs. As explained previously, they were extracted with and without adding context:

- with context: all the feature matrices share the same $108 \times 26$ dimension,
- without context: the feature matrices have a variable time length according to each word: $N \times 26$.

With the different machine learning algorithms used for the classification task, we show in Table 1 that using a bit of context leads to better results in accuracy. The FCNN with 3 hidden layers with 200 units in each layer, using the ReLU activation function, obtained the best performance with a 93.4% accuracy. The variations in performance indicated in the table correspond to the variations according to the five folds used for cross-validation.

**Table 1.** Accuracy comparison between different classifier types.

| With context | No | Yes |
|---|---|---|
| FCNN (1 layer) | $81.1 \pm 1.0\%$ | $89.9 \pm 1.7\%$ |
| FCNN (3 layers) | $81.3 \pm 5.9\%$ | $\mathbf{93.4 \pm 3.3\%}$ |
| CNN | – | $90.2 \pm 1.8\%$ |
| SVM | $81.5 \pm 1.0\%$ | $92.9 \pm 2.3\%$ |
| LDA | $76.8 \pm 2.4\%$ | $89.0 \pm 3.6\%$ |

In a second experiment, we tested the use of a CNN model. As we showed in Table 1, using context allowed better performance on this task. Consequently, we used the F-BANK images with context as input to a CNN (matrices of shape $108 \times 26$, which represent the extracted F-BANKs features over 1 s of speech signal).

In order to train a CNN model, we needed a validation subset so that we used 70% for training, 10% for validation, and 20% for testing and always with 5 folds. During training, we noticed a clear overfitting of the model on the validation subset so that we used $L^2$ regularization to overcome this issue. The averaged performance on the five folds was 90.2% ($\pm$1.8%), which is not as good as the SVM (92.9%) and the FCNN (93.4%).

### 3.3   Error Analysis

In this subsection, we analyze the errors made by the best classifier, the FCNN.

A first interesting cue concerns the influence of word duration on performance. The mean duration of the wrong predicted words is about 200 ms. The false positives (emphasized word predicted as neutral) predominantly concern short words such as *"moi"* (me), *"un"* (a), *"pas"* (not), *"cela"* (that), and their mean duration is about 140 ms. On the contrary, the false negatives are longest word mostly, such as *"historien"* (historian), *"constamment"* (constantly). Mean duration of the false negatives is around 400 ms.

By listening to some word utterances incorrectly predicted as neutral, we noticed that the relative focus on these words was as obvious as other emphasized realizations. Smaller intensity values can also be observed in their corresponding spectrograms.

We also explored if there were any relation between the word positions in sentences and the detection errors. Figure 3 shows histograms counting the errors (in black the false positives, in orange the false negatives) according to the word position: at the beginning, middle, or end of a sentence. No clear impact of word position can be observed. Nevertheless, it seems that more false negatives (emphasized words predicted as neutral) occur at the beginning and end of sentences.

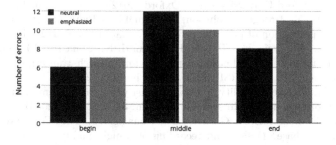

**Fig. 3.** Number of incorrect predictions according to the word position in sentences. (Color figure online)

## 4    Conclusions

In this paper, we presented an approach to detect emphasis/neutral intonation at word level specific to the French language. As a first step, a word alignment is carried out, which automatically aligns the expected text to the audio speech signal. Then, F-BANK coefficients are extracted and fed to a binary classifier that takes decisions on the emphasized/neutral decision at word-level.

Evaluation was conducted on SIWIS, a publicly available speech database, that provides read speech material in French with a sub-part manually annotated in terms of emphasis.

Several types of classifiers were tested and the best performance was obtained with a neural network comprised of three fully-connected layers of 200 units each.

As future work, we plan to exploit this system to attempt to improve our keyword extraction module applied to speech transcripts in spoken French. Additionally, we would like to use sequence modelling approach to carry out the detection of emphasized words through entire sentences.

## References

1. Abadi, M., Agarwal, A., Barham, P., Brevdo, E., Chen, Z., Citro, C., Corrado, G.S., Davis, A., Dean, J., Devin, M., et al.: Tensorflow: large-scale machine learning on heterogeneous distributed systems. arXiv preprint arXiv:1603.04467 (2016)
2. Bisani, M., Ney, H.: Joint-sequence models for grapheme-to-phoneme conversion. Speech Commun. **50**(5), 434–451 (2008)
3. Campbell, N.: Loudness, spectral tilt, and perceived prominence in dialogues. In: Proceedings ICPhS, vol. 95, pp. 676–679 (1995)
4. Campbell, N.: On the use of nonverbal speech sounds in human communication. In: Esposito, A., Faundez-Zanuy, M., Keller, E., Marinaro, M. (eds.) Verbal and Nonverbal Communication Behaviours. LNCS, vol. 4775, pp. 117–128. Springer, Heidelberg (2007). doi:10.1007/978-3-540-76442-7_11
5. Campbell, W.N.: Prosodic encoding of English speech. In: Second International Conference on Spoken Language Processing (1992)
6. Cohn, A.C., Fougeron, C., Huffman, M.K.: The Oxford Handbook of Laboratory Phonology. Oxford University Press, Oxford (2012). Sect. 6.2, pp. 103–114
7. Cole, J., Mo, Y., Hasegawa-Johnson, M.: Signal-based and expectation-based factors in the perception of prosodic prominence. Lab. Phonol. **1**(2), 425–452 (2010)
8. Galliano, S., Geoffrois, E., Mostefa, D., Choukri, K., Bonastre, J.F., Gravier, G.: The ESTER phase II evaluation campaign for the rich transcription of French broadcast news. In: INTERSPEECH, pp. 1149–1152 (2005)
9. Heldner, M.: On the reliability of overall intensity and spectral emphasis as acoustic correlates of focal accents in swedish. J. Phon. **31**(1), 39–62 (2003)
10. Honnet, P.E., Lazaridis, A., Garner, P.N., Yamagishi, J.: The SIWIS French speech synthesis database? Design and recording of a high quality French database for speech synthesis. Technical report, Idiap (2017)
11. Kingma, D., Ba, J.: Adam: a method for stochastic optimization. arXiv preprint arXiv:1412.6980 (2014)
12. Li, K., Meng, H.: Automatic lexical stress and pitch accent detection for L2 English speech using multi-distribution deep neural networks. Speech Commun. (2016)

13. Li, K., Zhang, S., Li, M., Lo, W.K., Meng, H.M.: Prominence model for prosodic features in automatic lexical stress and pitch accent detection. In: INTERSPEECH, pp. 2009–2012 (2011)
14. Narupiyakul, L., Keselj, V., Cercone, N., Sirinaovakul, B.: Focus to emphasize tone analysis for prosodic generation. Comput. Math. Appl. **55**(8), 1735–1753 (2008)
15. Noth, E., Batliner, A., Kießling, A., Kompe, R., Niemann, H.: Verbmobil: the use of prosody in the linguistic components of a speech understanding system. IEEE Trans. Speech Audio Process. **8**(5), 519–532 (2000)
16. Povey, D., Ghoshal, A., Boulianne, G., Burget, L., Glembek, O., Goel, N., Hannemann, M., Motlicek, P., Qian, Y., Schwarz, P., et al.: The kaldi speech recognition toolkit. In: IEEE 2011 Workshop on Automatic Speech Recognition and Understanding, No. EPFL-CONF-192584. IEEE Signal Processing Society (2011)
17. Shriberg, E., Stolcke, A., Hakkani-Tür, D., Tür, G.: Prosody-based automatic segmentation of speech into sentences and topics. Speech Commun. **32**(1), 127–154 (2000)
18. Sluijter, A.M., Shattuck-Hufnagel, S., Stevens, K.N., Van Heuven, V., et al.: Supralaryngeal resonance and glottal pulse shape as correlates of prosodic stress and accent in American English (1995)
19. Sluijter, A.M., Van Heuven, V.J.: Spectral balance as an acoustic correlate of linguistic stress. J. Acoust. Soc. Am. **100**(4), 2471–2485 (1996)
20. Streefkerk, B.M., Pols, L.C., Ten Bosch, L., et al.: Automatic detection of prominence (as defined by listeners' judgements) in read aloud Dutch sentences. In: ICSLP (1998)
21. Tepperman, J., Narayanan, S.: Automatic syllable stress detection using prosodic features for pronunciation evaluation of language learners. In: IEEE International Conference on Proceedings of the Acoustics, Speech, and Signal Processing (ICASSP 2005), vol. 1, pp. I–937. IEEE (2005)
22. Van Kuijk, D., Boves, L.: Acoustic characteristics of lexical stress in continuous telephone speech. Speech Commun. **27**(2), 95–111 (1999)
23. Wheatley, B., Doddington, G., Hemphill, C., Godfrey, J., Holliman, E., McDaniel, J., Fisher, D.: Robust automatic time alignment of orthographic transcriptions with unconstrained speech. In: 1992 IEEE International Conference on Acoustics, Speech, and Signal Processing, ICASSP-1992, vol. 1, pp. 533–536. IEEE (1992)
24. Wightman, C.W., Ostendorf, M.: Automatic labeling of prosodic patterns. IEEE Trans. Speech Audio Process. **2**(4), 469–481 (1994)
25. Yu, K., Mairesse, F., Young, S.: Word-level emphasis modelling in HMM-based speech synthesis. In: 2010 IEEE International Conference on Acoustics Speech and Signal Processing (ICASSP), pp. 4238–4241. IEEE (2010)
26. Zeiler, M.D., Ranzato, M., Monga, R., Mao, M., Yang, K., Le, Q.V., Nguyen, P., Senior, A., Vanhoucke, V., Dean, J., et al.: On rectified linear units for speech processing. In: 2013 IEEE International Conference on Acoustics, Speech and Signal Processing (ICASSP), pp. 3517–3521. IEEE (2013)
27. Zhao, J., Yuan, H., Liu, J., Xia, S.: Automatic lexical stress detection using acoustic features for computer assisted language learning. In: Proceedings of the APSIPA ASC, pp. 247–251 (2011)
28. Zhu, Y., Liu, J., Liu, R.: Automatic lexical stress detection for English learning. In: Proceedings of the 2003 International Conference on Natural Language Processing and Knowledge Engineering, pp. 728–733. IEEE (2003)

# Speaker Change Detection Using Binary Key Modelling with Contextual Information

Jose Patino$^{(\boxtimes)}$, Héctor Delgado, and Nicholas Evans

Department of Digital Security, EURECOM, Sophia Antipolis, France
{patino,delgado,evans}@eurecom.fr

**Abstract.** Speaker change detection can be of benefit to a number of different speech processing tasks such as speaker diarization, recognition and detection. Current solutions rely either on highly localized data or on training with large quantities of background data. While efficient, the former tend to over-segment. While more stable, the latter are less efficient and need adaptation to mis-matching data. Building on previous work in speaker recognition and diarization, this paper reports a new binary key (BK) modelling approach to speaker change detection which aims to strike a balance between efficiency and segmentation accuracy. The BK approach benefits from training using a controllable degree of contextual data, rather than relying on external background data, and is efficient in terms of computation and speaker discrimination. Experiments on a subset of the standard ETAPE database show that the new approach outperforms the current state-of-the-art methods for speaker change detection and gives an average relative improvement in segment coverage and purity of 18.71% and 4.51% respectively.

**Keywords:** Speaker identification and verification · Speaker change detection · Binary keys · Speaker diarization

## 1 Introduction and Related Work

Speaker change detection (SCD), also known as speaker turn detection and more simply speaker segmentation, aims to segment an audio stream into speaker-homogeneous segments. SCD is often a critical pre-processing step or enabling technology before other tasks such as speaker recognition, detection or diarization.

The literature shows two general approaches. On the one hand, metric-based approaches aim to determine speaker-change points by computing distances between two adjacent, sliding windows. Peaks in the resulting distance curve are thresholded in order to identify speaker changes. Bayesian information criterion (BIC) [8] and Gaussian divergence (GD) [4] are the most popular metric-based approaches. On the other hand, model-based approaches generally use off-line training using potentially large quantities of external data. Example model-based approaches utilise Gaussian mixture models (GMMs) [20], universal background

© Springer International Publishing AG 2017
N. Camelin et al. (Eds.): SLSP 2017, LNAI 10583, pp. 250–261, 2017.
DOI: 10.1007/978-3-319-68456-7_21

models (UBMs) [24], or more recent techniques such as those based on the i-vector paradigm [21,25] or deep learning [6,17,18,23].

Despite significant research effort, SCD remains challenging, with high error rates being common, particularly for short speaker turns. Since they use only small quantities of data within the local context, metric-based approaches are more efficient and domain-independent, though they tend to produce a substantial number of false alarms. This over-segmentation stems from the intra-speaker variability in short speech segments. Model-based approaches, while more stable than purely metric-based approaches, depend on external training data and hence may not generalise well in the face of out-of-domain data.

The work reported in this paper has sought to combine the merits of metric- and model-based approaches. The use of external data is avoided in order to promote domain-independence. Instead, the approach to SCD reported here uses variable quantities of contextual information for modelling, i.e. intervals of the audio recording itself. These intervals range from the whole recording to shorter intervals surrounding a hypothesized speaker change point.

The novelty of the approach lies in the use of an efficient and discriminative approach to modelling using binary keys (BKs) which have been reported previously in the context of speaker recognition [1,5], emotion recognition [3,19], speech activity detection [15] and, more extensively, speaker diarization [2,11–15,22]. In all of this work, segmentation consists in a straightforward partition of the audio stream into what are consequently non-heterogeneous speaker segments. In this case, speaker segmentation is only done implicitly at best; none of this work has investigated the discriminability of the BK approach for the task of explicit SCD. The novel contribution of this paper includes two BK-based approaches to explicit SCD. They support the flexible use of contextual information and are both shown to outperform a state-of-the-art baseline approach to SCD based on the Bayesian information criterion (BIC).

The remainder of the paper is organised as follows. Section 2 describes binary key modelling. Its application to speaker change detection is the subject of Sect. 3. Section 4 describes the experimental setup including databases, system configuration and evaluation metrics. Section 5 reports experimental results and discussion. Conclusions are presented in Sect. 6.

## 2    Binary Key Modelling

This section presents an overview of the binary key modelling technique. The material is based upon original work in speaker diarization [2] and recent enhancements introduced in [22].

### 2.1    Binary Key Background Model (KBM)

Binary key representations of acoustic features are obtained using a binary key background model (KBM). The KBM plays a similar role to the conventional

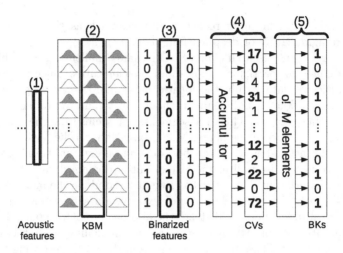

**Fig. 1.** An illustration of the BK extraction procedure based upon the comparison of acoustic features and the KBM. (Color figure online)

UBM. Just like a UBM, the KBM can be estimated using either external data [1] or the test data itself [2].

The KBM training procedure involves the selection of discriminative Gaussians from a large pool of candidates. The pool is learned from standard acoustic features extracted with a conventional frame-blocking approach and spectral analysis. The selection process uses a single-linkage criterion and a cosine distance metric to select the most discriminant and complementary Gaussians. This approach favours the selection of dissimilar candidates thereby resulting in a set of Gaussians with broad coverage of the acoustic space; closely related, redundant candidates (which likely stem from the over-sampling of homogeneous audio segments) are eliminated. The process is applied to select an arbitrary number of $N$ Gaussian candidates which then constitute the KBM. As in [22], the size of the KBM is expressed not in terms of a fixed number of components, but is defined adaptively according to the quantity of available data. The resulting KBM size is hence defined as a percentage $\alpha$ of the number of Gaussians in the original pool. The KBM is hence a decimated version of a conventional UBM where Gaussian components are selected so as to be representative of the full acoustic space in terms of coverage rather than density. Full details of the KBM training approach can be found in [13].

## 2.2   Feature Binarisation

Binarised features are obtained from the comparison of conventional acoustic features to the KBM. The process is illustrated in Fig. 1. A sequence of $n_f$ acoustic features is transformed into a binary key (BK) whose dimension $N$ is dictated by the number of components in the KBM. For each acoustic feature vector in the input sequence (labelled 1 in Fig. 1), the likelihood given each of

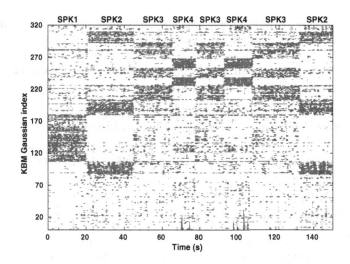

**Fig. 2.** A matrix of BKs from an arbitrary 2.5-min speech fragment from the ETAPE database. Each column of the matrix is an individual BK with $N = 320$ elements extracted according to the procedure illustrated in Fig. 1. Distinguishable BK patterns indicate distinct speakers whereas differences between them indicate speaker change points. (Color figure online)

the $N$ KBM components is computed and stored in a vector which is sorted by Gaussian index. The top $N_G$ Gaussians defined as those with the $N_G$ highest likelihoods (2 - illustrated in solid blue) are then selected and used to create binarised versions of the acoustic features (3).

This process is repeated for each frame of acoustic features thereby resulting in a binary matrix of $n_f \times N$, each column of which has $N_G$ values equal to binary 1. A row-wise addition of this matrix is then used to determine a single cumulative vector (CV) which reflects the number of times each Gaussian in the KBM was selected as a top-Gaussian (4). The final BK is obtained from the $M$ positions with highest values in the CV (5). Corresponding elements in the BK are set to binary 1 whereas others are set to 0. The BK provides a sparse, fixed length representation of a speech segment based on similarities to the acoustic space modelled by the KBM. Full details of the feature binarisation approach are also available in [13].

### 2.3   An Illustrative Example

By way of illustrating the speaker-discriminability of the BK approach, Fig. 2 depicts a sequence of BKs extracted following the procedure described in Sect. 2.2 from an arbitrary speech fragment in the order of 2.5 min duration. Each column of the matrix is a BK computed from a 1 s window with a 0.1 s shift using a KBM of size $N = 320$. Speaker labels towards the top of the plot indicate the speaker which is active during each apparent segment. The vertical axis indicates the sorted KBM Gaussian indexes whereas the horizontal axis indicates time. The intra-speaker consistency of BKs is immediately evident, as are the inter-speaker differences which indicate speaker changes or turns. The apparent diagonal component towards the upper half of the figure stems from the sequential, temporal nature with which Gaussian candidates are added to the KBM pool.

## 3    BK-based Speaker Change Detection

This section describes the application of BK modelling to speaker change detection. Two such approaches are proposed.

### 3.1    KBM Estimation

The KBM can be learned using either one of the two approaches illustrated in Figs. 3 and 4. The first is a *global-context* approach whereby the KBM is learned with data from the entire test sequence. This approach follows the algorithm described in Sect. 2.1. The second is a variant referred to as a *local-context* approach whereby the KBM is learned from a shorter context window centred on the hypothesised speaker change point. Unlike the global-context approach, the local-context approach uses all the Gaussians contained in the defined context (no selection process is performed). This approach to KBM learning enables the flexible use of acoustic context information.

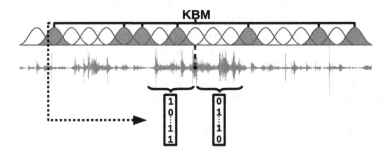

**Fig. 3.** Global-context KBM obtained through the selection of Gaussians from a global pool.

### 3.2    Speaker Change Detection

Speaker change detection (SCD) is performed using data from two smaller and non-overlapping windows, one either side of hypothesized speaker change points. BKs are extracted for each window and are compared using the Jaccard distance, defined as:

$$D_J(\mathrm{v}_a, \mathrm{v}_b) = 1 - S_J(v_a, v_b) \tag{1}$$

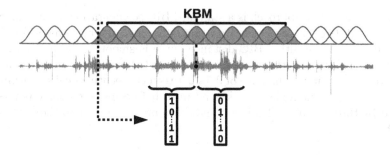

**Fig. 4.** Local-context KBM constructed using all Gaussians estimated from within a local context.

where $S_J(\mathrm{v}_a, \mathrm{v}_b)$ is the Jaccard similarity between two binary vectors $v_a$ and $v_b$ defined as:

$$S_J(\mathrm{v}_a, \mathrm{v}_b) = \frac{\sum_{i=1}^{N}(v_a[i] \wedge v_b[i])}{\sum_{i=1}^{N}(v_a[i] \vee v_b[i])} \tag{2}$$

where $N$ is the vector dimension, $\wedge$ is the boolean AND operator and $\vee$ is the boolean OR operator.

This procedure is applied sequentially to obtain a curve of window distances at regular intervals. Local peaks in this curve represent speaker change candidates. Speaker change decisions are then obtained by thresholding the distance curve using an empirically optimised threshold.

## 4   Experimental Setup

This section describes the database, the configuration of baseline and BK-based approaches to SCD and the evaluation metrics.

### 4.1   Database

In keeping with previous work on SCD, e.g. [6,17], this work was performed with the ETAPE database [16] which contains audio recordings of a set of French TV and radio shows. The TV show development partition used for all work reported here consists of 9 audio files containing excerpts of debate, entertainment and broadcast TV shows broadcast on a number of different French TV channels. Together the recordings contain in the order of 5.5 h of audio of which 3.9 h contain the speech of 61 speakers in 2304 speech segments.

### 4.2   Baseline System

Acoustic features comprise 19 static Mel-frequency cepstral coefficients (MFCCs) which are extracted from pre-emphasised audio signals using an analysis window of 25 ms with a time shift of 10 ms using a 20-channel Mel-scaled filterbank. No dynamic features are used.

The baseline SCD approach is a standard BIC segmentation algorithm [8]. It is applied with two windows of 1 s duration either side of a hypothesised speaker change point. The resulting BIC distance curve is smoothed by replacing each point with the average estimated over 1 s context. Local maxima are identified by enforcing a minimum distance of 0.5 s between consecutive peaks. Within any 0.5 s interval, only the highest peak is retained before speaker change points are selected by thresholding. This is a standard approach similar to those reported in [7,9,10].

## 4.3 Binary Key System

Acoustic features are the same as for the baseline system. Candidate Gaussians for the KBM pool are learned from windows of 2 s duration with a time shift of 1 s. The number of components in the final KBM is chosen adaptively according to a percentage $\alpha$ of the number in the initial pool. Reported below are a set of experiments used to optimise $\alpha$. The number of top Gaussians $N_G$ used for BK extraction (step 5 in Fig. 1) is set to 5 and the number of bits $M$ that are set to 1 is set to 20% of the number of KBM components $N$.

Two BKs are extracted every 0.1 s with sliding windows of 1 s duration positioned either side of the hypothesized change point. The distance between each pair of BKs is calculated using the Jaccard similarity (Sect. 3.2) and the distance curve is smoothed in the same way as for the baseline system. Speaker change points are again selected by thresholding.

## 4.4 Evaluation Metrics

SCD performance is evaluated using the approach used in [6], namely through estimates of segment coverage and purity. According to the work in [6], coverage is defined as:

$$\text{coverage}(\mathcal{R}, \mathcal{H}) = \frac{\sum_{r \in \mathcal{R}} \max_{h \in \mathcal{H}} |r \cap h|}{\sum_{r \in \mathcal{R}} |r|} \tag{3}$$

where $|r|$ is the duration of segment $r$ within the set of reference segments $\mathcal{R}$, and where $r \cap h$ is the intersection of segments $r$ and segments $h$ within the set of hypothesis segments $\mathcal{H}$. Purity is analogously defined with $\mathcal{R}$ and $\mathcal{H}$ in Eq. 3 being interchanged.

An over-segmented hypothesis (too many speaker changes) implies a high segment purity at the expense of low coverage (hypothesised segments *cover* a low percentage of reference segments). In contrast, an under-segmented hypothesis (too few speaker changes) implies the opposite, namely high coverage, but low purity. Purity and coverage are hence a classical trade-off, with the optimal algorithm configuration depending on the subsequent task.

In order to concentrate on the assessment of SCD alone, ground-truth annotations are used for speech activity detection (SAD). It is noted that the use of ground-truth SAD as a hypothesis with a single speaker delivers a ceiling coverage of 100% and a floor purity of 83%. These values can be taken as a performance reference.

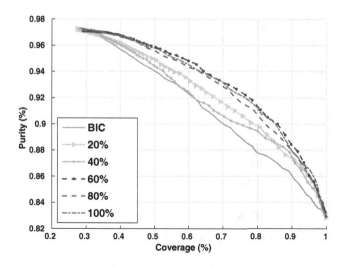

**Fig. 5.** Performance measured in segment purity and coverage using global-context KBM, obtained by varying the decision threshold $\theta$. (Color figure online)

## 5    Results

Figures 5 and 6 show plots of purity and coverage for global- and local-context KBMs respectively. Each profile shows the trade-off between the two metrics as the distance threshold $\theta$ is varied. Profiles are shown for KBMs whose size $\alpha$ is set to 20, 40, 60, 80 and 100% of the total number of original Gaussians. In both cases, the performance of the BIC baseline system is illustrated with a solid blue line.

The BK approach with global-context KBMs (Fig. 5) gives universally better performance than the baseline, even if the trend is somewhat inconsistent. This behaviour is due to the Gaussian selection process which can result in a selection of Gaussians that are not representative of certain audio segments. KBMs of larger size have inherently better potential to cover the full acoustic space and hence better potential to produce more discriminant BKs. Larger KBMs then give better performance, e.g. for $\alpha$ greater than 40%. The optimal $\alpha$ is 60%. Greater value of $\alpha$ do not necessarily give better performance. This is because the acoustic space is already fully covered and the introduction of additional Gaussians is largely redundant. The BK approach with local-context KBMs (Fig. 6) also outperforms the baseline. While the trend is consistent for lower values of coverage, across the range the optimal $\alpha$ varies between 60% and 100%.

Table 1 illustrates the variation in coverage against purity for BK-based SCD using global-context KBMs for $\alpha = 60\%$. The performance is compared to that obtained with the baseline system. The BK approach gives higher coverage at all operating points, especially for those with higher purity. Estimated using an area-under-the-curve (AUC) metric, the average relative improvement in coverage and

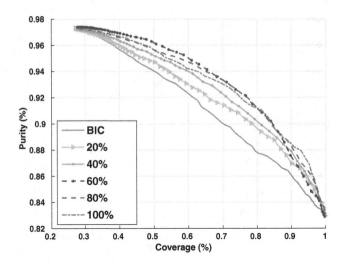

**Fig. 6.** Performance measured in segment purity and coverage using local-context KBM, obtained by varying the decision threshold $\theta$. (Color figure online)

purity across all operating points is 17.39% and 4.48%, for coverage and purity metrics respectively.

Table 2 illustrates the same analysis for BK-based SCD using local-context KBMs for $\alpha = 70$. Using the same AUC metric, the average relative improvement in coverage and purity is 18.71% and 4.51% respectively. These improvements are similar to results achieved by more advanced deep learning-based solutions such as that in [6].

It is of interest to compare the two proposed methods not only in terms of performance, but also in terms of efficiency and practical application. Even if the local-context approach slightly outperforms the global-context one, each approach can be better suited for different application modes. On one hand, in the case of offline processing (when the entire input stream is available in advance), the global-context approach is more efficient since the KBM is fix for all the process, hence allowing to compute frame-wise likelihoods only once and then reuse them for subsequent operations. However, in the local-context approach, the KBM changes over time (by using Gaussian components estimated on the contextual window around the current time point). This forces to recompute frame-wise likelihoods every time the window is shifted, therefore adding an extra computation cost. On the other hand, in online processing scenarios, the global-context approach cannot be used since the complete input stream is required in advance to train the KBM. However, the local-context approach is well suited for online applications since it only utilises local information around the current time point. In the latter case, system latency is proportional to the amount of contextual data considered.

**Table 1.** Comparative performance measured in coverage for several fixed purity values using the **global-context KBM** approach.

| Purity (%) | | 84 | 88 | 92 | 96 |
|---|---|---|---|---|---|
| Coverage (%) | BIC | 96.48 | 79.54 | 60.92 | 37.90 |
| | BK | 98.88 | 91.71 | 78.46 | 48.99 |

**Table 2.** Comparative performance measured in coverage for several fixed purity values using the **local-context KBM** approach.

| Purity (%) | | 84 | 88 | 92 | 96 |
|---|---|---|---|---|---|
| Coverage (%) | BIC | 96.48 | 79.54 | 60.92 | 37.90 |
| | BK | 98.99 | 92.86 | 77.45 | 51.51 |

## 6 Conclusions

This paper introduces a binary key (BK) solution to speaker change detection (SCD). The algorithm uses traditional acoustic features and a configurable quantity of contextual information captured through a binary key background model (KBM). Speaker-discriminative BKs are then extracted from the comparison of acoustic features to the KBM. The binarisation of acoustic features resembles a form of quantisation which helps to reduce noise and hence improve the robustness of subsequent SCD. The latter is performed by thresholding the distance between BKs extracted from two adjacent windows either side of hypothesized speaker change points. While not requiring the use of external data, two variants of the novel BK SCD algorithm are shown to outperform a baseline approach based on the classical Bayesian information criterion. Results obtained using a standard dataset show average relative improvements which compare favourably to results reported recently for more computationally demanding deep learning solutions.

**Acknowledgements.** This work was supported through funding from the Agence Nationale de la Recherche (French research funding agency) in the context of the ODESSA project (ANR-15-CE39-0010). The authors acknowledge Hervé Bredin's help in the evaluation of speaker change detection.

## References

1. Anguera, X., Bonastre, J.F.: A novel speaker binary key derived from anchor models. In: Proceedings of the INTERSPEECH, pp. 2118–2121 (2010)
2. Anguera, X., Bonastre, J.F.: Fast speaker diarization based on binary keys. In: Proceedings of the IEEE International Conference on Acoustics, Speech and Signal Processing (ICASSP), pp. 4428–4431. IEEE (2011)
3. Anguera, X., Movellan, E., Ferrarons, M.: Emotions recognition using binary fingerprints. In: Proceedings of the IberSPEECH (2012)

4. Barras, C., Zhu, X., Meignier, S., Gauvain, J.L.: Multistage speaker diarization of broadcast news. IEEE Trans. Audio Speech Lang. Process. **14**(5), 1505–1512 (2006)
5. Bonastre, J.F., Miró, X.A., Sierra, G.H., Bousquet, P.M.: Speaker modeling using local binary decisions. In: Proceedings of the INTERSPEECH, pp. 13–16 (2011)
6. Bredin, H.: Tristounet: triplet loss for speaker turn embedding. In: Proceedings of the IEEE International Conference on Acoustics, Speech and Signal Processing (ICASSP), pp. 5430–5434. IEEE (2017)
7. Cettolo, M., Vescovi, M.: Efficient audio segmentation algorithms based on the BIC. In: Proceedings of the IEEE International Conference on Acoustics, Speech and Signal Processing (ICASSP), vol. 6, pp. VI–537. IEEE (2003)
8. Chen, S., Gopalakrishnan, P.: Speaker, environment and channel change detection and clustering via the Bayesian information criterion. In: Proceedings of the DARPA Broadcast News Transcription and Understanding Workshop, vol. 8, pp. 127–132 (1998)
9. Cheng, S.S., Wang, H.M., Fu, H.C.: BIC-based speaker segmentation using divide-and-conquer strategies with application to speaker diarization. IEEE Trans. Audio Speech Lang. Process. **18**(1), 141–157 (2010)
10. Delacourt, P., Wellekens, C.J.: DISTBIC: a speaker-based segmentation for audio data indexing. Speech Commun. **32**(1), 111–126 (2000)
11. Delgado, H., Anguera, X., Fredouille, C., Serrano, J.: Improved binary key speaker diarization system. In: Proceedings of the 23rd European Signal Processing Conference (EUSIPCO), pp. 2087–2091 (2015)
12. Delgado, H., Anguera, X., Fredouille, C., Serrano, J.: Global speaker clustering towards optimal stopping criterion in binary key speaker diarization. In: Navarro Mesa, J.L., Ortega, A., Teixeira, A., Hernández Pérez, E., Quintana Morales, P., Ravelo García, A., Guerra Moreno, I., Toledano, D.T. (eds.) IberSPEECH 2014. LNCS, vol. 8854, pp. 59–68. Springer, Cham (2014). doi:10.1007/978-3-319-13623-3_7
13. Delgado, H., Anguera, X., Fredouille, C., Serrano, J.: Fast single-and cross-show speaker diarization using binary key speaker modeling. IEEE Trans. Audio Speech Lang. Process. **23**(12), 2286–2297 (2015)
14. Delgado, H., Anguera, X., Fredouille, C., Serrano, J.: Novel clustering selection criterion for fast binary key speaker diarization. In: Proceedings of the INTERSPEECH, pp. 3091–3095, Dresden, Germany (2015)
15. Delgado, H., Fredouille, C., Serrano, J.: Towards a complete binary key system for the speaker diarization task. In: Proceedings of the INTERSPEECH, pp. 572–576 (2014)
16. Gravier, G., Adda, G., Paulson, N., Carré, M., Giraudel, A., Galibert, O.: The ETAPE corpus for the evaluation of speech-based TV content processing in the French language. In: LREC-Eighth International Conference on Language Resources and Evaluation, p. na (2012)
17. Gupta, V.: Speaker change point detection using deep neural nets. In: Proceedings of the IEEE International Conference on Acoustics, Speech and Signal Processing (ICASSP), pp. 4420–4424. IEEE (2015)
18. Hrúz, M., Zajíc, Z.: Convolutional neural network for speaker change detection in telephone speaker diarization system. In: Proceedings of the IEEE International Conference on Acoustics, Speech and Signal Processing (ICASSP), pp. 4945–4949. IEEE (2017)

19. Luque, J., Anguera, X.: On the modeling of natural vocal emotion expressions through binary key. In: Proceedings of the 22nd European Signal Processing Conference (EUSIPCO), pp. 1562–1566 (2014)

20. Malegaonkar, A.S., Ariyaeeinia, A.M., Sivakumaran, P.: Efficient speaker change detection using adapted Gaussian mixture models. IEEE Trans. Audio Speech Lang. Process. **15**(6), 1859–1869 (2007)

21. Neri, L.V., Pinheiro, H.N., Ren, T.I., Cavalcanti, G.D.D.C., Adami, A.G.: Speaker segmentation using i-vector in meetings domain. In: Proceedings of the IEEE International Conference on Acoustics, Speech and Signal Processing (ICASSP), pp. 5455–5459. IEEE (2017)

22. Patino, J., Delgado, H., Evans, N., Anguera, X.: EURECOM submission to the Albayzin 2016 speaker diarization evaluation. In: Proceedings of the IberSPEECH (2016)

23. Wang, R., Gu, M., Li, L., Xu, M., Zheng, T.F.: Speaker segmentation using deep speaker vectors for fast speaker change scenarios. In: Proceedings of the IEEE International Conference on Acoustics, Speech and Signal Processing (ICASSP), pp. 5420–5424. IEEE (2017)

24. Wu, T.Y., Lu, L., Chen, K., Zhang, H.: Universal background models for real-time speaker change detection. In: MMM, pp. 135–149 (2003)

25. Zajíc, Z., Kunešová, M., Radová, V.: Investigation of segmentation in i-vector based speaker diarization of telephone speech. In: Ronzhin, A., Potapova, R., Németh, G. (eds.) SPECOM 2016. LNCS, vol. 9811, pp. 411–418. Springer, Cham (2016). doi:10.1007/978-3-319-43958-7_49

# Perception of Expressivity in TTS: Linguistics, Phonetics or Prosody?

Marie Tahon[(✉)], Gwénolé Lecorvé, Damien Lolive, and Raheel Qader

IRISA/University of Rennes 1, 6 Rue de Kérampont, 22300 Lannion, France
{marie.tahon,gwenole.lecorve,damien.lolive,raheel.qader}@irisa.fr
https://www-expression.irisa.fr/

**Abstract.** Actually a lot of work on expressive speech focus on acoustic models and prosody variations. However, in expressive Text-to-Speech (TTS) systems, prosody generation strongly relies on the sequence of phonemes to be expressed and also to the words below these phonemes. Consequently, linguistic and phonetic cues play a significant role in the perception of expressivity. In previous works, we proposed a statistical corpus-specific framework which adapts phonemes derived from an automatic phonetizer to the phonemes as labelled in the TTS speech corpus. This framework allows to synthesize good quality but neutral speech samples. The present study goes further in the generation of expressive speech by predicting not only corpus-specific but also expressive pronunciation. It also investigates the shared impacts of linguistics, phonetics and prosody, these impacts being evaluated through different French neutral and expressive speech collected with different speaking styles and linguistic content and expressed under diverse emotional states. Perception tests show that expressivity is more easily perceived when linguistics, phonetics and prosody are consistent. Linguistics seems to be the strongest cue in the perception of expressivity, but phonetics greatly improves expressiveness when combined with and adequate prosody.

**Keywords:** Expressive speech synthesis · Perception · Linguistics · Phonetics · Prosody · Pronunciation adaptation

## 1 Introduction

Speech synthesis usually consists of the conversion of a written text to a speech sound, also named as Text-To-Speech (TTS) process. While TTS has reached a fairly acceptable level of quality and intelligibility on neutral speech in the last decades, the lack of expressivity is often criticized, as it usually sounds different from spontaneous human conversations [17]. The shift of TTS from read to spontaneous and expressive speech would greatly help to reproduce situations where the synthetic voice talks with a user, for instance in the field of human-machine interactions. As a result, there is a crucial need not only for just intelligible speech carrying linguistic information, but also for expressive speech. The present study investigates affective speech for TTS and finds applications in

© Springer International Publishing AG 2017
N. Camelin et al. (Eds.): SLSP 2017, LNAI 10583, pp. 262–274, 2017.
DOI: 10.1007/978-3-319-68456-7_22

many domains such as education and entertainment. According to Campbell [5], the main challenge in expressive TTS is to find the adequation of affective states in the input and the realization of prosodic characteristics to express them in the output speech. Undoubtedly, prosody is an important cue in the perception of expressivity in speech. However, in the framework of expressive speech synthesis, prosody is highly related to the sequences of phonemes to be expressed and to the words below these phonemes. Therefore, lexical and phonetic cues also play a significant role in the perception of expressivity. Here expressivity is reduced to the expression of emotional states. The present work investigates the shared impacts of linguistics, phonetics and prosody in the perception of quality and expressivity of speech samples generated with a TTS system.

Three main data-driven approaches coexist for TTS [17], unit selection, statistical parametric and Deep Neural Networks systems, all of them require variable affective speech data of good audio quality. In that sense, there is a growing interest for audio books as shown by the Blizzard Challenge 2016 [13]. They are very interesting for TTS as they contain both a text of interest, with different characters, speaking styles and emotions, and the corresponding audio signal [6]. In the present study, three speech corpora with different levels of expressivity are used, one being collected from an audio book, another from high quality speech for synthesis, and the last from TV commentaries. A solution to introduce some flexibility in TTS consists in training acoustic models on speech produced with different speaking styles or in adapting models to specific voices or prosodic styles [7,12]. Expressivity can also be controlled in symbolic terms (diphone identity, position, etc.) [1] or in prosodic terms (fundamental frequency, energy, duration) [18]. Those elements are usually used in the speech synthesizer directly in the cost function or in the construction of the acoustic model [15]. In addition, voice transformation techniques can be applied to synthetic samples [9,21]. The TTS used in this paper is a neutral unit selection system [1], expressivity being controlled with different types of text, pronunciation and speech databases.

While a lot of work on expressive speech focus on acoustic models and prosody variations, very few of them deal with pronunciation. A perception study [4] showed that samples synthesized with the *realized* pronunciation were preferred to those synthesized with the pronunciation derived from an automatic phonetizer – the *canonical* pronunciation. In previous works, we proposed a statistical framework which adapts the *canonical* pronunciation to a target pronunciation. This framework allows to predict phoneme sequences by using Conditional Random Fields (CRF) models trained on lexical, phonological, articulatory and prosodic features. The framework was used to generate spontaneous English pronunciations and the results show that a trade-off between quality and intelligibility is necessary [16]. It was also used to predict a corpus-specific pronunciation, i.e. a pronunciation adapted to the TTS voice corpus, thus conducting to a significant improvement of the overall quality of synthesized speech [19,20]. In the work realized in [19], we manage to synthesize good quality speech samples on a neutral voice. The present study goes further in the generation of expressive speech samples by predicting not only a corpus-specific pronunciation but also an expressive pronunciation. We also investigate the shared impacts of linguistics, phonetics and prosody on the perception of expressivity, as well as the best configuration towards an expressive synthesis system. In the remainder, an overview of the general process is presented in Sect. 2. Speech, pronunciation and text databases are detailed in Sect. 3. Features and models are exposed in Sect. 4. Finally, Sect. 5 presents the perception test protocol and results.

## 2   General Overview

The process used in this study has been set up in order to study the impact of linguistic, phonetic and prosodic expressive variations on the perception of expressivity. Expressive variations of linguistic and prosodic features are easily managed through different corpora, whereas expressive pronunciation variants need to be generated with a pronunciation adaptation framework as illustrated in Fig. 1. As detailed in [19], the goal of pronunciation adaptation is to reduce the differences between phonemes derived from a phonetizer (*canonical*) and phonemes as labelled in the pronunciation corpus (*realized*). To do so, the proposed method is to train CRFs phoneme-to-phoneme (P2P) models which predict adapted phonemes from *canonical* ones. To go further towards expressive pronunciation generation, this study combines two P2P models. The voice-specific pronunciation P2P model is trained on the TTS speech corpus with *canonical* phonemes and predicts neutral voice-specific (VoSpe) phonemes. The expressive

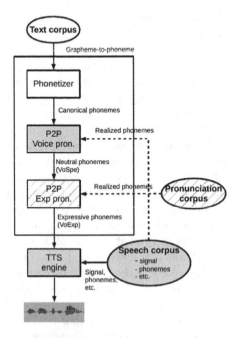

**Fig. 1.** General overview. Databases are symbolized with ellipses.

pronunciation P2P model is trained on the pronunciation corpus with VoSpe phonemes and predicts voice-specific and expressive (VoExp) phonemes. One could argue that adaptation could have been realized without any voice-specific adaptation. Such a method could probably improve the expressiveness of the synthesized speech samples, but inconsistencies between speech and pronunciation corpora would remain, thus lowering the TTS quality. Overcoming the disadvantages of the aforementioned method, the protocol illustrated in Fig. 1 was designed to generate expressive speech samples of good quality. Adapted VoSpe and VoExp pronunciations are evaluated with expressive and with neutral utterances. Such a protocol is of interest in evaluating the influence of words in the perception of expressivity. Finally, three different speech corpora are used to create TTS voices, each one having its own prosodic characteristics.

# 3   Databases

This section presents the databases used in the following experiments, which characteristics are given in Table 1. Three speech corpora are used for voice-specific pronunciation modelling and in the TTS voice creation. An emotional pronunciation corpus is used for expressive pronunciation modelling. Finally, utterances from two subcorpora of the aforementioned databases are used to evaluate the influence of linguistics.

**Table 1.** Characteristics of each database. Mean (standard deviations) of fundamental frequency $(F_0)$ in semitone and speech rate $(SR)$ in syllable per seconds are given.

| Corpus | Expressivity | # utt. | Dur. | # phon. | $F_0$(st) | $SR$ |
|---|---|---|---|---|---|---|
| Speech corpora | | | | | | |
| *Telecom* - train 70% | Neutral | 5044 | 4h51' | 151,945 | 89 (2.7) | 4.7 (2.1) |
| *Audiobook* | Moderate | 3339 | 10h45' | 379,897 | 77 (3.2) | 6.3 (1.2) |
| *Commentary* | Expressive | 1631 | 5h25' | 173,858 | 85 (5.0) | 6.0 (1.7) |
| Pronunciation corpus | | | | | | |
| *Expressive* | Expressive | 6 × 47 | 0h41' | 16,248 | 84 (7.1) | 6.3 (1.8) |
| Text corpora | | | | | | |
| *Telecom* - eval 30% | Neutral | 2162 | 2h04' | 64,960 | | |
| *Expressive* | Expressive | 6 × 47 | 0h41' | 16,248 | | |

## 3.1   Speech Corpora

Speech corpora are used to train voice-specific pronunciation P2P models. They are also used to create TTS voices.

**Telecom corpus** features a French speech corpus dedicated to interactive vocal servers. As such, this corpus covers all diphonemes present in French. Phonemes and non speech sounds have been manually corrected. The *Telecom* corpus has been randomly split in two subsets. 70% are left for training purposes and the remaining 30% are kept for evaluations. This corpus comprises most words used in the telecommunication field. Utterances are neutral such as: *"On nous invite à visiter les églises de onze heures à trois heures."* ("We are pleased to visit the churches from 11 a.m. to 3 p.m."). It features a neutral female voice which pitch is normal (170 Hz, 89 st) and pitch standard deviation is quite small. The speech rate is in the normal range according to [10]. According to these prosodic characteristics, this corpus can be considered as little expressive.

**Audiobook corpus** is extracted from a French audio book [11]. The reader is a professional male French actor with a very low voice (91 Hz, 77 st). The book "Albertine disparue" was written by the French author Marcel Proust. Data was automatically segmented and annotated (phonemes, syllables, non-speech sounds and syntactic features) using the process described in [3]. Since the main topic is an analysis of love pain, the tone is mainly serious, as this example suggests: *"Alors je pleurais ce que je voyais si bien et qui, la veille, n'était pour moi que néant."* ("Then I was crying what I was seeing so well, and what, before, was for me just a void"). Compared to the *Telecom* corpus, pitch variations are more important, speech rate is also faster. This corpus is then considered as moderately expressive.

**Commentary corpus** is extracted from commentaries which precede science -fiction French series. The male speaker presents the synopsis of each episode in a very expressive way. Data was also automatically annotated using the process described in [3]. The commentator often calls out to the audience, and gets it interested in viewing the episode. For example, he says: *"Qu'avez-vous pensé de ce géant qui s'avère être une tour humaine formée par trois acrobates ? Réalité, ou fiction ?"* ("What did you think of this giant who turns out to be a human tour made of three acrobats? Reality or fiction?"). In this corpus, the global pitch is quite high (136 Hz, 85 st) for a male speaker, and the variations are important, revealing a large diversity in prosody. The speech rate and its variations are at the same level as in the *Audiobook* corpus. For these reasons, this corpus is the most expressive.

### 3.2 Pronunciation Corpus

The pronunciation expressive corpus is used to train expressive pronunciation models for each emotion.

**Expressive corpus** has been collected for expressive synthesis purposes. A male speaker recorded French sentences in various emotion styles with a high activation degree: anger, disgust, joy, fear, surprise and sadness. The speech material has been aligned to the corresponding text for prosodic analysis and alignment has been manually corrected [2]. The linguistic content of the sentences is informal and emotionally coloured language, as for example in the expression of anger: *"Oh! Merde! Il y a un bouchon, c'est pas vrai, on va encore louper le début du film !"* ("Oh! Shit! There is traffic, I can't believe it, we are going to miss the beginning of the film again!"). The choice of such sentences greatly helps the speaker to simulate an emotion while reading. Each of 6 expressive pronunciation model will be trained and evaluated in cross-validation using the 47 available utterances per emotion. Unsurprisingly, pitch and energy are highly variable throughout the corpus and the speech rate is as fast as in *Audiobook*. The *Expressive* corpus offers the opportunity to study expressed pronunciations for different emotional states, this aspect being left for further studies.

### 3.3   Text Corpora

120 utterances were randomly selected from *Telecom-eval* and *Expressive* corpora by sub-sampling the corpus according to the Phoneme Error Rate (PER) between *canonical* and *realized* pronunciations. These utterances will be used as neutral and expressive input text to evaluate the models. The 60 utterances selected from *Telecom-eval* differ from the utterances used to train the voice-specific pronunciation model and to create the TTS voice. On the contrary, due to the small size of the corpus, the 60 utterances selected from *Expressive* corpus are also used to train the expressive pronunciation model. Therefore, this corpus has been split in 5 folds and managed under cross-validation conditions.

## 4   P2P Models

Voice-specific and expressive phoneme sequences are predicted using CRFs as pronunciation models. This section describes the features, then voice-specific and expressive pronunciation CRF models.

### 4.1   Features

CRFs are trained using the Wapiti toolkit [14] with the default BFGS algorithm on a speech or pronunciation corpus with different features. Precisely, as detailed in [19], four groups of features were investigated: 26 lexical, 17 phonological, 9 articulatory and 8 prosodic features. Relevant features for pronunciation adaptation are then automatically selected according to a cross-validation protocol. Prosodic features are extracted in an oracle way, i.e., directly from the recorded utterances of the speech corpus. In the future, a prosodic model could be included in the synthesizer, thus making prosodic features available. Such a protocol allows to know to what extent prosody affects pronunciation models.

### 4.2   P2P Voice-Specific Pronunciation Model

The voice pronunciation model adapts *canonical* phonemes to phonemes as *realized* in the *speech corpus*. In previous work [19,20], we have presented the training process of a P2P voice-specific model with the corpus *Telecom*. Table 2 shows the distribution of selected features within groups. Feature selection performed on the voice-specific model (VoSpe) excludes articulatory features. In the end,

**Table 2.** Number of selected features within groups with a $W_0$ phoneme window. Feature selection results are presented for adaptation to the voice pronunciation on *Telecom* (VoSpe) then to the expressive pronunciations on the *Expressive* corpus for each emotion (VoExp).

| Feature group (# feat.) | VoSpe | VoExp | | | | | |
|---|---|---|---|---|---|---|---|
| | | Anger | Disgust | Joy | Fear | Surprise | Sadness |
| Lexical (26) | 2 | 3 | 5 | 2 | 4 | 5 | 3 |
| Phonological (17) | 7 | 5 | 7 | 6 | 6 | 3 | 3 |
| Articulatory (9) | 0 | 3 | 4 | 1 | 1 | 2 | 2 |
| Prosodic (8) | 0 (removed) | 6 | 6 | 7 | 5 | 7 | 8 |
| Total (52) | 9 | 17 | 22 | 16 | 16 | 17 | 16 |

a set of 15 features including lexical, phonological and prosodic features with a 5-phoneme window (two phonemes surrounding the current phoneme, named as $W_2$) were automatically selected. An optimal PER of 2.7% (baseline 11.2%) was reached when training models on the data. However, a perception test has shown that speech samples generated with the 15-feature set were perceived with the same or a lower quality than samples generated with a 9-feature set excluding prosodic features. Since prosodic features are not generated from text yet but are estimated in an oracle way, only the selected lexical and phonological 9-feature set is used. For the same reason, a 5-phoneme window ($W_2$) is applied to train voice-specific P2P models. The corpora used for training voice-specific pronunciation models are the three speech corpora described in Sect. 3.1.

### 4.3    P2P Expressive Pronunciation Model

The expressive pronunciation model adapts VoSpe phonemes which are predicted with the voice-specific pronunciation model described before, to phonemes as labelled in the *Expressive* pronunciation corpus. More precisely, 6 pronunciation models are trained for each emotion contained in the *Expressive* corpus. A greedy feature selection process is performed in 5-folds cross-validation conditions for each emotion separately starting from at least VoSpe phonemes and target *realized* expressive phonemes, then adding features one by one. Features are selected separately in the four groups and with three window sizes: $W_0$, $W_1$ and $W_2$. The window $W_0$ has shown to reach the best PER.

The number of selected features and its distribution within groups differ across emotions, as reported in Table 2, while applying $W_0$ on the phoneme sequence. According to Table 2, whatever the emotion, most of the prosodic features seem to be highly relevant for expressive pronunciation modeling, while articulatory features are not. Very few lexical features were selected. Among them, word and stem are often selected, while word (disgust) and Part-of-Speech (fear) context and frequency (surprise) were selected for some emotions only. The case of sadness is interesting as all prosodic features were selected, and very few features from other groups are included in the final subset.

**Table 3.** Average [standard deviation] PER (%) over emotions between realized and predicted phonemes, with neutral and expressive text.

| Speech corpus → | Telecom | | Audiobook | | Comment. | |
|---|---|---|---|---|---|---|
| Text corpus → | Neu. | Exp. | Neu. | Exp. | Neu. | Exp. |
| VoSpe | 3.0 [0] | 16.0 [0.6] | 6.9 [0] | 15.6 [1.1] | 6.4 [0] | 16.3 [1.1] |
| VoExp ($W_0$) | 8.0 [0.8] | 12.5 [2.3] | 10.0 [0.5] | 12.7 [0.7] | 9.9 [0.5] | 13.1 [1.5] |
| VoExp ($W_0$ + sel. feat.) | 9.0 [0.3] | 5.1 [1.2] | 10.0 [0.6] | 6.0 [1.0] | 9.9 [0.5] | 5.1 [0.8] |

## 4.4 Objective Evaluation of the Models

*Canonical* phonemes extracted automatically from neutral (in cross-validation conditions) and expressive (in cross-corpus conditions) sentences are used as inputs to evaluate the models. These are evaluated in terms of PER between realized expressive or neutral phonemes and canonical or predicted phonemes. The results are reported in Table 3. In the case of neutral utterances in input, no realized expressive pronunciation is available since the corpus *Telecom* was designed for neutral speech only. On the contrary, in the case of expressive utterances in input, no realized neutral pronunciation is available since the corpus *Expressive* was designed for emotional data collection. Therefore, with neutral text in input, the PER obtained with VoSpe is smaller than the one obtained with VoExp, and with expressive text in input, the PER obtained with VoSpe is higher than the one obtained with VoExp. Both results being obtained whatever the *Voice* corpus, as shown in Table 3.

The combination of voice and expressive pronunciation models – which outputs VoExp phonemes – helps in reducing phonemic differences between the predicted and the realized expressive sequences when text is expressive. Furthermore, the addition of selected features is not of significant interest when the input text is neutral, but is when text is expressive. Average PER improvement reaches 6.7 pp. with *Audiobook*, 7.4 pp. with *Telecom* and 8.0 pp. with *Commentary* with the $W_0$+ selected features. A perception test will be able to evaluate the models in a similar way for both expressive and neutral text in input.

## 4.5 Example

Table 4 illustrates some differences which occur between a neutral and an expressive pronunciation. In this example, the *realized* pronunciation comes from *Expressive* corpus. Canonical pronunciation is adapted to the pronunciation of the speech corpus (VoSpe) then to the emotional pronunciation corpus (VoExp). Some deletions appear to be characteristic of an expressive pronunciation in French, for example deletion of the vowel /ø/ or the liquid /l/. Also the liaison /z/ is missing in the three adapted VoExp pronunciations as well as in the *realized* pronunciation. This example also presents an interesting case: the *canonical* pronunciation /ʒ ø n/ is subsituted by /ʒ/ in expressive pronunciations (see also /i n ø/. This is a regular case in French: first the deletion of /ø/ gives /ʒ n/ and the deletion of the negative *ne* gives the final pronunciation.

**Table 4.** Example of pronunciation adaptations. The input text is *Je suis dégoûtée, ils ne m'ont pas embauchée parce que je n'ai pas le diplôme.* "I am gutted, they did not hire me because I do not have the diploma:"

| | | |
|---|---|---|
| Cano | ʒ ø sɥidegutei lnø mɔ̃pa z ɑ̃boʃepaʁs ə køʒ ø n ɛ pal ø diplom ə |
| Real | ʒ - sɥidegutei - - - mɔ̃pa - ɑ̃boʃepaʁs - køʒ - - e pal - diplom ə |
| *Telecom* | |
| VoSpe | ʒ ø sɥidegutei lnø mɔ̃pa z ɑ̃boʃepaʁs - køʒ ø n e pal ø diplom - |
| VoExp | ʒ - sɥidegutei - - - mɔ̃pa - ɑ̃boʃepaʁs - køʒ - - e pal - diplom - |
| *Audiobook* | |
| VoSpe | ʒ ø sɥidegutei ln- mɔ̃pa z ɑ̃boʃepaʁs - køʒ - n ɛ pal ø diplom - |
| VoExp | ʒ - sɥidegutei - - - mɔ̃pa - ɑ̃boʃepaʁs - køʒ - - - pal - diplom - |
| *Commentary* | |
| VoSpe | ʒ ø sɥidegutei lnø mɔ̃pa z ɑ̃boʃepaʁs - køʒ ø n ɛ pal ø diplom - |
| VoExp | ʒ - sɥidegutei - - - mɔ̃pa - ɑ̃boʃepaʁs - køʒ - - e pal - diplom - |

## 5  Perception Test Results

In this section, we present perception tests which evaluate the respective influences of linguistics, phonetics and prosody in terms of quality and expressivity.

### 5.1  Experimental Set-Up

Six perception AB tests were conducted independently. Each test combines a TTS voice built on one of the 3 speech corpora (*Telecom*, *Audiobook* or *Commentary*) and either neutral or expressive input text. Within a test, AB pairs combine 3 different pronunciations: *canonical* (Cano), adapted neutral voice-specific (VoSpe) and adapted expressive (VoExp). For each test, 11 participants were asked to judge 30 utterances per AB pair, consequently each AB pair was evaluated more than 300 times. The listeners had to answer to the questions reported in Fig. 2 for the 30 utterances presented randomly. Speech samples were synthesized directly from the phoneme sequence (*canonical* or predicted with one of the 2 pronunciation models) derived from the tested input text. A TTS voice was created with the corpus-based unit selection TTS system described in [1] for each of the 3 speech corpora.

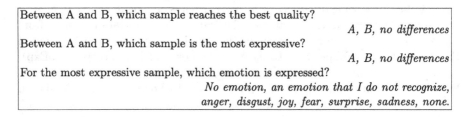

Fig. 2. Perception test design.

## 5.2   Results

The results concerning quality are reported in Table 5, the ones concerning expressivity are reported in Table 6. The number of preferred samples in % is given for each pronunciation of the AB pair. Cases for which the two pronunciations are judged as similar are not reported. Significant preferences are annotated with a * according to the confidence interval used in [8]. VoSpe is globally preferred to Cano whatever the voice and the input text. This neutral adapted pronunciation is also judged with a better quality than the expressive adapted pronunciation (VoExp). Moreover, it seems that VoExp reaches a better quality than Cano when input text is expressive rather than when the input text is neutral with *Telecom* and *Commentary* voices.

Interestingly, in cases where VoExp's quality is preferred to Cano's quality, expressivity of VoExp is also preferred: with *Telecom* voice whatever the text and with *Commentary* voice and expressive text, thus underlying the required adequation between expressivity and audio quality. There is no differences in the perception of expressivity between VoSpe and VoExp while input text is neutral. With the *Audiobook* voice, Cano is preferred to VoExp both in terms of quality and expressivity. This result was expected since the phonetizer was tuned with *Audiobook* speech data. Obtained results show us that the expressive pronunciation adaptation framework improves the perception of expressivity especially

**Table 5.** Preferred samples (%) in terms of quality.

| Text | Sample | Telecom | | Audiobook | | Comment. | |
|---|---|---|---|---|---|---|---|
| | A/B | A | B | A | B | A | B |
| Neutral | Cano/VoSpe | 5.0 | **65*** | 20 | **27** | 8.1 | **27*** |
| | Cano/VoExp | 32 | **48*** | **47*** | 8.3 | **35*** | 18 |
| | VoExp/VoSpe | 13 | **41*** | 12 | **34*** | 14 | **35*** |
| Expressive | Cano/VoSpe | 9.2 | **54*** | 14 | **26*** | 15 | **28*** |
| | Cano/VoExp | 30 | **39** | **46*** | 22 | 25 | **32** |
| | VoExp/VoSpe | 28 | **47*** | 14 | **44*** | 22 | **25** |

**Table 6.** Preferred samples (%) in terms of expressivity.

| Text | Sample | Telecom | | Audiobook | | Comment. | |
|---|---|---|---|---|---|---|---|
| | A/B | A | B | A | B | A | B |
| Neutral | Cano/VoSpe | 22 | **30** | **22** | 13 | 15 | **26** |
| | Cano/VoExp | 20 | **30** | **23** | 15 | **23** | 14 |
| | VoExp/VoSpe | 15 | 15 | 13 | **18** | **17** | 14 |
| Expressive | Cano/VoSpe | **28** | 22 | 24 | **30** | **29** | 21 |
| | Cano/VoExp | 27 | **43*** | **34** | 28 | 24 | **29** |
| | VoExp/VoSpe | **41*** | 24 | 26 | **33** | **26** | 24 |

as the speech corpus contains neutral speech (such as *Telecom*). Should the speech corpus be already expressive, the voice-specific pronunciation adaptation improves the global perceived quality, while expressive pronunciation adaptation does not achieve to improve the perceived expressivity, probably because expressivity is already contained in the prosody of the voice. Therefore, we show that the perception of expressivity relies on the adequation of phonetics and prosody.

With neutral text, whatever the prosody (i.e. the voice speech corpus), participants are not able to recognize any emotion (correctly recognized emotions <5% over all pronunciations). However, they do when linguistic content is expressive (correctly recognized emotions >30% over all pronunciations), whatever the prosody, thus meaning that emotion perception is strongly linked with linguistic content. Whilst *Commentary* has been characterized as the most expressive, the moderate voice *Audiobook* reaches the best recognition rate: average $F_1$ measure is 82% with *Audiobook* and only 67% with *Commentary*. Precisely, it seems that *Commentary* voice is not suitable for the expression of sadness ($F_1^{sad} = 34\%$), while *Audiobook* is ($F_1^{sad} = 74\%$). We have mentioned in Sect. 4.3 that prosodic features were the most selected features in the expressive pronunciation model. Same prosodic features can be used to model a sad pronunciation and can be reached by the TTS *Commentary* voice. However, even if the *Commentary* corpus is expressive, sadness is probably under-represented thus introducing the observed mismatch.

## 6    Conclusions

The present work evaluates the respective influence of linguistics, phonetics and prosody on the perception of quality and expressivity of synthetic speech samples. Neutral and expressive input texts, pronunciations and synthetic voices are used in a TTS system to evaluate the shared influences of these factors. The experiments confirm the interest of voice-specific adaptation for the perceived quality of TTS with different voices. Perception tests show that expressivity is better perceived when synthetic samples also have a good quality. While the perception of expressivity mainly relies on the adequation of phonetics and prosody, the perception of emotions is strongly linked with linguistics. The presented results open new perspectives in emotional data collection. In the framework of expressive speech synthesis, the use of a moderately expressive voice is of interest for the expression of affect and also for the quality of synthetic speech samples. In the future, prosodic features could be predicted directly from text, thus allowing to select appropriate speech units in the TTS voice. Further experiments are needed to label speech units according to their expressiveness, for instance with emotion recognition frameworks or speaking styles models.

**Acknowledgments.** This study has been realized under the ANR (French National Research Agency) project SynPaFlex ANR-15-CE23-0015.

# References

1. Alain, P., Chevelu, J., Guennec, D., Lecorvé, G., Lolive, D.: The IRISA Text-to-Speech system for the Blizzard Challenge 2016. In: Blizzard Challenge (Satellite of Interspeech) (2016)
2. Bartkova, K., Jouvet, D., Delais-Roussarie, E.: Prosodic parameters and prosodic structures of French emotional data. In: Speech Prosody, Shanghai, China (2016)
3. Boeffard, O., Charonnat, L., Maguer, S.L., Lolive, D., Vidal, G.: Towards fully automatic annotation of audiobooks for TTS. In: LREC, Istanbul, Turkey (2012). http://www.lrec-conf.org/proceedings/lrec2012/pdf/632_Paper.pdf
4. Brognaux, S., Picart, B., Drugman, T.: Speech synthesis in various communicative situations: impact of pronunciation variations. In: Interspeech, pp. 1524–1528, September 2014
5. Campbell, N.: Expressive/Affective Speech Synthesis, pp. 505–518. Springer, Heidelberg (2008)
6. Charfuelan, M., Steiner, I.: Expressive speech synthesis in MARY TTS using audiobook data and EmotionML. In: Interspeech, Lyon, France, August 2013
7. Chen, Y.Y., Wu, C.H., Huang, Y.F.: Generation of emotion control vector using MDS-based space transformation for expressive speech synthesis. In: Interspeech, San Fransisco, USA, pp. 3176–3180, September 2016
8. Chollet, G., Montacié, C.: Evaluating speech recognizers and databases. Recent Adv. Speech Understand. Dialog Syst. NATO ASI F: Comput. Syst. Sci. **46**, 345–348 (1988)
9. Feugère, L., d'Alessandro, C., Delalez, S., Ardaillon, L., Roebel, A.: Evaluation of singing synthesis: methodology and case study with concatenative and performative systems. In: Interspeech, San Fransisco, USA, pp. 1245–1249, September 2016
10. Goldman-Eisler, F.: The significance of changes in the rate of articulation. Lang. Speech **4**(4), 171–174 (1961)
11. Guennec, D., Lolive, D.: Unit selection cost function exploration using an A* based Text-to-Speech system. In: Sojka, P., Horák, A., Kopeček, I., Pala, K. (eds.) TSD 2014. LNCS, vol. 8655, pp. 432–440. Springer, Cham (2014). doi:10.1007/978-3-319-10816-2_52
12. Kanagawa, H., Nose, T., Kobayashi, T.: Speaker-independent style conversion for HMM-based expressive speech synthesis. In: IEEE International Conference on Acoustics, Speech and Signal Processing, pp. 7864–7868, May 2013
13. King, S., Karaiskos, V.: The Blizzard Challenge 2016. In: Blizzard Challenge (Satellite of Interspeech) (2016)
14. Lavergne, T., Cappé, O., Yvon, F.: Practical very large scale CRFs. In: Proceedings the 48th Annual Meeting of the Association for Computational Linguistics (ACL), Uppsala, Sweden, pp. 504–513 (2010)
15. Pammi, S., Charfuelan, M.: HMM-based sCost quality control for unit selection speech synthesis. In: ISCA Speech Synthesis Workshop, Barcelona, Spain, pp. 53–57, September 2013
16. Qader, R., Lecorvé, G., Lolive, D., Tahon, M., Sébillot, P.: Statistical pronunciation adaptation for spontaneous speech synthesis. In: TSD, Pragua, Czech Republic (2017)
17. Schröder, M.: Expressive speech synthesis: past, present, and possible futures. In: Tao, J., Tan, T. (eds.) Affective Information Processing, pp. 111–126. Springer, London (2009). doi:10.1007/978-1-84800-306-4_7

18. Steiner, I., Schröder, M., Charfuelan, M., Klepp, A.: Symbolic vs. acoustics-based style control for expressive unit selection. In: ISCA Speech Synthesis Workshop (SSW7), Kyoto, Japan (2010)
19. Tahon, M., Qader, R., Lecorvé, G., Lolive, D.: Improving TTS with corpus-specific pronunciation adaptation. In: Interspeech, San Fransisco, USA (2016)
20. Tahon, M., Qader, R., Lecorvé, G., Lolive, D.: Optimal feature set and minimal training size for pronunciation adaptation in TTS. In: Král, P., Martín-Vide, C. (eds.) SLSP 2016. LNCS, vol. 9918, pp. 108–119. Springer, Cham (2016). doi:10. 1007/978-3-319-45925-7_9
21. Turk, O., Schröder, M.: Evaluation of expressive speech synthesis with voice conversion and copy resynthesis techniques. IEEE Trans. Audio Speech Lang. Process. **18**(5), 965–973 (2010)

# Author Index